Sea Stories

SEA STORIES

28 Thrilling Tales of the Deep

EDITED BY TOM McCARTHY

LYONS
PRESS

Guilford, Connecticut

An imprint of Globe Pequot, the trade division of The Rowman & Littlefield Publishing Group, Inc.
4501 Forbes Boulevard, Suite 200, Lanham, Maryland 20706
www.rowman.com

Distributed by NATIONAL BOOK NETWORK

British Library Cataloguing in Publication Information Available

Library of Congress Cataloging-in-Publication Data
Names: McCarthy, Tom, 1952– editor.
Title: Sea stories : 28 thrilling tales of the deep / edited by Tom McCarthy.
Description: Guilford, Connecticut : Lyons Press, an imprint of The Rowman & Littlefield Publishing Group, Inc., [2021]
Identifiers: LCCN 2021016538 (print) | LCCN 2021016539 (ebook) | ISBN 9781493060030 (cloth) | ISBN 9781493060047 (epub)
Subjects: LCSH: Sea stories. | Ocean travel.
Classification: LCC G525 .S436 2021 (print) | LCC G525 (ebook) | DDC 910.4/5—dc23
LC record available at https://lccn.loc.gov/2021016538
LC ebook record available at https://lccn.loc.gov/2021016539

♾™ The paper used in this publication meets the minimum requirements of American National Standard for Information Sciences—Permanence of Paper for Printed Library Materials, ANSI/NISO Z39.48-1992.

Contents

Contents

Contents

Introduction

Sᴇᴀ sᴛᴏʀɪᴇs ᴀʀᴇ, ᴘʟᴀɪɴ ᴀɴᴅ sɪᴍᴘʟᴇ, ᴡᴏɴᴅᴇʀꜰᴜʟ sᴘʀɪɴɢʙᴏᴀʀᴅs ꜰᴏʀ vicarious adventure.

There is nothing like a sea story to entertain, thrill, move, shock, or inspire a reader, and this collection will do just that.

What is it about the sea that lends itself to so many indelibly classic stories? The sea is a wonderful stage on which to unroll a dramatic narrative or introduce a heroic character. It's no wonder so many masterpieces are set on the seas of the world.

From sublime moments gunkholing with Erskine Childers in "An Introduction to Informality" to sheer terror with the ill-fated men among sharks in Raymond B. Lech "*The Loss of* the *Indianapolis*" to astounding respect for the endurance of Ernest Shackleton and his storm-tossed men in "Escape from the Ice," there is simply nothing that can compare to what awaits in this collection of twenty-eight thrilling stories. Many, having withstood the test of time and the vagaries of popular culture are classics Classic or not, the stories in this collection are good reading--breathtaking, and entertaining. They offer unexpected pleasures.

You'll soon find there is much surviving among these stories. What internal chemistry is required to spend five days drifting in the Pacific under a broiling sun while shipmates floating beside you are ripped apart by sharks? How does one maintain any sort of sanity? What degree of strength is needed to live for seventy days in a small boat as you watch your boatmates voluntarily drop into the sea to end it all? For that matter, at what point does one decide not to mourn over the body of a crewmate but rather to eat him?

For anyone sitting adrift in a small, fragile boat while a ship sinks rapidly nearby, the prospect of what lies ahead can hold only terror.

Everyone who has experienced this has been afraid. In these stories, those who controlled their fear—the stronger ones who somehow used it to motivate themselves—made it to shore. Those who were overcome by fear soon enough perished.

Read on.

The stories in this collection will provide a master class for readers seeking to understand the singular psychic strength that can persist under the direst of conditions. On a simpler level, these tales offer an astounding look at how one can stare down death and survive—with courage and strength and amazing patience. There is among the survivors a certain arrogance toward death.

These tales are a resounding affirmation of the power of hope.

There is much more to this collection than mere survival. Here is history, and drama, intrigue, well-paced tales, and colorful characters—all available to enjoy as you sit back in the comfort of a warm and cozy (and I hope, dry) reading chair anchored to your living room floor.

That's the magic of a good sea story tale and why some of the best writers chose the sea as a backdrop.

The accounts in this collection are stronger and more dramatic for their total lack of affectation, their frankness, and their absence of ego.

The stories here are frank and—one must admit—inspiring stories of adventure from authors who by luck, more often than not, bad luck, found themselves at sea, alone on a boat in devastating circumstances.

Read the first person accounts of William Lay and Cyrus M. Hussey on board the ship *Globe* out of Nantucket, who witnessed the bloody carnage of a mutiny. Or perhaps Lewis Holmes's account of a whaler's frozen voyage. Or Owen Chase's understated but eloquent tale of survival.

Here also are giants among authors, not only of seafaring classics but of literature, Herman Melville, Robert Louis Stevenson. Joseph Conrad, James Fenimore Cooper, Jack London--who celebrated the uniqueness of the American character so eloquently that their writing has never faded from public view. But here also are other writers who produced hidden jewels that simply slipped away quietly. It's time to revive them all.

Few people would want to test their mettle in an ice-encrusted boat with Ernest Shackleton, sail the Straits of Magellan with Joshua Slocum, or watch with Owen Chase as an angry whale sends his ship to the bottom thousands of miles from the nearest land.

That is why it is best to simply *read* about them.

"The Capture," by Aaron Smith is a true account of the author's abduction by Cuban pirates while en route from Jamaica to England in 1822. Smith subsequently gained his freedom, but his bad karma continued when he was later arrested as a pirate in Havana, returned to England in chains, and put on trial for his life.

Another first-hand account of the travails of life at sea is Owen Chase's chilling retelling of watching his whaler, the Essex out of Nantucket, being rammed by a white whale and sunk. Chase was only one of eight crew to survive an ordeal that included spending ninety three days adrift in a whaleboat after the ship was sunk. Chase's 1821 narrative inspired Melville to write Moby Dick, a portion of which appears here. Of Chase's story Melville wrote, "The reading of this wondrous story upon the landless sea, and very close to the latitude of the shipwreck has a surprising effect on me." Melville, the icon, appears here also with "Rounding Cape Horn," from his 1850 work, White-Jacket. Melville's contemporary Richard Henry Dana presents an interesting glimpse into the highly superstitious life of a sailor in "Loss of a Man—Superstition," from Two Years Before the Mast.

No collection of sea stories, classic or not, would be complete without Joseph Conrad. "Dirty Weather," from Typhoon, contains the requisite storm, but Conrad's portrayal of Captain MacWhirr is masterful and rich and unforgettable.

It is impossible to read Erskine Childers's 1903 novel The Riddle of the Sands and not want to chuck it all, find a small boat, and go gunkholing for months on end. "An Introduction to Informality," in which the stiff, vain, excessively formal Curutthers meets up with his unglamorous acquaintance Arthur Davies, is taken from The Riddle, a book The Irish Times called "one of the most famous of all thrillers and one of the best of all sea stories." It was the only work of fiction Childers produced. Turning his attention to the vast difficulties of Irish politics and independence, Childers was executed by the British in 1922.

Joshua Slocum, who at one point commanded some of the finest tall ships of his era, became in 1898 the first man to sail around the world single-handedly, in his small sloop Spray—a 46,000mile trip he began three years earlier at age 51. "Yammerschooner" is taken from his incomparable Sailing Alone Around the World.

What can one say about legendary Sir Ernest Shackleton? Leader of perhaps the greatest, or at least the most remarkable survival feats ever, Shackleton was an unflappable character whose story would be deemed unbelievable if it had been presented as fiction. "Escape from the Ice" is taken from his account, South.

Three Men in a Boat, from which "Mal de Mer" is excerpted, is considered British writer Jerome K. Jerome's great comic masterpiece. Though humor rarely travels or ages well, Three Men in a Boat is, more than one hundred years after Jerome wrote it, still funny.

It might strike some as unusual to see a piece by James Fenimore Cooper—an author better known for the landlocked Natty Bumpo and The Last of the Mohicans, in a collection of sea stories. But Cooper was a longtime aficionado of the American Navy. His account of the early days of the USS Constitution appears in "Young Ironsides."

I hope readers find this collection broad, balanced, and interesting.

ONE

An American Sealer in the Russian Sea

Jack London

"BUT THEY WON'T TAKE EXCUSES. YOU'RE ACROSS THE LINE, AND THAT'S enough. They'll take you. In you go, Siberia and the salt mines. And as for Uncle Sam, why, what's he to know about it? Never a word will get back to the States. 'The *Mary Thomas*,' the papers will say, 'the *Mary Thomas* lost with all hands. Probably in a typhoon in the Japanese seas.' That's what the papers will say, and people, too. In you go, Siberia and the salt mines. Dead to the world and kith and kin, though you live fifty years."

In such manner John Lewis, commonly known as the "sea-lawyer," settled the matter out of hand.

It was a serious moment in the forecastle of the *Mary Thomas*. No sooner had the watch below begun to talk the trouble over, than the watch on deck came down and joined them. As there was no wind, every hand could be spared with the exception of the man at the wheel, and he remained only for the sake of discipline. Even "Bub" Russell, the cabin-boy, had crept forward to hear what was going on.

However, it was a serious moment, as the grave faces of the sailors bore witness. For the three preceding months the *Mary Thomas* sealing schooner, had hunted the seal pack along the coast of Japan and north to Bering Sea. Here, on the Asiatic side of the sea, they were forced to give over the chase, or rather, to go no farther; for beyond, the Russian cruisers patrolled forbidden ground, where the seals might breed in peace.

A week before she had fallen into a heavy fog accompanied by calm. Since then the fog-bank had not lifted, and the only wind had been light airs and catspaws. This in itself was not so bad, for the sealing schooners are never in a hurry so long as they are in the midst of the seals; but the trouble lay in the fact that the current at this point bore heavily to the north. Thus the *Mary Thomas* had unwittingly drifted across the line, and every hour she was penetrating, unwillingly, farther and farther into the dangerous waters where the Russian bear kept guard.

How far she had drifted no man knew. The sun had not been visible for a week, nor the stars, and the captain had been unable to take observations in order to determine his position. At any moment a cruiser might swoop down and hale the crew away to Siberia. The fate of other poaching seal-hunters was too well known to the men of the *Mary Thomas*, and there was cause for grave faces.

"Mine friends," spoke up a German boat-steerer, "it vas a pad piziness. Shust as ve make a big catch, und all honest, somedings go wrong, und der Russians nab us, dake our skins and our schooner, und send us mit der anarchists to Siberia. Ach! a pretty pad piziness!"

"Yes, that's where it hurts," the sea-lawyer went on. "Fifteen hundred skins in the salt piles, and all honest, a big pay-day coming to every man Jack of us, and then to be captured and lose it all! It'd be different if we'd been poaching, but it's all honest work in open water."

"But if we haven't done anything wrong, they can't do anything to us, can they?" Bub queried.

"It strikes me as 'ow it ain't the proper thing for a boy o' your age shovin' in when 'is elders is talkin'," protested an English sailor, from over the edge of his bunk.

"Oh, that's all right, Jack," answered the sea-lawyer. "He's a perfect right to. Ain't he just as liable to lose his wages as the rest of us?"

"Wouldn't give thruppence for them!" Jack sniffed back. He had been planning to go home and see his family in Chelsea when he was paid off, and he was now feeling rather blue over the highly possible loss, not only of his pay, but of his liberty.

"How are they to know?" the sea-lawyer asked in answer to Bub's previous question. "Here we are in forbidden water. How do they know but what we came here of our own accord? Here we are, fifteen hundred skins in the hold. How do they know whether we got them in open water or in the closed sea? Don't you see, Bub, the evidence is all against us. If you caught a man with his pockets full of apples like those which grow on your tree, and if you caught him in your tree besides, what'd you think if he told you he couldn't help it, and had just been sort of blown there, and that anyway those apples came from some other tree—what'd you think, eh?"

Bub saw it clearly when put in that light, and shook his head despondently.

"You'd rather be dead than go to Siberia," one of the boat-pullers said. "They put you into the salt mines and work you till you die. Never see daylight again. Why, I've heard tell of one fellow that was chained to his mate, and that mate died. And they were both chained together! And if they send you to the quicksilver mines you get salivated. I'd rather be hung than salivated."

"Wot's salivated?" Jack asked, suddenly sitting up in his bunk at the hint of fresh misfortunes.

"Why, the quicksilver gets into your blood; I think that's the way. And your gums all swell like you had the scurvy, only worse, and your teeth get loose in your jaws. And big ulcers forms, and then you die horrible. The strongest man can't last long a-mining quicksilver."

"A pad piziness," the boat-steerer reiterated, dolorously, in the silence which followed. "A pad piziness. I vish I vas in Yokohama. Eh? Vot vas dot?"

The vessel had suddenly heeled over. The decks were aslant. A tin pannikin rolled down the inclined plane, rattling and banging. From above came the slapping of canvas and the quivering rat-tat-tat of the after leech of the loosely stretched foresail. Then the mate's voice sang down the hatch, "All hands on deck and make sail!"

Never had such summons been answered with more enthusiasm. The calm had broken. The wind had come which was to carry them south into safety. With a wild cheer all sprang on deck. Working with mad haste, they flung out topsails, flying jibs and staysails. As they worked, the fog-bank lifted and the black vault of heaven, bespangled with the old familiar stars, rushed into view. When all was shipshape, the *Mary Thomas* was lying gallantly over on her side to a beam wind and plunging ahead due south.

"Steamer's lights ahead on the port bow, sir!" cried the lookout from his station on the forecastle-head. There was excitement in the man's voice.

The captain sent Bub below for his night-glasses. Everybody crowded to the lee-rail to gaze at the suspicious stranger, which already began to loom up vague and indistinct. In those unfrequented waters the chance was one in a thousand that it could be anything else than a Russian patrol. The captain was still anxiously gazing through the glasses, when a flash of flame left the stranger's side, followed by the loud report of a cannon. The worst fears were confirmed. It was a patrol, evidently firing across the bows of the *Mary Thomas* in order to make her heave to.

"Hard down with your helm!" the captain commanded the steersman, all the life gone out of his voice. Then to the crew, "Back over the jib and foresail! Run down the flying jib! Clew up the foretopsail! And aft here and swing on to the main-sheet!"

The *Mary Thomas* ran into the eye of the wind, lost headway, and fell to curtsying gravely to the long seas rolling up from the west.

The cruiser steamed a little nearer and lowered a boat. The sealers watched in heartbroken silence. They could see the white bulk of the boat as it was slacked away to the water, and its crew sliding aboard. They could hear the creaking of the davits and the commands of the officers. Then the boat sprang away under the impulse of the oars, and came toward them. The wind had been rising, and already the sea was too rough to permit the frail craft to lie alongside the tossing schooner; but watching their chance, and taking advantage of the boarding ropes thrown to them, an officer and a couple of men clambered aboard. The boat then sheered off into safety and lay to its oars, a young midshipman, sitting in the stern and holding the yoke-lines, in charge.

The officer, whose uniform disclosed his rank as that of second lieutenant in the Russian navy went below with the captain of the *Mary Thomas* to look at the ship's papers. A few minutes later he emerged, and upon his sailors removing the hatch-covers, passed down into the hold with a lantern to inspect the salt piles. It was a goodly heap which confronted him—fifteen hundred fresh skins, the season's catch; and under the circumstances he could have had but one conclusion.

"I am very sorry," he said, in broken English to the sealing captain, when he again came on deck, "but it is my duty, in the name of the tsar, to seize your vessel as a poacher caught with fresh skins in the closed sea. The penalty, as you may know, is confiscation and imprisonment."

The captain of the *Mary Thomas* shrugged his shoulders in seeming indifference, and turned away. Although they may restrain all outward show, strong men, under unmerited misfortune, are sometimes very close to tears. Just then the vision of his little California home, and of the wife and two yellow-haired boys, was strong upon him, and there was a strange, choking sensation in his throat, which made him afraid that if he attempted to speak he would sob instead.

And also there was upon him the duty he owed his men. No weakness before them, for he must be a tower of strength to sustain them in misfortune. He had already explained to the second lieutenant, and knew the hopelessness of the situation. As the sea-lawyer had said, the evidence was all against him. So he turned aft, and fell to pacing up and down the poop of the vessel over which he was no longer commander.

The Russian officer now took temporary charge. He ordered more of his men aboard, and had all the canvas clewed up and furled snugly away. While this was being done, the boat plied back and forth between the two vessels, passing a heavy hawser, which was made fast to the great towing-bitts on the schooner's forecastle-head. During all this work the sealers stood about in sullen groups. It was madness to think of resisting, with the guns of a man-of-war not a biscuit-toss away; but they refused to lend a hand, preferring instead to maintain a gloomy silence.

Having accomplished his task, the lieutenant ordered all but four of his men back into the boat. Then the midshipman, a lad of sixteen, looking strangely mature and dignified in his uniform and sword, came aboard to take command of the captured sealer. Just as the lieutenant prepared to depart his eye chanced to alight upon Bub. Without a word of warning, he seized him by the arm and dropped him over the rail into the waiting boat; and then, with a parting wave of his hand, he followed him.

It was only natural that Bub should be frightened at this unexpected happening. All the terrible stories he had heard of the Russians served to make him fear them, and now returned to his mind with double force. To be captured by them was bad enough, but to be carried off by them, away from his comrades, was a fate of which he had not dreamed.

"Be a good boy, Bub," the captain called to him, as the boat drew away from the *Mary Thomas*'s side, "and tell the truth!"

"Aye, aye, sir!" he answered, bravely enough by all outward appearance. He felt a certain pride of race, and was ashamed to be a coward before these strange enemies, these wild Russian bears.

"Und be politeful!" the German boat-steerer added, his rough voice lifting across the water like a fog-horn.

Bub waved his hand in farewell, and his mates clustered along the rail as they answered with a cheering shout. He found room in the stern-sheets, where he fell to regarding the lieutenant. He didn't look so wild or bearish after all—very much like other men, Bub concluded, and the sailors were much the same as all other man-of-war's men he had ever known. Nevertheless, as his feet struck the steel deck of the cruiser, he felt as if he had entered the portals of a prison.

For a few minutes he was left unheeded. The sailors hoisted the boat up, and swung it in on the davits. Then great clouds of black smoke poured out of the funnels, and they were under way—to Siberia, Bub could not help but think. He saw the *Mary Thomas* swing abruptly into line as she took the pressure from the hawser, and her side-lights, red and green, rose and fell as she was towed through the sea.

Bub's eyes dimmed at the melancholy sight, but—but just then the lieutenant came to take him down to the commander, and he straightened up and set his lips firmly, as if this were a very commonplace affair and he were used to being sent to Siberia every day in the week. The cabin in which the commander sat was like a palace compared to the humble fittings of the *Mary Thomas*, and the commander himself, in gold lace and dignity, was a most august personage, quite unlike the simple man who navigated his schooner on the trail of the seal pack.

Bub now quickly learned why he had been brought aboard, and in the prolonged questioning which followed, told nothing but the plain truth. The truth was harmless; only a lie could have injured his cause. He did not know much, except that they had been sealing far to the south in open water, and that when the calm and fog came down upon them, being close to the line, they had drifted across. Again and again he insisted that they had not lowered a boat or shot a seal in the week they had been drifting about in the forbidden sea; but the commander chose to consider all that he said to be a tissue of falsehoods, and adopted a bullying tone in an effort to frighten the boy. He threatened and cajoled by turns, but failed in the slightest to shake Bub's statements, and at last ordered him out of his presence.

By some oversight, Bub was not put in anybody's charge, and wandered up on deck unobserved. Sometimes the sailors, in passing, bent curious glances upon him, but otherwise he was left strictly alone. Nor could he have attracted much attention, for he was small, the night dark, and the watch on deck intent on its own business. Stumbling over the strange decks, he made his way aft where he could look upon the side-lights of the *Mary Thomas*, following steadily in the rear.

For a long while he watched, and then lay down in the darkness close to where the hawser passed over the stern to the captured schooner. Once an officer came up and examined the straining rope to see if it were chafing, but Bub cowered away in the shadow undiscovered. This, however, gave him an idea which concerned the lives and liberties of twenty-two men, and which was to avert crushing sorrow from more than one happy home many thousand miles away.

In the first place, he reasoned, the crew were all guiltless of any crime, and yet were being carried relentlessly away to imprisonment in Siberia—a living death, he had heard, and he believed it implicitly. In the second place, he was a prisoner, hard and fast, with no chance to escape. In the third, it was possible for the twenty-two men on the *Mary Thomas* to escape. The only thing which bound them was a four-inch hawser. They dared not cut it at their end, for a watch was sure to be maintained upon it by their Russian captors; but at this end, ah! at his end—

Bub did not stop to reason further. Wriggling close to the hawser, he opened his jack-knife and went to work. The blade was not very sharp, and he sawed away, rope-yarn by rope-yarn, the awful picture of the solitary Siberian exile he must endure growing clearer and more terrible at every stroke. Such a fate was bad enough to undergo with one's comrades, but to face it alone seemed frightful. And besides, the very act he was performing was sure to bring greater punishment upon him.

In the midst of such somber thoughts, he heard footsteps approaching. He wriggled away into the shadow. An officer stopped where he had been working, half-stooped to examine the hawser, then changed his mind and straightened up. For a few minutes he stood there, gazing at the lights of the captured schooner, and then went forward again.

Now was the time! Bub crept back and went on sawing. Now two parts were severed. Now three. But one remained. The tension upon this was so great that it readily yielded. Splash the freed end went overboard. He lay quietly, his heart in his mouth, listening. No one on the cruiser but himself had heard.

He saw the red and green lights of the *Mary Thomas* grow dimmer and dimmer. Then a faint hallo came over the water from the Russian prize crew. Still nobody heard. The smoke continued to pour out of the cruiser's funnels, and her propellers throbbed as mightily as ever.

What was happening on the *Mary Thomas*? Bub could only surmise; but of one thing he was certain: his comrades would assert themselves and overpower the four sailors and the midshipman. A few minutes later he saw a small flash, and straining his ears heard the very faint report of a pistol. Then, oh joy! both the red and green lights suddenly disappeared. The *Mary Thomas* was retaken!

Just as an officer came aft, Bub crept forward, and hid away in one of the boats. Not an instant too soon. The alarm was given. Loud voices rose in command. The cruiser altered her course. An electric search-light began to throw its white rays across the sea, here, there, everywhere; but in its flashing path no tossing schooner was revealed.

Bub went to sleep soon after that, nor did he wake till the gray of dawn. The engines were pulsing monotonously, and the water, splashing noisily, told him the decks were being washed down. One sweeping glance, and he saw that they were alone on the expanse of ocean. The *Mary Thomas* had escaped. As he lifted his head, a roar of laughter went up from the sailors. Even the officer, who ordered him taken below and locked up, could not quite conceal the laughter in his eyes. Bub thought often in the days of confinement which followed that they were not very angry with him for what he had done.

He was not far from right. There is a certain innate nobility deep down in the hearts of all men, which forces them to admire a brave act, even if it is performed by an enemy. The Russians were in nowise different from other men. True, a boy had outwitted them; but they could not blame him, and they were sore puzzled as to what to do with him. It would never do to take a little mite like him in to represent all that remained of the lost poacher.

So, two weeks later, a United States man-of-war, steaming out of the Russian port of Vladivostok, was signaled by a Russian cruiser. A boat passed between the two ships, and a small boy dropped over the rail upon the deck of the American vessel. A week later he was put ashore at Hakodate, and after some telegraphing, his fare was paid on the railroad to Yokohama.

From the depot he hurried through the quaint Japanese streets to the harbor, and hired a sampan boatman to put him aboard a certain vessel whose familiar rigging had quickly caught his eye. Her gaskets were off, her sails unfurled; she was just starting back to the United States. As he came closer, a crowd of sailors sprang upon the forecastle-head, and the windlass-bars rose and fell as the anchor was torn from its muddy bottom.

"'Yankee ship come down the ribber!'" the sea-lawyer's voice rolled out as he led the anchor song.

"'Pull, my bully boys, pull!'" roared back the old familiar chorus, the men's bodies lifting and bending to the rhythm.

Bub Russell paid the boatman and stepped on deck. The anchor was forgotten. A mighty cheer went up from the men, and almost before he could catch his breath he was on the shoulders of the captain, surrounded by his mates, and endeavoring to answer twenty questions to the second.

The next day a schooner hove to off a Japanese fishing village, sent ashore four sailors and a little midshipman, and sailed away. These men did not talk English, but they had money and quickly made their way to Yokohama. From that day the Japanese village folk never heard anything more about them, and they are still a much-talked-of mystery. As the Russian government never said anything about the incident, the United States is still ignorant of the whereabouts of the lost poacher, nor has she ever heard, officially, of the way in which some of her citizens "shang-haied" five subjects of the tsar. Even nations have secrets sometimes.

Two

The Capture: A True Event

Aaron Smith

AT TWO O'CLOCK P.M. WHILE WALKING THE DECK IN CONVERSATION with Captain Cowper, I discovered a schooner standing out towards us from the land; she bore a very suspicious appearance, and I immediately went up aloft with my telescope to examine her more closely. I was instantly convinced that she was a pirate, and mentioned it to Captain Cowper, who coincided with me, and we deemed it proper to call Mr. Lumsden from below & inform him.

When he came on deck we pointed out the schooner and stated our suspicions, recommending him to alter his course and avoid her. We were at this moment about six leagues from Cape Roman, which bore S.E. by E. Never did ignorance, with its concomitant obstinacy, betray itself more strongly than on this occasion; he rejected our advice and refused to alter his course, and was infatuated enough to suppose, that, because he bore the English flag, no one would dare to molest him.

To this obstinacy and infatuation I must attribute all my subsequent misfortunes—the unparalleled cruelties which I have suffered—the persecutions and prosecutions which I have undergone—the mean and wanton insults which have been heaped upon me—and the villainy and dishonesty to which I have been exposed from the author of them all; who, not satisfied with having occasioned my sufferings, would have basely taken advantage of them to defraud my friends of what little of my property had escaped the general plunder.

In about half an hour after this conversation, we began to discover that the deck of the schooner was full of men, and that she was beginning to hoist out her boats. This circumstance greatly alarmed Mr. Lumsden, and he ordered the course to be altered two points, but it was then too late, for the stranger was within gunshot. In a short time she was within hail, and, in English, ordered us to lower our stern boat and send the captain on board of her. Mr. Lumsden either did not understand the order, or pretended not to do so, and the corsair, for such she now proved to be, fired a volley of musketry.

This increased his terror, which he expressed in hurried exclamations of Aye, aye! Oh, Lord God! and then gave orders to lay the main yard aback. A boat from the pirate now boarded the *Zephyr*, containing nine or ten men, of a most ferocious aspect, armed with muskets, knives, and cutlasses, who immediately took charge of the brig, and ordered Captain Cowper, Mr. Lumsden, the ship's carpenter, and myself, to go on board the pirate, hastening our departure by repeated blows with the flat part of their cutlasses over our backs, and threatening to shoot us.

The rapidity of our movements did not give us much time for consideration; and, while we were rowing towards the corsair, Mr. Lumsden remarked that he had been very careless in leaving the books, which contained the account of all the money on board, on the cabin table. The captain of the pirate ordered us on deck immediately on our arrival. He was a man of most uncouth and savage appearance, about five feet six inches in height, stout in proportion, with an aquiline nose, high cheek bones, a large mouth, and very large full eyes. His complexion was sallow, and his hair black, and he appeared to be about two and thirty years of age.

In his appearance he very much resembled an Indian, and I was afterwards informed that his father was a Spaniard and his mother a Yucatan Squaw. He first addressed Mr. Lumsden, & inquired in broken English what the vessels were that he saw ahead. On being informed that they were French merchantmen, he gave orders for all hands to go in chase. The *Zephyr* was observed in the mean time to make sail and stand in the direction of Cape Roman.

The captain now addressed himself to Mr. Lumsden on the subject of his cargo, which he was informed consisted of sugars, rum, coffee, arrow root, dye woods, & co. He then severally inquired who and what we were; and then whether we had spoken any vessel on our passage. On being informed of the schooner from New Brunswick, he asked if we thought she had specie on board. We told him that those vessels in general sold their cargoes for cash and he seemed very anxious to learn whether she was ahead or astern of us, and whether she was armed. Mr. Lumsden now entreated the captain to make a signal to the *Zephyr* not to stand nearer to the land, as he was apprehensive of her going on shore, and was told that he need not be under any alarm, as there was a very experienced pilot on board of her. He was, however, dissatisfied with this reply, and repeated his entreaty, when the other, in a menacing tone, enjoined silence, and went forward. In a short time he returned and questioned Mr. Lumsden as to what money he had on board, and when told that there was none, he replied, 'Do not imagine that I am a fool, Sir; I know that all vessels going to Europe have specie on board; & if you will give up what you have, you shall proceed on your voyage without further molestation.' Mr. Lumsden repeated his answer, and the pirate declared that if the money was not produced, he would detain the *Zephyr*, throw her cargo overboard, and, if any was found concealed, he would burn her with every soul belonging to her. He then asked whether there were any candles, wine, or porter, on board; and Mr. Lumsden foolishly replied, not any that he could spare, without appearing to consider that we were in his power, and that he could if he pleased possess himself of any thing he might wish without consulting his convenience.

The night was at this time fast approaching, and the breeze had begun to die away. The captain appeared to despair of coming up with his chase, which we could not clearly perceive to be a ship and a brig, and asked Captain Cowper and myself whether he should be able to overtake them before dark. We replied in the negative, and he then gave orders to shorten sail and stand towards the *Zephyr*. The pirates then began to prepare for supper, and were very liberal in serving out spirits to our boat's crew, and also offered us a share, or wine if we preferred it, but we declined both.

The captain now turned to me and said, that, as he was in a bad state of health and none of his ship's company understood navigation, he should detain me for the purpose of navigating the schooner. I tried as much as possible to conceal my emotions at this intimation, and endeavoured to work upon his feelings by telling him that I was married and had three children—that they, together with my wife and aged parents, were anxiously expecting me at home; and represented, in as pathetic language as I could, the misery & distraction which it would cause them, beseeching him to spare my wife and children, and not bring down the grey hairs of my unfortunate parents with sorrow to the grave. But I appealed to a monster, devoid of all feeling, inured to crime, and hardened in iniquity. Mr. Lumsden in the mean time interfered, and hoped that he would not deprive him of my services: but he savagely told him, 'If I do not keep him, I shall keep you.' This threat evidently alarmed and agitated him, and he seemed to regret the part he had taken.

A few minutes, however, displayed the unfeeling & selfish character of this man in the strongest light. 'Mr. Smith,' said he, turning to me, 'for God's sake do not importune the captain, or he will certainly take me: you are a single man, but I have a large family dependent upon me, who will become orphans and be utterly destitute. The moment I am liberated, I shall proceed to the Havannah, and despatch a man of war in search of the corsair, and at the same time publish to the world the manner in which you have been forcibly detained. Nay, I will represent the whole affair at Lloyd's; and, should the pirate be captured hereafter, and you found on board, no harm shall befall you. Whatever property you have shall be safely delivered to your family, and mine will for ever bless you for the kind and generous act.' During this address he was much affected, and the tears streamed from his eyes. I sympathized in his feelings, and replied, that I hoped that neither of us would be detained; but if the lot must fall upon one, under these circumstances, and on these conditions, I could consent to become the victim. This declaration calmed his agitated spirits; but little did I think of the treachery and duplicity that had been masked beneath them, and which subsequent events have too clearly demonstrated.

Supper having been prepared, the captain and his officers, six or seven in number, sat down to it, and invited us to join them, which, for fear of giving offense and exciting their brutality, we did. Our supper consisted of garlic and onions chopped fine and mixed up with bread in a bowl, for which there was a general scramble, every one helping himself as he pleased, either with his fingers or any instrument with which he happened to be supplied.

During supper Mr. Lumsden begged to be allowed to go on board the *Zephyr* to the children, as he was fearful that they would be alarmed at our absence and the presence of strangers, in which request I joined; but he replied, that no one would injure them, and that, as soon as the two vessels came to an anchor, he would accompany us on board.

The corsair was at this time fast approaching the *Zephyr*, when the captain ordered a musket to be fired, and then tacked in shore; the signal was immediately answered, & the brig followed our movements. One of our boat's crew was then ordered to the lead, with directions to give notice the moment he found soundings, and the captain then inquired if we had any Americans on board as seamen. He expressed himself very warmly against them, and declared he would kill all belonging to that nation in revenge for the injuries that he had sustained at their hands, one of his vessels having been lately taken and destroyed by them; adding, at the same time, that if he discovered that we had concealed the fact from him, he would punish us equally. To the Americans he said that he should never give quarter; but as all nations were hostile to Spain, he would attack all.

The man at the lead, during this conversation, gave notice of soundings in fourteen fathoms, and the captain ordered the boat down, and told Mr. Lumsden he would accompany him on board his vessel. The men we had brought were ordered into the boat, but Captain Cowper, the carpenter, and myself, were not allowed to go into her. The boat then proceeded towards the *Zephyr*, with Mr. Lumsden and the captain of the corsair, and shortly after returned with some of the men whom the pirate had put on board; who brought with them Captain Cowper's watch, the ship's spyglass, and my telescope, together with some of my clothes, and a goat. The goat had no sooner reached the deck, than one of these inhuman wretches cut its throat, and proceeded to flay it while it was yet alive, telling us at the same time, that we should all be served in the same manner if no money was found on board. The corsair had then got into four fathoms' water and came to an anchor, as also did the *Zephyr*, about fifty yards from her, and the pirates that were on board began hailing their companions, & congratulating one another at their success.

The watch on board the corsair was now set, and Captain Cowper, the carpenter, and myself, were ordered to sleep on the Companion. Thither we repaired; but to sleep was impossible. The carpenter then took an opportunity of informing us that there was specie on board, and expressed his apprehension, that, if discovered, the cruel threat would be put into execution. Captain Cowper and myself, however, were ignorant of the circumstance, and felt rather inclined to believe that the carpenter was mistaken, but he assured us that such was the case, and that Mr. Lumsden had consulted him a day or two before about a place for its concealment. The expression which had dropped from him in the boat then occurred to us; but we still felt inclined to believe that it was some private money of his own. The whole night was passed in giving way to various conjectures, and hope and fear and the dread of assassination completely drove sleep away. Each reflected on what might be his future fate, and imparted his hopes or his apprehensions to his fellow-sufferers.

At daylight we perceived the pirates on board beating the *Zephyr*'s crew with their cutlasses, & began to tremble for our own safety. After this we perceived the sailors at work hoisting out her boats, and hauling a rope cable from the afterhatchway and coiling it on deck, as if preparing to take out the brig's cargo. The crew of the corsair meanwhile began to take their coffee, and the officers invited us to partake of some, which we willingly did, & found it very refreshing after a night spent in sleepless apprehension.

At seven o'clock the captain hailed his crew from the *Zephyr*, where he had passed the night, and ordered the boat to be sent, in which he returned in a short time, with some curiosities belonging to myself. On his arrival, he approached me, and, brandishing a cutlass over my head, told me to go on board the *Zephyr*, & bring every thing necessary for the purposes of navigation, as it was his determination to keep me. To this mandate I made no reply; so, brandishing his cutlass again, he asked me, with an oath, if I heard him. I replied that I did, when, with a ferocious air, he said, 'Mind and obey me then, or I will take off your skin.' At this threat I went into the boat, and pulled towards the *Zephyr*, and on my arrival found Mr. Lumsden at the gangway. I told him the nature of my visit, at which he expressed his sorrow, but advised me not to oppose the pirate, lest it might produce bad usage, as he seemed bent upon detaining me. He then informed me that they had taken possession of every thing, and that he himself had narrowly escaped assassination on account of his watch.

On entering my cabin, I found my chest broken to pieces, & contents taken away, with two diamond rings and some articles of value. From a seaman, I received my gold watch, sextant, and some other valuable things, which I had previously given to him to conceal; and with these I returned to my own stateroom, and proceeded to pack up what few clothes had been left by the plunderers. My books, parrot, and various other articles, I gave in charge to Mr. Lumsden, who engaged to deliver them safely into the hands of my friends, should he reach England.

The corsair had, during the interim, weighed anchor and hauled alongside of the *Zephyr*; and, having made fast, the crew had commenced moving all the trunks on board of her. Among these was the desk of Captain Cowper, containing all his papers and vouchers, which he begged me to claim as mine and recover for him, for which purpose he gave me the key. Well aware of the serious loss he would sustain, I undertook the dangerous task; and, passing into the corsair, informed the captain that my desk had been taken, and begged, as it only contained papers, which were of importance to my family, that it might be restored. He ordered me to open it, and, having examined the interior, he granted my request; and I had the pleasure of obliging Captain Cowper, who, in return, promised to represent my case to the underwriters at Lloyd's.

The pirates next commenced taking out the *Zephyr*'s cargo, at which Mr. Lumsden & myself were compelled to assist; but the former was soon after removed on board the schooner, in consequence of the crying of the children, who, the captain said, had been instigated by him to do so. There he was employed in striking the cargo into the hatchways. The pirates in the mean time became intoxicated, and gave way to the most violent excesses. All subordination was at an end, and quality seemed to be the order of the day. Mr. Lumsden was now called on board the

Zephyr, and questioned as to the cargo down the main hatchway, when he read & explained the manifesto. Orders were immediately given to four sailors and myself to prepare for hoisting the cargo up, and to clear away the dye wood that was in our way. Mr. Lumsden directed us to throw it overboard, which we commenced doing, and threw some over; but this was prevented by the captain, who said that he only wanted to throw the ship's cargo overboard, in order to say that it was taken from him, and defraud the underwriters. We continued our occupation until we had hoisted up two scroons of indigo, a quantity of arrow root, and as much coffee as they thought sufficient. The seamen were then ordered to send down the foretop gallant mast and yard, both of which were taken on board the pirate, with whatever spars they thought would be of utility. Even the playful innocence of the children could not protect them from the barbarity of these ruffians; their earrings were taken out of their ears, and they were left without a bed to lie upon or a blanket to cover them.

They next commenced taking out the ship's stores, with all the live stock and some water; and Mr. Lumsden and Captain Cowper were then ordered on the quarter deck, and told that if they did not either produce the money or tell where it was concealed, the *Zephyr* should be burned and they with her. On this occasion the same answer was given as before, and the inhuman wretch instantly prepared to put his threat into execution, by sending the children on board the schooner, and ordering those two gentlemen to be taken below decks and to be locked to the pumps. The mandate was no sooner issued than it was obeyed by his fiendlike myrmidons, who even commenced piling combustibles round them. The apparent certainty of their fate extorted a confession from Lumsden, who was released and taken on deck, where he went to the round house and produced a small box of doubloons, which the pirate exhibited with an air of exultation to the crew. He then insisted that there was more; and, notwithstanding that the other made the most solemn asseverations to the contrary, and that even what he had given was not his own, he was again lashed to the pumps. The question was then applied to Captain Cowper, and fire was ordered to be put to the combustibles piled round him. Seeing his fate inevitable, he offered to surrender all he had, and, being released, he gave them about nine doubloons, declaring that what he had produced was all he had, and had been entrusted to his care for a poor woman, who, for aught he knew, might at this moment be in a state of starvation. 'Do not speak to me of poor people,' exclaimed the fiend, 'I am poor, and your countrymen and the Americans have made me so; I know there is more money, and will either have it, or burn you and the vessel.' The unfortunate man was then once more ordered below and fire directed to be applied. In vain did they protest that he had got all; he persisted in his cruelty. The flames now began to approach their persons, and their cries were heart-rending, while they implored him to turn them adrift in a boat, at the mercy of the waves, rather than torture them thus, and keep the *Zephyr*; when, if there was money, he would surely find it.

Finding that no further confession was extorted, he began to believe the truth of these protestations, & ordered his men to throw water below and quench the flames. The unfortunate sufferers were then released and taken into the round house, and the seamen, children, and myself, allowed to go on board the brig. There we were left for a while at liberty, while the pirates caroused and exulted over their booty.

When they had finished their meal, the captain told them that it was now time to return to their own vessel, and ordered me to accompany them. I hesitated at first to obey; but he was not to be thwarted, and, drawing his knife, threatened with an oath to cut my head off if I did not move instanter. I thought it best to pretend ignorance of his order, and said that I had not heard him at first, and hoped, as I had some accounts to settle with Mr. Lumsden, that he would give me time to do so before he took me away. He complied, with some difficulty. I then requested Mr. Lumsden to sign a written bill of lading for the two tierces of coffee belonging to me on board, consigned to Mr. Watson, ship-chandler, in London, and also a promissory note for eighteen pounds ten shillings, payable to the same person, on account of monies due to me.

Having made these arrangements, I returned on board the schooner, and the captain asked me if I had my watch. I answered in the affirmative; he took it from me and looked at it, and, admiring it, gave very strong hints that he should like to have it. I took no notice of the hint, and said that it was a gift from my aged mother, whom I never expected to see again, and should like to send it to her by Mr. Lumsden; but I was afraid that his people would take it away from that person, if I gave it into his hands. 'Your people have a very bad opinion of us,' he replied; 'but I will convince you that we are not so bad as we are represented to be; come along with me, and your watch shall go safely home.' Saying this, he took me on board the *Zephyr*, with the watch in his hand, and gave it to Mr. Lumsden, and desired his people not to take it from him on any account. I then asked to remain awhile, & bid farewell to the children and crew; which he, with some difficulty, allowed me to do. In this interval, the owner's son, who was on board learning navigation, said that his quadrant and all his clothes had been taken away, and begged me to recover them for him, if possible. I promised him I would do my best, and represented to the pirate that he was the son of poor parents, and could ill afford the loss, and begged that they might be restored. He sulkily replied, that I was presuming too much, but this must be my last request; and the lad might have his things, with the exception of checked shirts, which must be left for his crew. Having performed this good office, the captain became impatient for my return, and I was obliged to take a hasty but affecting leave of my former messmates, to become one of a desperate banditti. The inhuman wretch thought even this affecting ceremony too long, and drove me on board at the point of his knife.

When I had reached the deck of the corsair, he asked me if I had got every instrument necessary to the purposes of navigation; and if not, to go and get them, for he would have no excuses by and by, and if I made any he would kill me. I answered, that I had got all that was necessary, and he then gave orders to cast loose from the *Zephyr*, and told Mr. Lumsden he might proceed on his voyage, but on no account to steer for the Havannah; for, if he overtook him on that course, he would destroy him and his vessel together. He promised that he would not touch there, and the vessels were accordingly cast loose, in a short time afterwards. Mr. Lumsden, Captain Cowper, & the children, stood on the gangway and bade me adieu, and my heart sunk within me as the two vessels parted.

The horrors of my situation now rushed upon my mind; I looked upon myself as a wretch, upon whom the world was closed for ever: exposed to the brutality of a ferocious and remorseless horde of mis-creants—doomed to destruction and death, and, perhaps, to worse, to disgrace and ignominy; to become the partaker of their enormities, and be compelled, I knew not how soon, to embrue my hands in the blood of a fellow-creature, and, perhaps, a fellow-countryman. The distraction, grief, and painful apprehensions of my parents, and of one to whom I was under the tenderest of all engagements, filled my mind with terror. I could no longer bear to look upon the scene my fancy presented to me, and I would have sought a refuge from my own miserable thoughts, in self-destruction; but my movements were watched, and I was secured, and a guard set over me. The captain then addressed me, and told me that if I made a second attempt I should be lashed to a gun, and there left to die through hunger; and, for the sake of security, ordered me below; but, at my earnest entreaty, I was allowed to remain on deck till it was dark.

THREE

The Proud Narragansett's Escape

James Barnes

"TWENTY OF THOSE CONFOUNDED YANKEES GIVE ME MORE TROUBLE than three decks full of Frenchmen," remarked Captain Brower of the prison-ship *Spartan*, one of the fleet of dismantled battle-ships that thronged the harbor of Plymouth, England.

Lieutenant Barnard, commanding the neat little sloop of war *Sparrow*, then on the guard station, laughed.

"They are troublesome beggars, sure enough," he said; "but the funny thing is that they behave almost exactly the way our fellows do, or at least would under the same circumstances; that I verily believe."

"Well, such insolence and impudence I never saw in my life," returned Brower. "I shall be glad when I get rid of this last batch and will rest easy when they have been sent ashore to Dartmoor. You should have seen the way they behaved about two weeks ago. Let me see, it was the evening of the fourth, I believe. In fact the whole day through they were at it—skylarking and speech-making and singing."

It was July, 1814. Many vessels in the government service of Great Britain, returning from America, or from the high seas, brought into Plymouth crews of American vessels, and not a few of the troops captured about the Lakes and on the Canadian frontier had been brought over also. They were usually kept on board one of the prison hulks for three or four months; sometimes it was a year or more before they were transferred to the military prisons, the largest of which was situated at

Dartmoor, and the second in size at Stapleton, not far from the town of Gloucester. Although the prison-ships and the prisons themselves were crowded with Frenchmen, the Yankees were three or four times as much trouble to control and to command. When they were not planning to escape, they were generally bothering the sentinels, drawing up petitions, or having some row or other, if only for the fun of turning out the guard.

"I wish somebody else had this position," grumbled Captain Brower, pouring out a glass of port. "I don't think that I was made for it. When I am left alone, I am liable to become too lenient, and when I am angered, perhaps I may be too hasty. . . . At any rate, I wish some one else was here in my place. . . . I had to laugh the other day, though; you know old Bagwigge of the *Germanicus,* here alongside, what a hot-tempered, testy old fellow he is?

"Well, the other day he was walking up and down his old quarter-deck, and about fourscore of my Yankee prisoners were up on deck for air and exercise. Suddenly they began singing. Now, I don't object to that; if they'd never do anything worse, I'd be happy. They've only cut four holes through different parts of this ship, and once well-nigh scuttled her; but never mind; to go on: Bagwigge, he walks to the side and shouts across to my vessel: 'Hi, there! you confounded Yankees! avast that everlasting row.' I didn't see that it was any of his business, as it was on my own ship; but the Yankees—I wish you had seen them, Barnard, upon my soul."

"What did they do? Slanged him, I suppose, terrible."

"Well, you see," continued Captain Brower, "the potatoes had just been given out for the use of the prison mess cooks, and three big baskets of them lay there on the deck. One of the Yankees threw a potato that caught old Captain B. fair and square on the side of his head, capsizing his hat and nearly fetching away his ear. 'You insolent villains!' he cried, almost jumping up on the rail, 'I'll make you sweat your blood for this.' Well, ha, ha, not only one potato was thrown this time, but about half a bushel. I' faith, but those rascals were good shots. Old Bagwigge, he was raked fore and aft. Turning, he ran for it, and dove in the cabin."

The younger man laughed. The officer about whom the tale had been told was not popular in the service. He had had no Americans on board his prison hulk, and the Frenchmen who were temporarily his guests trembled at his frown and cringed at his gesture. He was an overbearing, hot-tempered martinet, and was hated accordingly. But this was not the end of Captain Brower's story, and as soon as the Lieutenant had stopped laughing, he resumed:—

"Let me go on, for I haven't finished yet. When Bagwigge returned, he had with him a file of marines. Up he marches 'em, and the Yankees greeted them with a cheer, and then seeing that the Captain was going to speak to them, they desisted to let him talk.

"'Now,' he said, 'you impudent scoundrels, below with you; every mother's son of you, or I'll—' He hadn't got any farther than that when the same fellow who threw the first potato hit him again. He was only about forty feet away, you know, and with such force was the vegetable thrown that it nearly took his head off his shoulders. 'Fire!' he roared. 'Fire at them!' I doubt whether the marines could have taken aim, they were so busy dodging potatoes, and as for Bagwigge himself, he was jumping, bubbling, and sizzling like a blob of butter in a skillet. I rushed forward and jumped on to the forecastle rail.

"'If you dare fire, Captain Bagwigge,' I cried, 'you'll swing for it!' At this, he dove down the companionway again, with his marines after him. I turned to the prisoners and ordered them below, where they went readily enough. As to Bagwigge, I don't suppose that I'll hear from him again; I hope that he will attend to his own vessel and leave mine alone."

All this conversation, or at least the relation of Captain Brower's story, had taken place in the *Spartan*'s cabin, and when the two officers left, a detail of the prisoners was on the deck, walking briskly back and forth under the eyes of armed sentries, who guarded the gangways and patrolled narrow board walks, raised some two or three feet above the hammock-nettings.

"Do you see that tall, brown fellow, there?" asked Captain Brower, pointing. "He is the one who did such sharp shooting with the potatoes."

"A strange-looking creature, surely," responded the Commander of the *Sparrow*. "He looks a half-tamed man. Well, I wish you less trouble and all success. Good day to you; I have to return to my ship."

Brower turned and went back into his cabin. Although he did not know it, and would have denied it if he had been told the truth, he was exactly the man for the position, for he was just and painstaking, humane and careful. Although there had been all sorts of attempts to escape formulated among the Yankees, and almost carried into successful execution, Brower had not lost a single prisoner, and his presence among them could restore order and quell a disturbance better than the parading of a file of soldiers.

They were a strange lot, these captives. They came from all walks of life, and from every sort of place. Raw militiamen, who had been surrendered by Hull (the army Hull, mark you, not the brave Commodore), privateersmen, captured in all sorts of crafts and dressed in all fashions, but now principally in rags, and men-of-warsmen who had given themselves up while serving on board English ships rather than fight against their country. These last held themselves rather aloof from the others and messed by themselves. Poor devils, they had never had the satisfaction, even, of having struck a blow. They had turned from one kind of slavery to another; that was all.

The tall, odd-looking figure that Captain Brower had pointed out, belonged to the wildest mess on the orlop deck. His appearance might, perhaps, be called startling; he was far from ill-looking, with straight aquiline features, deep-set and quick black eyes that could laugh or look cruel almost at the same moment. His teeth were beautifully white and even, and although he was not heavy or compact looking, he was as strong almost as any two other men on board the ship. He spoke English without an accent, but with an odd form and phrasing that would have attracted attention to him anywhere. His clear skin was the color of new copper sheathing, and his straight black hair that was gathered sailor fashion into a queue was as coarse as a horse's mane. The grandson of a chief he was, a descendant of the line of kings that had ruled the Narragansett tribes—a full-blooded Indian. But he rejoiced in no fine name.

A sailor before the mast he had been since his sixteenth year, and he had appeared on the books of the privateer brig *Teaser* as John Vance, A.B. It is a wrong supposition that an Indian will never laugh or that he is not a fun-maker. John Vance was constantly skylarking, and he was a leader in that, as he was in almost all the games of skill or strength. Every one liked him, and to a certain extent he was feared, for a tale was told in which John and a knife figured extensively. The flash that would come into his eye gave warning often when the danger limit was being approached, yet he was popular, and even the detested marine guard treated him with some deference. In the last attempt to escape, the Narragansett had been captured after he had swum half-way to the shore and had dived more than twenty times to escape musket-balls from the guard-ships. Suddenly the order came "Prisoners below"—and the ship-bell struck eight sonorous strokes. As the last four or five men left the deck, the Indian touched one of them upon the shoulder.

"Watch me," he said, "and say nothing."

There was a narrow door in a bulkhead close to the companionway, but out of reach unless there was something like a box or barrel on which to stand. It was closed by a padlock thrust through two iron staples. As John descended, he caught the combing of the hatch and drew himself up to a level with his chin. Holding himself there with one arm, he reached forward and caught the padlock in his brown, sinewy fingers. Slowly he turned his hand. The iron bent and gave a little. A grin crossed his face. Swinging himself forward, he landed on a man's shoulders beneath him, and with a wild warwhoop he tumbled a half-dozen down the rest of the ladder, and they sprawled in a heap on the deck. Disdaining to notice the half-humorous curses, he sprang to his feet. Three other men who belonged to his mess followed him.

"Can you do it, Red?" asked one.

"Yes, surely," John replied. "So I can to-night."

The whole of the gun-deck forward of the forecastle hatch had been divided, by a strong partition, into a sort of storeroom. There was one entrance into it from above from the topgallant forecastle, where part of the marine guard were stationed, and the other opening onto the hatch-way, to be used in case of emergency.

It was just past the midnight watch when four stealthy figures crept out from the shadows into the light of the dingy lantern that hung at the foot of the companionway. At night there was only one sentry stationed there, and he generally sat half-way up the ladder, and it was impossible for the prisoners to tell without crossing the dead-line that was drawn at night whether he was asleep or not. This was the risk that had to be undertaken; for if the man should see any one pass beneath that old rope that was drawn across the deck, he would have a right to fire. If the fellow was asleep, yet to gain the deck above, the venturesome prisoner would have to pass within arm's length of him.

Perhaps John Vance had inherited from his long line of red ancestors the peculiar knack of moving without sound, the art of crawling on his belly like a snake, perhaps he had a acquired it by constant practice since he had been a prisoner. For it was his boast, and one that had been proved to be true, that contrary to rules he had visited every part of the ship, and after hours; as has been told, he had been retaken a number of times when just on the point of making good his escape.

The three seamen who accompanied him on this occasion could see the legs of the sentry from the knee down, as he sat on the steps of the ladder leading to the berth-deck above. They could also see the butt of his musket as it rested beside him. Vance had disappeared in the black shadow that lay along the starboard side, and now the watchers saw a curious thing take place. The sentry's musket suddenly tilted forward, as if of its own volition, and then disappeared backward into the darkness, without a sound, much in the manner of a vanishing slide in a magic lantern. The man's legs did not move.

"He is asleep," whispered Ned Thornton to Bill Pratt.

"He's asleep," reiterated Bill Pratt to Gabe Sackett, who made the fourth one of the "constant plotters," as they were termed by the other prisoners.

But in one minute that sentry was seen to be very wide awake indeed. That is, if movement signified wakefulness. His legs shot out in two vicious and sudden kicks. A hand, with wide-spread, reaching fingers, stretched out as if searching for the missing musket. The man wriggled from one side to another and floundered helplessly, with his body half-way off the edge of the ladder. But not one sound did he utter!

"Red's got hold of him," croaked Thornton, and with the assurance of hunters who had watched their quarry step into the trap that held him fast, they stepped forward without fear or caution.

It was as Thornton had said. The poor sentry's head was wedged against the steps. Around his throat were clasped the fingers of two sinewy, bronze-colored hands that held the victim as closely and in as deadly a clasp as might the strap of the Spanish garrote. The scene was really horrible. Sackett leaned about the edge of the ladder, and then he saw what a wonderful thing the Narragansett had done. The combing of the hatchway was fully six feet from where the sentry sat. Below yawned the black abyss into the mid-hold. Across this Vance had been forced to lean, balancing himself with one hand when he relieved the sentry of his musket, and then springing forward he had caught him from behind, about the throat. There the Indian hung as a man might hang over the mouth of a well. No wonder the unfortunate marine had been unable to cry out!

"Let go of him, Red," whispered Gabe. "You've choked him enough." The Indian stretched out one of his feet and hooked it over the hatch combing. With a supple movement and without a stumble, he stood erect upon the deck. The sentry would have plunged over into the hold, had not the two others grasped him firmly by the shoulders. They carried him to one side and laid him in the deep shadow against a bulkhead. He was breathing, but insensible.

The rest of the escape can be told in a few words: The lock of the door leading into the storeroom was wrenched away, and noiselessly the four entered, closing it behind them. They had been just in time, for they could hear, on the deck above, the new watch coming on. A port on one side of the storeroom was guarded by three flimsy iron bars. There was enough light outside from the young moon to show the direction of the opening.

Vance bent the irons double at the first attempt. They were almost twenty feet above the water, for the old hulk floated high. But everything seemed working for the furtherance of their plan. There was a new coil of rope on the deck, and looking out of the port right beneath them, they could see a ship's dingy with the oars in it. Sackett slid down first; the other two followed, and Vance remained until the last. No sooner had he made the boat in safety than a great hubbub and confusion sounded through the ship. There came a sharp blare of a bugle, the rolling of the alarm drum, and they could hear the slamming of the heavy hatches that prevented communication from one part of the vessel to the other. The prisoners, cooped up below, knew what it all meant. Some one was out, and there in the pitch darkness they fell to cheering.

But to return to the "constant plotters," in the dingy: they had made but a dozen boat's-lengths when they were discovered, for there was light enough to see objects a long distance across the water. There came a quick hail, followed by a spurt of flame.

"Lord!" Pratt, who was pulling stroke oar with Sackett alongside of him, groaned; "I caught that in the shoulder." One of his arms drooped helplessly, but he continued rowing with the other.

"Let go," grunted Sackett; "I can work it alone—lie down in the stern sheets."

There were three or four vessels, mostly prison or sheer hulks, to be passed before they gained the shore. From each one there came a volley. Poor Sackett received a ball through his lungs and fell into the bottom of the boat, bleeding badly. And now the boats were after them!

Vance and Thornton pulled lustily at the oars; but the others gained a foot in every four. The dingy was splintered by the hail of musket-balls. One of the prison hulks—the last they had to pass—let go a carronade loaded with grape. It awoke the echoes of the old town. So close was the charge delivered that it had hardly time to scatter, and churned the water into foam just astern of the little boat as if some one had dumped a bushel of gravel stones into the waters of the harbor. Not three hundred feet ahead of the foremost pursuing boat, the dingy's keel grated on the shingle.

The Narragansett sprang out, Thornton after him. Sackett could not be raised. Pratt, holding his wounded and disabled arm, staggered up the incline towards some stone steps leading to the roadway above. But he had hardly reached the foot when there came another shot. He fell face downward and made no attempt to rise. Sackett and he would join in no more plots; but Vance and Thornton were now running down a side street.

They dodged about a corner into an alley; crossed a small common, and just as they reached the other side they ran, bows on, into a heavy cloaked figure, who, seeing their haste, hailed them peremptorily, and sprang a huge rattle, making much the same noise that a small boy does when he runs down a picket fence with a stick. Thornton was laboring ahead like a wherry in a tideway. But the Indian was striding along like a racehorse, with the easy, springing gait inherited from his own father, "Chief Fleetfoot," who, if the story told be true, could run down a red deer in the woods. He turned to assist his comrade by taking hold of him and giving him a tow. But as he did so, Thornton's foot struck a round stone and he fell forward, and lay there groaning.

"Run on, Red! run on!" he cried breathlessly. "I've broken a leg; something's carried away in my pins; on with you!"

"Come you with me too," answered the Narragansett, pulling Thornton to his feet with one hand; but the poor lad groaned and fell again.

"Run ahead, curse you!" he said. "Don't stay here and be taken!"

The watchman's rattle had attracted the notice of the people in the houses. Windows were opened and heads were thrust forth, and from about a corner came another cloaked figure carrying a lantern, and a big pike was in his hand.

There was nothing else to do, and, obeying Thornton's angry order, the Indian struck out again into his long distance-covering gait. Which way he ran it made little matter to him. He did not know the country; he had no plans; but the feel of the springy earth beneath his feet was good to him. The sight of the stars shining through the branches of the trees overhead—for he had soon reached the open country and left the town behind him—made him breathe the air in long, deep breaths, and tempted him to shout. It was freedom; liberty!

The dim moonlight softened everything, and to his mind he seemed to be flying. He passed by great stone archways leading to private parks and great estates. Twice he had avoided little hamlets of thatched cottages. Once he had run full speed through the streets of a little village, and had been hailed by the watchman, who sprang his harmless rattle. But it was growing light. He must find some place to hide, for travel during the daytime he knew he could not. Leaping a fence, he made his way into an adjoining field and lay down, panting, beneath some bushes.

Soon cocks began to crow; daylight widened; a bell in an ivy-covered tower tolled musically. Insects commenced their morning hum; birds twittered, and people moved out to their toil. From his hiding-place the

Narragansett watched the unusual sight. In a field below him—for he lay at the top of a small hill—he could see some men and women working in a field of grain. One of the girls had placed a basket beneath the shade of a bush. The Indian was hungry. It required little trouble to snake himself through the grass and secure the contents of the little hamper, a loaf of bread and a large piece of cheese. Then he carefully replaced the cover and stole back to his former hiding-place. Soon he observed, in the road below him, a man riding along at a fast gait; he pulled in his horse and shouted something to the workers in the field. This done, he rode at top speed into the village. Very soon another horseman appeared, and soon quite a little band of them, among whom was a mounted soldier or two, and three or four in the pink coats of the hunting-field.

But near footsteps sounded. A man in leather gaiters, with a fowling-piece over his shoulder, was coming down a little path from some deep woods on the right. A setter dog played in front of him. The man was reading a freshly printed notice. The ink was smeared from handling. The man spelled it out aloud. "Escaped from the hulks; a dangerous prisoner; a wild American Indian; ten pounds reward," and much more of it.

All of a sudden the dog stopped; then with a short bark, he sprang forward. At the same instant the gamekeeper dropped the printed notice that had been handed to him but a minute previously by a horseman on the road. Surely he could not be mistaken, something had dodged down behind yonder hedge; and as the setter sprang forward, barking viciously, a strange figure arose, a man with a copper-colored face, and streaming, unkempt, black locks; he wore big gold ear-rings, and he was clad in a torn canvas shirt and trousers, with a sailor's neckerchief around his throat.

The dog was bounding forward when suddenly the figure raised its arm. No cricketer that ever played on the village green could throw with such unerring force. A large stone struck the dog and took the fight out of him. Yelping, he sneaked back to his master's heels. The startled gamekeeper raised his gun and fired. Whether it was because of his sudden fright or the quickness with which the agile figure dropped at the flash, the charge whistled harmlessly through the leaves. But the sound of the shot had attracted the attention of the people in the fields. A cry arose, as a weird figure broke from the bushes and dashed down the hill, making for the woods.

"Gone away! gone away! whoop, hi!"—the view hallo of the huntsman.

A man in a red coat had sighted the chase. He leaped a fence, and four or five other horsemen followed. Soon there came the shrill yelping of the dogs as they found the plain trail of the barefoot man running for his life.

It was a great run, that man-hunt, and one remembered to this day. Over fence and hedge, across ditch and stream, the Narragansett led them. No trained hurdler that ever ran across country in the county of Devonshire could have held the pace that Vance kept up. Twice he threw them off the scent by running up a stream and doubling on his tracks. But the whole countryside was out and after him. The dogs were gaining on him swiftly, and at last at the foot of a great oak they had him cornered. He fought them off with a broken branch, and soon the pack surrounded him in a yelping circle, not daring to come nearer.

Up came the huntsmen. They halted at some distance and talked among themselves. Who among them was brave enough to go up and lay hold of this strange wild man? They called off the dogs and waited for the soldiers. Eight or ten yokels and some farmer folks joined the gaping crowd. Five men appeared with muskets, and one with a long coil of rope. But all this time the Narragansett had stood there with his back against an oak tree, with a sneer on his thin lips. They talked aloud as to how they should capture him. Some were for shooting him down at once; but as yet no one had addressed a word to him direct. Surely, he must speak an outlandish foreign tongue! Suddenly, the fugitive took a step forward and raised his hand.

"Englishmen," he said, "listen to me."

All started back in astonishment. Why, this wild man spoke their own language!

"Who is the chief here? Who is the captain?" Every one looked at a middle-aged man astride a sturdy brown cob. He was the Squire, and magistrate of the neighborhood.

"Well, upon my soul," he began, "I suppose—"

But the Narragansett interrupted him. "To you I give myself," he said, advancing. He glanced at the others with supreme contempt. As he came forward, he held out his hand, and involuntarily the man on horseback stretched forth his. It was a strange sight, that greeting. The crowd gave way a little, and three or four mounted dragoons came tearing up hill. They stopped in astonishment.

"You gave us a good run," said the Squire, with some embarrassment, not knowing what to say.

"You are too many; I am your prisoner," was the answer.

No one laid hands on him. Walking beside the Squire's horse down to the road, followed by the gaping, gabbling crowd, who still, however, kept aloof, the Narragansett walked proudly erect. When he reached the highway, he turned. There was a cart standing there. The Squire dismounted from his horse and spoke a few words to the driver. Then he mounted to the seat. John Vance sprang up beside him. At a brisk pace they started down the road towards Portsmouth, the soldiers and the horsemen trailing on behind them. At the landing where the boat from the old *Spartan* met them—for a horseman had ridden on with the news—was waiting a sergeant of marines. He advanced with a pair of handcuffs.

"None of that!" exclaimed the Squire. "This man has given me his word."

"The word of a chief's son," put in the Narragansett. The two men shook hands again; then proudly John Vance stepped into the boat, and unmanacled sat there in the stern sheets.

In twenty minutes he was once more down in the close, foul-smelling 'tween decks.

The only notice taken of the Narragansett's break for liberty was the fact that he was numbered among the next detail bound for Dartmoor; but the tradition of the man-hunt of Squire Knowlton's hounds, and its curious ending, lives in Devonshire to-day.

FOUR

The Loss of the *Indianapolis*

Raymond B. Lech

FIVE MINUTES PAST MIDNIGHT ON JULY 30, 1945, THE FIRST TORPEDO smashed into the starboard bow of the United States heavy cruiser *Indianapolis*, and an ear-shattering explosion rocked the ship. Three seconds later, the second torpedo found its mark directly under the bridge and blew up. The vessel lifted slightly out of the water, quivered, then promptly settled back down. At the same time, from the bridge to the bow on the starboard side, water was sent soaring into the midnight sky; flame, steam, and smoke belched out of her forward stack, and an enormous ball of fire swept through the entire forward half of the ship. Within seconds, the fire died away. Once again the *Indianapolis* was level and riding high, but now with the bow gone and two huge gaping holes in her right side.

From midships forward, the cruiser was a complete disaster; no light, no power, no communication, no pressure. Although the rear half of the vessel was untouched, the tons of water that gushed into the forward part of the cruiser sealed the fate of the *Indianapolis*.

IN THE WATER THE FIRST DAY: MONDAY, JULY 30, 1945

Quartermaster 3rd Class Vincent Allard found himself with six or seven other men, all desperately hanging onto a coiled floater net. One of them had a bared knife and was busy cutting the tangles in the net so that it would uncoil and spread out. While this was going on, Allard heard a cry for help. He quickly swam toward the sound and in a few seconds found a sailor floating on a pontoon from one of the ship's planes. He guided the man back to the group clustered around the net, but no sooner did he return when again he heard cries for help. Off he went once more and soon spotted two men holding onto a potato crate. One of the boys could swim, but the other could not and was very scared. Telling the swimmer to stick close, Allard began helping the nonswimmer to the safety of the net. On the way toward the group, he heard someone yell that he had a raft. Since it seemed that the raft was closer than the net, Allard changed course and headed for the sound. The voice called again, and Allard thought he recognized it as the *Indianapolis*'s skipper, Captain Charles Butler McVay III. Allard called out to ask if it were the captain calling, and Captain McVay replied that it was and to come aboard. They swam a short distance and reached the rafts.

The man who could swim climbed unassisted into the empty second raft, McVay and Allard helped the other sailor in, and then Allard joined McVay in his raft. The two men in the second raft had swallowed an enormous amount of water, and at first Captain McVay thought they were both dying. But after a while, they came around. Just before sunrise, they met up with five men on another raft that had a floater net tied to it. They lashed this raft to theirs, and, at first light Monday morning, the group consisted of three rafts, one net, and nine men. Captain McVay was the only officer.

An inspection of the rafts turned up two canoe paddles, a box of cigarettes, fishing gear, signaling mirrors, and a tin container that held twelve Very (star) shells and a pistol. They also found a canvas bag holding a first-aid kit and matches, but it was soaked and everything inside was useless except for some sealed tubes of ointment. During the day, a water breaker holding three gallons of water floated by. This was given to McVay to be tasted, but salt water had leaked into the archaic wooden container and the water was undrinkable. So as not to create unnecessary fear, the captain didn't pass on the bad news but told everyone it would be rationed out when he thought it was "absolutely necessary that they have a drink."

No food was found on any of the rafts, but fortunately, sometime during the day, an emergency ration can drifted by. Upon opening, they found it was dry inside, and they pulled out a number of cans of Spam and small tins of malted-milk tablets and biscuits. The skipper told the other eight men that one twelve-ounce tin of Spam would be opened daily and divided equally. In addition, everyone would daily receive two biscuits and two malted-milk tablets. Under this quota, he figured they had rations to last ten days.

When the rafts crashed into the sea, their gratings had broken. Nevertheless the men made themselves as comfortable as possible and hung on while they were tossed about by the heavy swells of the unending ocean. At one moment they would be deep in a valley of waves and the next moment on top, looking down into that same valley. While on this unwanted roller-coaster ride, resting momentarily on the crest of a wave, they spotted two other rafts also on the crest of their waves. One raft was about 1,500 yards away and appeared to have one man on it who was calling for help. The other raft was much farther away and looked like it held a group of men who seemed to be in good condition. At this time though, McVay's group was too exhausted to paddle over to the near raft, and any investigation had to be held off until the next day.

During this first day, a monstrous shark decided to investigate the raft and its edible cargo. The shark kept swimming under the raft. The dorsal fin was "almost as white as a sheet of paper," while the body was a darker color. The shark could therefore always be spotted because of the visibility of its white fin in the water. The frightened men attempted to catch the pilot fish by knocking them off with canoe paddles, but this was an exercise in futility. They also tried hitting the shark with paddles, but when they occasionally did manage to do so he swam away and returned a few minutes later. In the days to follow, this unwanted nuisance was to become a real menace.

After spotting the two distant rafts, McVay and the others assumed that they were the only survivors of the ship and, all in all, figured no more than twenty-five or thirty men, including themselves, made it off. What they didn't know at the time was that they had drifted seven to ten miles north of the main groups.

Stranded in the middle of the deep and seemingly never-ending Philippine Sea, the captain understandably became very depressed. He daydreamed about taking a bath, drinking a cocktail, and relaxing in comfort, and in the midst of such thoughts he wished to live, but soon reality broke in upon his fantasies.

He dreaded the idea of seeing again the wives of his now dead officers. While at Mare Island, he and Mrs. McVay had gotten to know these women, and now "I knew there was nothing I could say to them. . . ." His mind drifted back to Guam. He remembered the moment when he was told no escort was needed, and he cursed the people there for not having one available; if there had been an escort, it could have radioed for help and picked up survivors. His final, and unfortunately most nagging, thought was of his personal responsibility: he was the captain, like it or not.

Two hours prior to the close of their first day, a plane flew overhead, its red-and-green running lights clearly visible. McVay fired one of the star shells skyward, but it went unnoticed. The container holding the shells had sixteen fillers but only twelve shells, which was the standard issue for this type of raft. It irked McVay to see four empty slots. Why couldn't they just fill the entire thing up and be done with it?

As the day drew to an end, however, spirits were high in anticipation of the morrow's rescue. The *Indianapolis* was due in Leyte Gulf in the morning, and when the heavy cruiser didn't show up questions would be asked, a search made, and rescue would be on the way.

* * *

After narrowly escaping from his after engine room, Lieutenant Richard Redmayne swam from the starboard side of the *Indianapolis*. Within five minutes, he found a kapok life preserver, which he put on, and for about a half hour he rested in the water alone. Then he spotted a life raft with men on it and joined them. During the remaining dark morning hours, two more rafts and two floater nets joined the group. The three rafts and two nets were lashed together, and they continued to drift, picking up water breakers, floating food containers, and other men.

Surveying the area at first light, they found the hostile sea covered with a heavy oil slick, five-inch powder cans, and an assortment of junk. Many of the men were terribly sick from swallowing sea water and oil, and the ones who had passed out in the water were being held up by their shipmates. A head count was attempted, and they discovered that their group consisted of approximately 150 men, including four officers and five chiefs. Lieutenant Redmayne, as the senior officer, took charge.

In addition to the three rafts and two nets, about 90 percent of the people in the water in this group were wearing life jackets; the ones who didn't have any held onto the side of the rafts or onto men who had jackets, or they hugged empty ammo cans. The rafts themselves were very overcrowded, each one averaging fifteen to twenty men, and the sailors who had been put on the rafts were the ones the officers and others in charge thought to be in the worst condition.

On Monday, nothing much happened. The large group floated, drifted, survived. They spotted the same two afternoon planes McVay had seen and also fired flares at the one plane that evening, with no success.

Certain early signs of insubordination surfaced. One of the men on the floater net was Petty Officer F. Giulio. Because of his particular job aboard ship, he was well known among the crew. On this first day, he kept complaining that he should be put on a raft since the life jacket kept slipping around his legs and he had a hard time keeping afloat. Giulio was the senior ranking man on that net and therefore their natural leader.

Distributed among the rafts and nets were four water casks and about nine or ten emergency tins of food, which contained malted-milk tablets, biscuits, and Spam. During the late afternoon, Giulio and some of his followers broke into the rations and began to eat. A short distance away, Chief Petty Officer Clarence Benton spotted them and immediately ordered them to stop, since all rations were to be divided equally. For the time being, Giulio and his small group obeyed the order.

During the evening Lieutenant Redmayne allowed a small amount of food to be rationed equally to all the men in the group.

* * *

At approximately 1:30 a.m., Quartermaster 1st Class Robert Gause spotted a fin. By estimating the distance between the dorsal and tail, he guessed the shark to be about twelve feet long.

Quite a few sailors in his group were critically wounded. There were a large number of severe flash burns of the face, arms, and body, and some men had compound fractures of one sort or another. There were no medical supplies of any kind for the frustrated Doctor Lewis Haynes, and many of the men with fractures and burns died from shock during the first few hours. After removing their life jackets, the dead were allowed to slip away. Before the boiling sun rose over the distant horizon on Monday morning, about fifty of the original four hundred were dead.

By daybreak, this mass of floating humanity had split into three subgroups. The largest group contained about two hundred men, the second one hundred, and the smallest about fifty. These subgroups were separated from each other by a distance of only several hundred yards, at most. Leader of the group of two hundred men was Captain Edward Parke, commanding officer of the Marine Detachment and holder of the Bronze Star for bravery on Guadalcanal. Strong and athletic, he was superb in his energy, leadership, and self-sacrifice. Dr. Haynes remembered him as the typical Marine, one who was very strict with the group and had the situation well in hand.

The main objective was for everyone to stay together. Captain Parke found a cork life ring with about one hundred feet of attached line. To prevent drifting, he strung the line out and each man grabbed a piece of it and took up the slack. In this way, they formed a long line of men which began to curl on itself, as a wagon train would circle against attack. The wounded were brought into the middle and tied to the life ring itself by the strings on their jackets. There was no confusion, and the men stayed well grouped together. If someone did drift off the line, Parke swam over to the man and herded him back in. On several occasions, he gave his jacket to a man without one and swam unsupported until he could find another preserver.

Bravery in this enormous group of "swimmers" was everywhere. Commander Lipski, the ship's gunnery officer, who had been very badly burned, was cheerfully supported all day Monday by Airman 1st Class

Anthony Maday. Lieutenant Commander Coleman, who came aboard in Guam, was the leader of a group, and he worked unceasingly to keep them together. Time after time, he swam out to bring in stragglers. Ultimately, Commander Coleman became so weak that he died from exhaustion. And there was Ensign Moynelo, who organized a large group of men. For three days, he kept the group together, rounded up drifters, and took off his own jacket many times and gave it to those without until he could find another. Finally he, too, collapsed and died.

* * *

Shortly after dawn on Monday, Lieutenant Commander Moss W. Flannery, commanding officer of VPB-133 based on Tinian, climbed into his Ventura bomber and headed out over the Philippine Sea on routine anti-submarine patrol. Visibility was unlimited and in order to obtain better horizon shots for navigation, instead of flying at his normal 5,000 feet, he dropped down and flew between 1,500 and 2,000 feet. At 9:20 a.m., he flew directly over Dr. Haynes and his group of 350 men. In the water, the men saw this plane coming directly at them, the sun reflecting off its front window, and they began splashing the water with their hands and feet to draw attention. Ensign Park, one of the ship's aviators, had some green marker dye in his jacket and spread it in the water. They all firmly believed that they had been seen and estimated that within five hours seaplanes from Guam would be landing in their midst.

Flannery, however, couldn't see a thing. The best way to spot something as small as a head in the ocean is not to look out at an angle but straight down, and at a height of 500 to 800 feet, not 1,500 feet. Flannery was looking out his side window, and his biggest problem was the glassy sea.

By 10:00 a.m., the sun was reflecting so sharply off the sea that everyone began to suffer from intense photophobia, an intolerance to light. Dr. Haynes was very concerned, since he considered this far worse than snow blindness. It caused severe pain, which was relieved only when the sun went down. Closing the eyelids did not help since the sun burned right through. In order to somewhat ease the discomfort, the men ripped their clothing and blindfolded themselves. Fortunately, their bodies did not burn; they were all covered by fuel oil, which the searing rays of the sun could not penetrate.

For the remainder of the first day, there was constant change among the three subgroups. They would merge for a short time then break apart again. The wounded stayed in fairly good shape, and only a few men died. In order to determine death, Dr. Haynes would place his finger on the pupil of an eye and if there was no reflex it was assumed the man was dead. The jacket would be removed and the body allowed to drift away. In the background, some of the men would recite the Lord's Prayer.

By noontime, the sea became choppy again, with large swells. Practically everyone by this time had swallowed some of the oil-soaked water, and they were all throwing up. Thirst was beginning to get to the men, and Haynes, while trying unsuccessfully to find some first-aid supplies, visited all three groups and cautioned them against drinking salt water. For the moment, all the men agreed not to drink from the sea.

The survivors were beginning to see sharks in the area, but, so far, there were no major attacks. Giles McCoy, of the Marine Detachment, saw a shark attack a dead man. He believed that because of the dead men in the water so much food was available that the sharks were not inclined to bother with those still alive.

That, however, had been in the morning and afternoon. By the time that the merciless sun began to set, large numbers of sharks had arrived on the scene, and the men were scared. Cuts were bleeding. When a shark approached a group, everyone would kick, punch, and create a general racket. This often worked, and the predator would leave. At other times, however, the shark "would have singled out his victim and no amount of shouts or pounding of the water would turn him away. There would be a piercing scream and the water would be churned red as the shark cut his victim to ribbons."

IN THE WATER THE SECOND DAY: TUESDAY, JULY 31, 1945

Yesterday they had been too exhausted to paddle over to the raft holding the one lone man, and this morning he was still calling to them. Thinking him hurt, the McVay group began the tremendous task of pulling nine men on three lashed rafts and a floater net to this isolated and scared soul.

Changing the two men paddling once every half hour, it took them four and a half hours to traverse the 1,600 yards separating them and their objective. Upon finally reaching the young man, they saw that, besides being lonely, there was nothing wrong with the new member, and McVay said, "As misery loves company, he wanted somebody to talk to."

There still remained the other group farther away that had been spotted the day before, but the men were now too exhausted to try to reach them. Besides, most of the men had blisters on their hands, and these were creating saltwater ulcers. The new man told the skipper he had seen no one else in the water, and the captain was convinced that his group, plus the small pack of men in the distance, were the sole survivors, even though it seemed incredible that no one else had escaped.

In the morning there was no wind, but the sea could still be described as rough. As the day wore on, the endless water calmed down. There were very long, sweeping swells, but they didn't break and no whitecaps could be seen. Considering the circumstances, the group was comfortable and in fairly good shape.

During the day, Vincent Allard took the large canvas bag that had held the matches, first-aid kits, etc., and fashioned out of the fabric a "cornucopia" cap for everyone. The men pulled the hats over their ears, and this, together with the fuel oil that covered them, saved them from the scorching rays of the sun. To further protect their hands from sunburn, they placed them under the oil-covered water sloshing around in the grating of the rafts.

The fishing kit they found on one of the rafts was a delight to any fisherman's eye, and both McVay and Allard were excellent fishermen. But it didn't help much since there were a number of sharks in the area, and the one big monster of the first day was still performing his merry-go-round act. They did manage to catch some black fish which McVay thought to be in the parrot family; although the meat was very white, he was afraid to let the men eat it. Instead, he used this flesh as bait, hoping to catch nearby schools of bonito and mackerel. However, every time they dropped the line, the shark took what they offered, and, after a while, they gave up the idea of fishing.

During this second twenty-four-hour period, two planes had been spotted; one at 1:00 a.m. and the second at 9:00 p.m. A pair of star shells were fired at both planes, but they weren't seen. The men griped about the shells, for once they reached their maximum height they burst like fireworks and then immediately died. The group wished parachutes were attached, which would float the light back and give the aviator more time to recognize the distress signal.

* * *

At dawn on the second day, the isolated Redmayne group had about sixty men on rafts and another sixty to eighty in the water. Meanwhile, during the dark morning hours, some of the more seriously injured men had died.

The water breakers turned out to be a disappointment. Some of the casks were empty while the others contained either salt or cruddy black water. Lieutenant Redmayne said, "It was dirty and tasted as though the salt content was about equal to the salt content of the seawater." These casks were made of wood, and when the rafts crashed into the sea the seams on the casks split, thereby allowing fresh water to escape and salt water to seep in. The casks were large, heavy, and difficult to handle, and in the standard life raft the water would probably become salty after the first use. Once the seal was broken to pour water, it couldn't properly be resealed, thus allowing salt water to seep in. Should the cup become lost, serving fresh water from the cask resulted in great wastage.

First-aid equipment was generally useless, since the containers were not watertight. Anything in tubes remained sealed, but there weren't enough remedies to go around for burns and eye troubles caused by salt water and fuel oil. The food stayed in good condition but, here again, there was a problem since the primary staple was Spam. Not only did this increase thirst because it was salty, but Spam draws sharks. The men discovered this when they opened a can of Spam and sharks gathered all around them.

The policy of the group was to put all men on rafts who were sick, injured, or didn't have life jackets or belts. The problem with this, however, was that men with belts or jackets began taking them off and allowing them to drift away in order to qualify for the relative safety of a raft. This necessitated keeping a close watch on the men. Giulio and his small band were now beginning to start trouble. Giulio, who was still on a floater net, kept insisting that he deserved some time on a raft. This request was not granted, and he continued to complain.

During the early part of this second day, some of the men swam over to Ensign Donald Blum and reported that the food had been broken into. Blum swam back with them to take a look and saw men eating and drinking. This was immediately reported to Redmayne, who then ordered that all food and water be placed on one raft and guarded at all times by the officers and chiefs. Later in the day there were reports that Giulio was again stealing food, but it was not clear whether food was being taken from the guarded raft or all the food had not been handed in. Ensign Harlan Twible, who was on a floater net about forty feet from Giulio, yelled out in a loud, clear voice, "The first man I see eating food not rationed I will report if we ever get in." He further told them that they were acting like a bunch of recruits and not seamen. As far as can be ascertained, there were no deaths in this group during the second day, and everyone appeared to be in fairly good shape. The only problem was Giulio and his gang. The next day would be a different story.

* * *

Even though total blackness surrounded them, because of the choppy sea the men were having a very difficult time sleeping. In this inky isolation, some of the weaker members of the crew, who could not face what they thought must be ahead of them, gave up all hope; they silently slipped out of their life jackets and committed suicide by drowning. Numerous deadly fights broke out over life jackets, and about twenty-five men were killed by their shipmates. At dawn, Dr. Haynes saw that the general condition of the men was not good, and the group appeared to be smaller. Haynes later recalled that basically two factors, other than lack of water, contributed greatly to the high mortality: the heat from the tropical sun and the ingestion of salt water. The drinking of salt water in his group was generally not deliberate but occurred during bouts of delirium or from the accidental swallowing of water in the choppy sea.

The constant breaking of waves over the men's heads the first two days, particularly when they tried to rest, caused most of them to develop a mechanical sinusitis. The swallowing of small amounts of seawater and fuel oil could not be avoided, and the sun caused intense headache and photophobia. The combination of these factors resulted in many deaths.

During the latter part of the day, the sea grew calmer. The men's thirst, however, had become overpowering as the placid water became very clear. As the day wore on, the men became more and more exhausted and complained of their thirst. Dr. Haynes noticed that the younger men, largely those without families, started to drink salt water first. As the hot sun continued to beat down on them, an increasing number of survivors were becoming delirious, talking incoherently, and drinking tremendous amounts of salt water.

They started becoming maniacal, thrashing around in the water and exhibiting considerable strength and energy compared to those who were exhausted but still sane. These spells would continue until the man either drowned or went into a coma. Several brave men, wearing rubber life belts, tried to support maniacal men and also drowned, for during the struggles the belts developed punctures or rips and deflated. Haynes kept swimming from one huge huddle of sailors to another, desperately trying to help. All during this time, people were getting discouraged and calling out for help, and he would be there to reassure and calm them down.

There were sharks in the area again. The clear water allowed the men to look down and see them. It seems that during this second day, however, the sharks were going after dead men, especially the bodies that were sinking down into the deeper ocean. They didn't seem to bother the men on the surface.

Things became progressively worse from sundown on the second day. The men's stories become mixed up, and some accounts are totally incoherent, making it difficult to piece together what actually happened. Haynes remembered that shortly after sundown they all experienced severe chills, which lasted for at least an hour. These were followed by high fever, as most of the group became delirious and got out of control. The men fought with one another, thinking there were Japanese in the group, and disorganization and disintegration occurred rapidly. Captain

Parke worked until he collapsed. Haynes was so exhausted that he drifted away from the group.

Some of the men attempted to help their shipmates. They swam outside the group, rounding up stragglers and towing them back in. The kapok jackets had a brass ring and also a snap on the back. At night, people who had these jackets on would form a circle and hook them all together. The rest of the men would get in the middle. The corrallers themselves were worried, however, since the jackets had lost so much buoyancy that the feeling of security they provided was rapidly ebbing.

By nightfall, more and more people were removing their preservers and throwing them away. Most of these men died. Haynes swam from one batch of crazed men to another, trying to calm them down. He would locate the groups by the screaming of the delirious men. From this night on, what happened in the water can only be described as a nightmare.

IN THE WATER THE THIRD DAY: WEDNESDAY, AUGUST 1, 1945

The captain and the men with him were continuing to fare relatively well. McVay still believed that his ship went down with all hands and that, at most, there could only be thirty survivors.

From the opening of this day, the central thought on the minds of the men was to kill the shark; it was big, it kept circling closer and closer, and they were frightened. This monster could easily rip the raft apart with one swift motion of his enormous jaws. But the only weapon they had was a knife from the fishing kit, with a one-inch blade, and there was no way they could tackle this massive creature with a blade that small. So the day passed with the men sitting and staring at the shark, annoyed that a larger weapon was not in the kit and further chafed that not one man had a sheath knife, an implement customarily carried by many of the sailors aboard ship.

Just before first light, a plane flew over, and two star shells were fired. Again at 1:00 p.m., a bomber, heading toward Leyte, passed above. They tried to attract this second plane with mirrors, yellow signal flags, and splashing, but to no avail.

* * *

Although the order had been given the day before to bring all food to the command raft, there was still a certain amount of hoarding going on. This morning, however, several more rafts handed their cached rations over to Redmayne. During the day, one cracker, a malted-milk tablet, and a few drops of precious water were allocated to each man. Some survivors tried their luck at fishing but, as with the McVay group, the numerous sharks in the area kept stealing the bait. Not everyone realized there was safety in numbers. Some men swam away. Attempts to stop them failed, and soon after leaving the security of the group these sailors were usually dragged beneath the surface by the sharks.

Toward late afternoon, some of the sailors started becoming delirious again. More and more men were drinking salt water. Chief Benton (Redmayne's assistant) attempted to talk to these half-crazed men in a calm, reassuring voice, but it wasn't much use. Fights broke out, men started swimming away, and people committed suicide by drowning themselves. A sailor yelled to Redmayne that things were getting very bad on his raft, and Ensign Eames was sent over to investigate. Upon returning, Eames reported that some of the men were making homosexual advances toward one of the other men. Upon hearing these reports, the chief engineer's reaction was to have the people around him recite the Lord's Prayer.

Giulio had been on a net for the previous two days, but this morning the pharmacist's mate decided to transfer him to a raft because Giulio complained that his eyes were bothering him. Shortly thereafter, it was noticed that Giulio and the people with him were eating and drinking. Upon checking the stored rations on the command raft, it was discovered that two of the four water breakers were missing, plus several cans of rations. The officers and chiefs ordered Giulio to return everything immediately, but he ignored them. Some of the senior people then swam over to the mutineers and tried to grab the food and water away, but they were unsuccessful since Giulio and his small band were much stronger than the tired officers. Throughout the day, he and his gang had themselves a veritable Roman feast while others suffered and died.

* * *

The early morning hours found Dr. Haynes with a large pack of swimmers headed by Captain Parke of the Marines who, through willpower, strength, and sheer determination, kept the group under control. Before dawn Haynes twice became delirious. At one point, he remembered, "The waves kept hitting me in the face, and I got the impression that people were splashing water in my face as a joke, and I pleaded with them that it wasn't funny and that I was sick. I begged them to stop and kept swimming furiously to make them stop, and then my head cleared."

Most of the men had become hysterical, and some were quickly going mad. A few of the sailors got the idea that people were trying to drown them and that there were Japanese in the group. The cry would circulate, "Get the Jap! Kill him!" Fights broke out, knives were drawn, and several men were brutally stabbed. Mass hysteria reigned.

The doctor did his best to calm them down but was unsuccessful and at one point he himself was held underwater by an insane crewman and had to fight his way back up. Captain Parke desperately tried to regain control but finally became delirious himself and eventually died. Once Parke was gone, the mass madness forced the subgroup to further dissolve, and the men scattered. They wanted to be alone, for no one trusted anyone else.

Under a cloudless sky and full moon, Haynes drifted, isolated but totally alert. A man floated by, and they instinctively backed away from each other. Everyone was crazy. Haynes hated being alone, however, and not very far away he heard the noises that the irrational members of another group were making and began swimming toward the sound. Only a few yards short of this band of men, his strength gave out, and he screamed for help. Breaking off from the pack, his chief pharmacist's mate, John Schmueck, grabbed him and towed him to the safety of their numbers.

Supported by Schmueck, who put his arm through the back of Haynes's jacket and lifted the doctor's body so that it rested on his own hip, Dr. Haynes fell asleep for a few hours. Schmueck himself was not in good shape and was having a difficult time with his rubber life ring. It was defective, and for two days—until he finally got a kapok jacket—he had had to hold his finger over the valve. When the ring would deflate too much, he would have to blow it up again and then hold his finger on it some more.

The new group was well organized and ably led by Ensign Moynelo. Someone in the group suggested using the leg straps on the kapok jackets to snap the men together. This worked very well and prevented them from drifting apart. By daybreak the sea was mirror calm, but the condition of the men was becoming critical. They had difficulty thinking clearly, and most of them talked incoherently and had hallucinations.

By this time, the kapok jackets just kept the men's heads out of the water. There was a great deal of anxiety within Moynelo's group concerning the buoyancy of the preservers since the Navy Manual stated that jackets would remain buoyant for only two days, and they were now well into their third. However, the kapok preservers maintained fair buoyancy, even after one hundred hours, and the mental distress that the men felt on this account turned out to have been uncalled for.

Preservers were, unfortunately, fairly easy to obtain. When a man died (and they were now dying en masse), Haynes would remove his jacket and add it to a pile in the middle of the group. This became their reserve when somebody's jacket went on the "fritz."

Sanity, as we know it, virtually disappeared on this third day. The few men who retained some semblance of sense tried to help their weaker shipmates, but it was a losing battle. Chief Gunner Harrison recalled that "Doctor Haynes's conduct throughout the time he was in the water was, in my opinion, above his normal call of duty. The comfort the men got from just talking to him seemed to quiet them down and relieve some of their worry."

Haynes felt that what kept him going was taking care of the men. They constantly asked him questions about whether the water was salty all the way down and when he thought the planes were coming.

Gunner Harrison remembered, "Early one morning somebody woke me up and wanted to know why we did not stop at an island that we passed. That story caused a great deal of trouble. Several of them believed that those islands were there—three islands. Lieutenant McKissick even dreamed he went to the island and there was a hotel there and they would not let him on the island. The first time I heard the story was, this kid woke me up and wanted to know why we did not stop there." All day long, small numbers of men broke off from the gathering and swam for the "island," never to be seen again.

Noticing a line of men stretching for some distance, Commander Haynes curiously swam to it and asked what was going on. He was told to be quiet for there was a hotel up ahead but it only had one room, and when it was your turn to get in you could sleep for fifteen minutes. Haynes turned and swam away from this procession of patient survivors. Stragglers were continually being rounded up and herded back to the group. Sometimes the job would take up to an hour but Haynes knew that they had to stay together in order to be found.

On this Wednesday afternoon, Ensign Moynelo disappeared with the group who were going to swim to Leyte. It all started out when some quartermaster claimed to have figured out the current and the wind, and how long it would take to swim to Leyte. Approximately twenty-five men joined him. They anticipated that it would take them a day and a half to reach the Philippines, based upon a two-knot current and swimming at one knot per hour. Once this large party disappeared from sight, it was never seen again. This was the largest single group of men lost during the days in the water. All of the strong leaders were now dead, except for Gunner Harrison and Commander Haynes. The doctor recalled that "Gunner Harrison and I were about the only ones left who were well enough to think, and he was just like the Rock of Gibraltar. He always had a smile and kept the group together. He used to say to the fellows, 'If that old broken-down Rickenbacker can stay out on the ocean for a week, we can stay for a month.'" Because of Harrison's leadership, "we managed to keep together. His morale was high, and his cheerful exhortations kept everyone united."

The doctor continued to pronounce men dead. He would remove their jackets, recite the Lord's Prayer, and release the bodies. The water was very clear, and Dr. Haynes remembered the bodies looking like small dolls sinking in the deep sea. He watched them until they faded from sight. A cloud of death hung over everyone, and rescue was no longer discussed. By early evening, all was calm—it was no longer a question of who would die, but when.

In the Water the Fourth Day: Thursday, August 2, 1945

With Lieutenant Redmayne delirious, Ensign Twible tried to command the group until he became totally exhausted and his effectiveness limited. Chief Benton was in a little better shape, however, and issued many orders on his own. During the morning, a man swam over to Twible's raft with cans of crackers and said Giulio sent them. No reason was given, and it is not known whether this was in response to a direct order or a limited act of charity.

More and more people were losing touch with their rational selves. For example, there were plenty of good kapok jackets available, but an insane sailor went up to a man wearing one of the rubber rings, ripped it off his body, and swam away. Unnecessary and foolish acts of this type were taking place throughout the groups. As Freud said, "The primitive stages can always be reestablished; the primitive mind is, in the fullest meaning of the word, imperishable."

The pharmacist's mate in this group, Harold Anthony, worked as hard as humanly possible to aid men in the water and became extremely fatigued. During the night he mentioned to one of his friends that he couldn't keep this pace up much longer and would probably be dead shortly. Twelve hours later, with the relentless Pacific sun beating down on this lonely spot of ocean, the lifeless body of the corpsman was permitted to drift away.

* * *

Doctor Haynes's group disbanded again. Small groups were continually forming and breaking up. The night had been particularly difficult, and most of the men suffered from chills, fever, and delirium. These lonely people were now dying in droves, and the helpless physician could only float and watch. By Thursday morning, August 2, the condition of most of the men was critical. Many were in coma and could be aroused only with exceptional effort. The group no longer existed, with the men drifting off and dying one by one. This isolation from the companionship of another human was cataclysmic.

* * *

At 9:00 a.m., on Thursday, August 2, securely strapped in the pilot's seat, Lieutenant (jg) Wilbur C. Gwinn pushed the throttles forward, brought the motors of his twin-engine Ventura bomber to an ear-splitting roar, and raced down the Peleliu runway. His mission was a regular day reconnaissance patrol of Sector 19V258. He was to report and attempt to sink any Japanese submarine in his area. The route for the outward leg of his journey just happened to have him flying directly over the heads of the dying men of the *Indianapolis*.

At the very rear of a Ventura is an antenna that trails behind the aircraft. It is used primarily for navigation. In order to keep the antenna from whipping around in the wind, which would make it useless, a weight (known as a "sock") is secured to the end. Once Gwinn gained enough speed to get airborne, he pulled back and the nose of the bomber pointed up toward the blue sky. At the same time, he lost the weight from his navigational antenna. With this "trailing antenna sock" gone, he had two choices: turn around and get it fixed, or continue on patrol and navigate by dead reckoning. Because the weather was excellent, Lieutenant Gwinn decided to go on, took the plane up to 3,000 feet, and over a glassy sea began looking for enemy submarines.

Dead reckoning navigation is not very accurate, and over the Pacific Ocean it is neither a very comfortable nor enviable position to be in. At 11:00 a.m., about an hour and forty-five minutes out of Peleliu, Gwinn figured that since caution is the better part of valor, the whipping antenna being pulled behind the plane should somehow be anchored down. Because the radioman was busy with something else and his co-pilot was concentrating on filling out a weather report, Gwinn resolved to repair it himself. Crawling through the after tunnel of the Ventura, he reached the narrow end and stared at the long, slender, thrashing piece of metal, wondering how to fix it. While attempting to come up with some creative solution to his problem, Gwinn happened to look down from his 3,000-foot perch into the Philippine Sea. At that precise moment, he saw it. The thin line of oil could only have come from a leaking submarine, and the startled pilot rushed back to his left-hand seat and began flying the airplane.

At 11:18 a.m., he changed his course so as to follow the snake-like slick. Not being able to see very well, he brought the bomber down to 900 feet. Mile after mile the slick continued, never seeming to reach an end. Five miles later, he suddenly saw them—thirty heads wrapped in a twenty-five-mile orbit of oil. Many were clinging to the sides of a raft, while others floated and feebly made motions to the plane. Who in the world could these people be? At 11:20 a.m., about two minutes after sighting what had looked like black balls on the water, the pilot dropped down to a wave-skimming 300 feet.

He ordered his radioman to get a message off, and at 11:25 a.m., the following transmission was sent:

SIGHTED 30 SURVIVORS

011-30 NORTH 133-30 EAST

DROPPED TRANSMITTER AND LIFEBOAT EMER-
GENCY

IFF ON 133-30

Now that he had positioned the thirty survivors, there was nothing more Gwinn could do so he decided to spread out his search. Following the slick on a northerly course, six miles farther on he found forty more men. Continuing on, four miles more had him pass over another fifty-five to seventy-five people—and still farther north, he found scattered groups of twos and threes. After an hour of flying and looking, Lieutenant Gwinn estimated that there were 150 men in the water.

The survivors were dispersed along a line about twenty miles long. He noticed a group so crowded on rafts that he was unable to tell the exact number of rafts they had. He could barely spot a lone oil-covered man, even at his low altitude, unless he was splashing the water.

Gwinn's antenna problem now had to be solved—quickly. The position he sent out in his first message was calculated by dead reckoning and couldn't possibly be accurate. He had to fix the whipping antenna, and once again he crawled through the dark tunnel to reach the end of the bomber. Once there, he put his hand out the tail, grabbed the long rod, and pulled it inside. Taking a rubber hose, he tied it around the tip of the antenna and pushed the length back out, hoping, while crawling back to the pilot's seat, that there would be enough weight to stop the shaking and get a decent fix. They tried, and it worked.

One hour and twenty minutes after sending his first message of thirty survivors, a second dispatch from the bomber was transmitted:

SEND RESCUE SHIP

11-15N 133-47E 150 SURVIVORS

IN LIFE BOAT AND JACKETS

DROPPED RED RAMROD

Gwinn received orders to stick around.

Dr. Haynes saw the thing and prayed it was real. Flying very low, the bomber zoomed over his head and as quickly as it came, it passed and soon was a dot on the opposite horizon. At that moment, Haynes knew he and his fellow survivors were dead men. Their last ounce of strength was giving out, and this plane was like all the others—blind to the living hell beneath it.

After scouting the area, there was no doubt in Gwinn's mind that these were American sailors below him. Turning the plane, he looked for a group which appeared to be alone and without rafts, and began dropping everything in the plane that floated.

When Dr. Haynes saw the distant dot suddenly reverse course and come back toward them, low over the water, he then knew that they had been sighted. Like a sudden tropical squall, things began falling from the sky. Two life rafts were dropped, together with cans of fresh water. The water cans ruptured on landing but the most important thing was that Gwinn saw them, and those fortunate enough to be still alive knew rescue was near.

Once there was nothing left to drop to the splashing, oil-covered men, Gwinn released dye markers and smoke bombs so as not to lose the position.

It would not be until the next day that the Navy finally discovered that these were survivors from the *Indianapolis*. By this time, the entire Pacific was curious as to who these people were. Ashore, many people thought that they had Japanese in the water and weren't in too big a rush to get things moving. A short time before, in this same area, escorts from a convoy had reported they had attacked a Japanese submarine.

However, after the second report citing "150 survivors" came in, all hell broke loose. Because submarines don't carry 150 men, Pacific Fleet knew they had a surface vessel to contend with, and if a Japanese warship had been sunk they would have known about it. It finally dawned on CinCPac that they might have an American ship down, and panic started to set in. Shortly after Gwinn's second message was received, CinCPac (now in a state of agitation) began radioing ships to report their positions.

For an hour after his second dispatch, Gwinn was all alone, attempting to comfort the dying men beneath him as best he could. Then another plane, on transport duty to the Philippines, appeared. It stayed with the Ventura for about an hour and dropped three of its rafts.

Back at Gwinn's base, the communications officer decoded the first message concerning the thirty survivors and quickly passed it on to his (and Gwinn's) boss, Lieutenant Commander George Atteberry, commanding officer of VPB-152. This was the Peleliu unit of the Search and Reconnaissance Command of Vice Admiral Murray, Commander, Marianas. The unit was under the command of Rear Admiral W. R. Greer.

Atteberry calculated the fuel supply of the lone, circling bomber and estimated that Gwinn would have to leave the scene by 3:30 p.m. in order to land with a small amount of reserve fuel. Not wanting to leave the survivors alone, Commander Atteberry started making some fast decisions.

Not far from the Ventura squadron was a squadron of seaplanes (Dumbos), and Atteberry picked up the phone and told the duty officer of VPB-23 to get a seaplane out to the area by 3:30 p.m. Not having intercepted Gwinn's message, "23" was skeptical about the whole thing and not eager to cooperate. Not liking this attitude, Atteberry drove over to their unit to ascertain the ready status personally. Once there, he decided they couldn't get a plane up in time to relieve Gwinn, so he quickly drove back to his own unit and ordered his plane and crew to get ready for takeoff. At exactly the same moment Gwinn's second message came in, Atteberry, whose call sign was "Gambler Leader," was lifting his bomber off the Peleliu runway.

During the hour-and-a-half flight out, "Leader" was in constant contact with his squadron office and was happy to hear that "23" finally had gotten airborne and on the way. At 2:15 p.m., Atteberry spotted Gwinn, together with the PBM, the large seaplane on transport duty, and immediately established voice contact with both. The commander was given a quick tour of the groups in order to size up the situation. Finally, so that the men in the water wouldn't think they were being deserted, the pilot of the PBM was ordered to circle the southwest half of the huge slick while "Gambler Leader" ranged the northeast portion.

Gwinn's fuel supply was running low, and twenty minutes after Atteberry arrived, he sent Gwinn on his way. Lieutenant Gwinn's third and final message read:

RELIEF BY 70V [Atteberry]

RETURNING TO BASE

The PBM also had to go, and for forty-five minutes Commander Atteberry was all alone, circling and comforting those below by his presence. Then out of his cockpit window, he saw the big, lumbering Dumbo waddling toward him from the distant southern horizon.

Patrol Bombing Squadron 23 was told that Atteberry and his planes were going to remain on the scene until "23" got one of their Catalinas out there. Lieutenant R. Adrian Marks happened to be the duty pilot at the time, and 1,400 gallons of gas were loaded into his seaplane. While this was taking place, Marks, together with his air combat intelligence officer, went to group operations to see if they could gather any more information than what Commander Atteberry had given them. Operations had nothing to offer and, unable to believe that there were so many men (i.e., thirty men as per Gwinn's first transmission) in the water, Marks assumed he was going out to pick up a ditched pilot. With a full tank of gas and extra air–sea rescue gear, Lieutenant Marks shoved his mammoth down the Peleliu runway and, once airborne, turned north. The time was 12:45 p.m.

On the way out, "Playmate 2" (Marks's call sign) received word that instead of thirty men in the sea there were now about 150. This was absolutely incomprehensible to Marks, and he assumed that the message must have been garbled in transmission. However, he "thought it would be a good idea to get to the scene as quickly as possible." At 3:03 p.m., he began picking up radio signals from Atteberry, and a little over three hours from takeoff, at 3:50 p.m., "Playmate 2" made visual contact and established communications with the commander.

Marks was dumbfounded—how did all these people get here? "Gambler Leader" instructed "Playmate 2" not to drop a single thing—there was much more than met the eye. For a half hour, Atteberry gave Marks the tour. Then the Dumbo dropped everything it had (saving only one small raft for itself), concentrating on those floaters who had only jackets.

With everything out of the plane, Marks wondered what he could do next. Looking down at the bobbing mass of humanity, he knew they were in horrible shape but also just as important—and maybe more so— he saw the sharks. Therefore, at "about 16:30 I decided a landing would be necessary to gather in the single ones. This decision was based partly on the number of single survivors, and the fact that they were bothered by sharks. We did observe bodies being eaten by sharks." Marks told "Gambler Leader" he was going in, and Atteberry notified his base that the Dumbo was landing and that he himself needed relief.

Preparations were made inside the Catalina for landing, while Marks looked for a spot where he thought the floating plane would do the most good. Never having made a landing at sea before, he was a little nervous. However, "at 17:15 a power stall was made into the wind. The wind was due north, swells about twelve feet high. The plane landed in three bounces, the first bounce being about fifteen feet high." "Playmate 2" was down safe—but not very sound.

The hull was intact, but rivets had sprung loose and seams ripped open from impact. While rivet holes were plugged with pencils and cotton shoved into the seams, the radio compartment was taking on water and was being bailed out at the rate of ten to twelve buckets per hour. In the meantime, the co-pilot went aft and began organizing the rescue effort. Because of the high swells, Marks couldn't see anything from his cockpit seat. Atteberry stayed in direct communication with him, however, and guided the Dumbo toward the survivors. Both pilots made the decision to stay away from men on rafts, since they appeared to be in better shape than those floating alone. There were problems, however, for although every effort was made to pick up the single ones it was necessary to avoid passing near the men on life rafts because they would jump onto the plane.

The side hatch had been opened, and the plane's ladder was hung out. Standing on the rungs was a crewman and, when they passed a swimmer, he would grab him and pull him aboard. This was very unsatisfactory though, because the people in the water were too weak to hang on. Furthermore, when a burned survivor, or one whose arm or leg was broken, was snatched, the pain was excruciating. They tried throwing out their remaining raft with a rope attached for a swimmer to grab (they were too frail to jump in). Then they would reel the raft back in. This proved to be impractical, because Marks continually kept the plane taxiing and anyone hanging on was dragged through the water. Finally, they settled on going up to a man, cutting the engines, bringing him aboard, and then starting up again and going to another swimmer. Once the engines were cut, silence enveloped the area except for the terrifying cries for help heard by the crew of "Playmate 2."

Before night fell, Marks had picked up thirty people and crammed them into the body of his leaking seaplane. All were in bad shape, and they were immediately given water and first aid. Naturally, as soon as the first man was plucked from the sea, Lieutenant Marks learned the *Indianapolis* had gone down. There was no way, however, that he was going to transmit this word in the clear and "I was too busy to code a message of this nature." So it would not be until Friday, August 3, that the U.S.

Navy finally learned that one of their heavy cruisers had been sunk just after midnight on July 30.

In the sky above the drifting Dumbo, Atteberry was busy directing Marks and telling other planes coming into the area where to drop their gear in order "to obtain the best possible distribution among them." Between the first sighting and midnight, planes continually flew in, and, at one point, there were eleven aircraft on the scene.

With night upon him, it was impossible for Marks to pick up any more individual swimmers, and he therefore taxied toward a large assembly of men who had had rafts dropped to them earlier in the day. This was Commander Haynes's group. Survivors were packed like sardines inside the hull of the Dumbo, so Marks ordered these men to be laid on top of the wings, covered with parachutes, and given water. This damaged the wing fabric, and it became doubtful whether the Catalina would ever fly again.

In the black of this Pacific night, things began to settle down; the stillness was interrupted only by the occasional pained moans of the *Indianapolis* crew. Marks couldn't move the plane for fear of running people down, so they drifted and waited for rescue. Just before midnight, a searchlight on the far horizon pierced the onyx sky, and at the same time a circling plane dropped a parachute flare over "Playmate 2." The ship changed course and steered toward the beat-up PBY and her precious cargo of fifty-six former *Indianapolis* crewmen.

* * *

It was 4:55 p.m. when 1st Lieutenant Richard Alcorn, U.S. Army Air Corps, 4th Emergency Rescue Squadron, forced his Catalina into the air over Palau. Two hours and twenty minutes later, he arrived, and after quickly surveying the situation tossed three of his eleven rafts out the door. He also saw Marks's plane already on the water picking up survivors. Noticing that the swimmers didn't have enough strength to pull themselves into the rubber boats, Alcorn decided not to throw any more out. Instead he landed at 7:30 p.m., bringing his plane down two miles north of Marks.

Within minutes his crew saw the first survivor and pulled him into the aircraft. Then they taxied a few feet, stopped; taxied again, stopped—and kept this up until darkness without seeing another living soul. When Alcorn stopped and searched, they found a tremendous amount of debris in the area, most of it having fallen from the sky during the day.

They also saw bodies, dead bodies everywhere. In the dark, they floated silently with their lone passenger. Soon they heard cries for help from a group of men and sergeants Needham and Higbee volunteered to take one of the rafts, pick them up and bring them back. Alcorn agreed, but with one provision—they could only go as far as the rope attached to raft and plane would take them. Unfortunately, the umbilical cord was not long enough, and the men returned disappointed.

Overhead, planes circled all night. Marks's Dumbo was totally out of commission, but Alcorn continued to signal to the flyers and they reassuringly flashed back to the two. By the end of the day, still no one on shore knew for certain who the people in the water were.

Yet after Gwinn's second frightful message was received, one of the largest rescue operations in U.S. naval history began. The *Cecil J. Doyle* (DE 368) was heading home after an unsuccessful submarine hunt, when she suddenly received orders from the Western Carolines Sub Area to reverse course and steam north to pick up survivors. This was immediately after Gwinn's first transmission. Once the second message came in, the destroyer escort increased speed to 22.5 knots.

At 2:35 p.m., *Doyle's* radio room made voice contact with Commander Atteberry, and they were kept informed of what was going on. The ship was asked to rush but replied that there was no way they could make it to the area until after midnight.

The destroyers *Ralph Talbot* and *Madison*, both on separate patrol off Ulithi, at 4:00 p.m. turned their sleek bows northward and hastened to the scene at thirty-two knots. It was 6:56 p.m. when the *Madison* made contact with the *Doyle* and pointed out that she wouldn't be able to help until 3:00 a.m. the next morning, and the *Talbot* announced that her ETA wasn't until 4:00 a.m.

At 9:49 p.m., *Doyle's* lookouts spotted their first star shell, and from that moment on flares were always visible. An hour later, the ship's giant twenty-four-inch searchlight was switched on and pointed skyward to give the guarding planes an idea of where she was. Instead of seeking individual people in the water, the destroyer escort headed straight for Marks's Dumbo and, shortly after midnight, the first survivor from the incredibly luckless *Indianapolis* was pulled aboard a rescue ship.

* * *

It was noon when they noticed the circling plane far to the south of them. An hour later, there was another, and as the day wore on the planes swarmed over the line separating sky from sea. Frantically the men signaled, but they were too small to be seen. They ripped the kapok out of jackets, threw the silky fiber into an empty 40-mm ammo can, and set it afire, hoping the rising smoke would draw attention to their plight. It didn't work.

Captain McVay was confused and couldn't imagine what was going on. If the men in his group were the only survivors of the ill-fated cruiser, what was going on ten miles to the south of them? They began to feel discouraged, for as darkness blanketed their isolated spot of ocean the search seemed to be moving farther away. McVay was almost certain they were not going to be found and ordered all rations cut in half.

Midnight saw them staring at the tiny pinprick of Doyle's light piercing the black sky, and now they were certain of other survivors. They were also certain, though, that the search area didn't extend north to their position and that it would be a long time, if ever, before they were found. No one slept, and, as the night wore on, this lonely group was very frightened.

* * *

The planes had no problem spotting the large Redmayne group and in the afternoon rafts, rations, and other emergency gear showered downward. With the security of sentinels circling above them, the men calmed down and patiently waited for rescue.

* * *

After Gwinn dropped the two rafts, they were quickly inflated, and, while the men held onto the side, Haynes was pushed in to investigate. The doctor ordered the sickest men put on the raft. He found an eleven-ounce can of water and doled it out in a plastic cup at the rate of one ounce per man. An enormous amount of equipment was dropped to this "swimmer" group, including a ten-man boat that soon had thirty people in it. But, during the day, it became so hot in the rafts that a great many men jumped back in the water to cool off.

Once the supplies were delivered, the group had almost everything they needed to keep them relatively comfortable until rescue ships arrived. Included in this bonanza were fresh water, rations, emergency medical supplies, and sun helmets. Dr. Haynes greatly appreciated the helmets for, when properly used, not only did they protect the wearer from the roasting sun but they also had a screen which dropped down in front of the face and prevented water from getting in the eyes and up the nose. As for the food, they found it impossible to eat the meat and crackers, but the malted-milk tablets and citrus candies went down easily.

Even though so much was dropped to them, the men's deteriorating physical condition made it essential that they be taken out of the water and given rudimentary first aid and medication; otherwise they wouldn't be alive when the ships came. Commander Haynes decided to swim for the plane. He told the group to stay where they were and explained what he was going to do. Then he swam toward Marks's plane and, after what seemed like two hours, finally reached it. His group still didn't have enough water, and he asked the crew of the plane to swing closer and give them some. They did so, and an emergency kit containing K-rations and a quart of water. Haynes treated burns and administered morphine to the more seriously wounded.

When nightfall came, they were in much better shape and had enough rafts so that all but four or five were out of the water. Fresh water was still a problem, but at sundown Haynes had found a saltwater converter in one of the rafts. He spent all night trying to make fresh water out of salt water. Because he was so exhausted, the directions didn't help and the effort was a failure. He eventually made two batches of water which tasted horrible, but which the men drank. They even asked for more, but it had taken almost four hours to make the first batch and Haynes had had it. The doctor, who had worked so hard over the last four days, finally surrendered. He took the converter, flung it into the hated sea, and began to cry.

IN THE WATER THE FIFTH DAY: FRIDAY, AUGUST 3, 1945

Ten minutes after midnight, in a rough sea with a north–northwest wind blowing between eight and ten miles per hour, the *Cecil J. Doyle* lowered her heavy-motor whaleboat. It headed directly for the closer of the two Dumbos. Twenty minutes later, it returned with eighteen former crewmen of the *Indianapolis*, taken from Marks's plane. As soon as the first man was lifted aboard, he was asked, "Who are you?" Minutes later, an urgent secret dispatch was sent to the Commander of the Western Carolines:

HAVE ARRIVED AREA X

AM PICKING UP SURVIVORS

FROM U.S.S. INDIANAPOLIS

(CA 35) TORPEODED [*sic*]

AND SUNK LAST SUNDAY NIGHT

Between 12:30 and 4:45 a.m., Doyle raised from the brutal sea ninety-three men, which included all survivors aboard Marks's plane and the lone man on Alcorn's. In addition about forty men were retrieved from the water and the rafts. While the whaleboat shuttled back and forth, the mother ship slowly cruised the area, sweeping the watery expanse with her huge searchlight and following the flares dropped from the circling planes. The crew of the whaleboat, meanwhile, had a tough time removing men from the plane and bringing them aboard ship. Transfer was difficult because of the condition of the survivors, some of whom were badly burned from the fires on board the ship, one of whom had a broken leg, and all of whom were terribly weak from thirst and exposure.

At 1:10 a.m., the *Doyle* saw a searchlight to the north and soon discovered it to be the high-speed transport *U.S.S. Bassett.* Two hours later, the destroyer escort *U.S.S. Dufilho* also appeared. Until dawn, the Doyle, Bassett and Dufilho worked independently, hoisting men to the safety of their steel decks. Sunup brought the two destroyers *Madison* and *Ralph Talbot* on the scene.

First light allowed Marks to inspect his Catalina, and he quickly determined that it would never fly again. At 6:00 a.m., *Doyle* sent her boat over to the Dumbo and transferred the crew and all salvageable gear to the ship.

Lieutenant Alcorn was relieved of his lone survivor by *Doyle* at 4:00 a.m. and, with the sun rising over the eastern horizon, he had to decide whether or not to take off. The sea was very rough and a heavy wind was blowing, but, fortunately, his Catalina was not nearly as beat up as Marks's. He resolved to try it, and at 7:30 a.m., with no trouble at all, he powered his way down the endless runway and lifted off. At almost the same time, *Doyle* poured eighty rounds of 40-mm gunfire into Marks's abandoned plane, and she sank in the same area as the ship whose men she had so valiantly rescued.

After sinking the seaplane, *Doyle* secured from general quarters, and all of her survivors were logged in, treated, and put to bed. The crew of the *Doyle* were extremely helpful to their fellow sailors who had so recently suffered through a living hell. Men moved out of their bunks to make room for the former crewmen of the *Indianapolis* and constantly hovered around them, waiting for the slightest request that they could fill. The men were all given baths, and the oil was removed from their tired bodies. Every thirty minutes, a half glass of water, hot soup, hot coffee, and fruit were served to them, and this continued throughout the night and into the next day. The *Doyle*'s doctor examined everyone and listed them all in medical condition ranging from serious to acute.

As it searched for the living, *Doyle* passed by the bodies of twenty-five to fifty dead sailors floating in life jackets. At 12:20 p.m., Madison ordered *Doyle* to take off for Peleliu, and this, the first ship on the scene, was now the first to leave, heading south at 22.5 knots.

* * *

All McVay and his isolated band could do was watch the distant search-lights, the falling flares, the circling planes. When the sun rose over the horizon, they were in despair. The entire morning was spent staring at the activity very far away. It did not seem to be coming closer. At 11:30 a.m., they spotted a plane making a box search. It was a very wide pattern, and on each leg it came closer. They found it extremely depressing, for the plane gave no recognition sign. Captain McVay contended that they were never spotted from the air. But they were, for this plane, flown by Marks's squadron leader, Lieutenant Commander M. V. Ricketts, saw them and reported that he sighted two rafts, with five survivors in one and four in the other. By voice radio, he directed the *U.S.S. Ringness* (APD 100) to pick them up. Like Bassett, *Ringness* was a high-speed transport sent by Philippine Sea Frontier, and it had just arrived. After receiving Ricketts's message, *Ringness* headed for the spot, and at 4,046 yards she picked McVay up on radar. On the rafts, the spell of isolation and despair was suddenly broken when somebody cried, "My God, look at this! There are two destroyers bearing down on us. Why, they're almost on top of us." The two destroyers were both transports, *Ringness* and the newly arrived *Register. Register* turned north to pick up another small group while *Ringness* headed for McVay.

Everyone made it aboard under his own power, and all were immedi-ately given first aid. They had lost about 14 percent of their body weight, and during the afternoon they were given ice cream, coffee, and as much water as they could drink. During the entire four and a half days on the rafts, no one in the group asked for a drink. This was surprising to McVay, since he had assumed people couldn't go that long without water—but they did.

* * *

While Doyle was taking care of the Haynes group, Bassett took care of Lieutenant Redmayne and his men. Lowering her four landing craft at 2:30 a.m., Bassett's boats picked up most of Redmayne's people. A head count was taken, and a little over eighty sailors were collected from the original group of 150. Bassett next sent a message to Frontier Headquarters:

SURVIVORS ARE FROM USS

INDIANAPOLIS (CA 35) WHICH

WAS TORPEDOED 29 JULY [*sic*] X

CONTINUING TO PICK UP

SURVIVORS X MANY BADLY

INJURED

Ralph Talbot picked up twenty-four survivors and then spent most of the afternoon sinking eight rafts and a small boat with her 20-mm guns. Later she transferred her survivors to *Register*. As soon as *Madison* arrived in the area, Bassett reported that she had 150 survivors aboard and desperately needed a doctor. Shortly thereafter, at 5:15 a.m., *Madison*'s physician, Lieutenant (jg) H. A. Stiles, was transferred to the transport. It was at the time the landing craft from Bassett came over to pick up Dr. Stiles that *Madison* first learned the survivors were from the *Indianapolis*.

During the day scouting lines were formed with the planes bird-dogging, but nothing was seen except for the dead, and they were generally left where they were. The unpleasant task of recovery and identification was postponed until the next day. The last living man plucked from the Philippine Sea was Captain McVay, who was the last man to enter it.

By the time the blazing Pacific sun reached its zenith on this day, not another living person from *Indianapolis* was to be found in that enormous ocean. She had sailed from San Francisco with 1,196 young men, was torpedoed, and about eight hundred of her crew escaped from the sinking ship. Of these eight hundred, 320 were rescued; two later died in the Philippines, and two on Peleliu. Because of complacency and carelessness, approximately five hundred U.S. sailors (no one will ever know the exact number) died in the waters of the Philippine Sea.

FIVE

Mutiny on Board the Ship *Globe* of Nantucket

William Lay and Cyrus M. Hussey

THE SHIP *GLOBE*, ON BOARD OF WHICH VESSEL OCCURRED THE HORRID transactions we are about to relate, belonged to the Island of Nantucket; she was owned by Messrs. C. Mitchell, & Co. and other merchants of that place; and commanded on this voyage by Thomas Worth, of Edgartown, Martha's Vineyard. William Beetle (mate), John Lumbert (2d mate), Nathaniel Fisher (3d mate), Gilbert Smith (boat-steerer), Samuel B. Comstock, do. Stephen Kidder, seaman, Peter C. Kidder, do. Columbus Worth, do. Rowland Jones, do. John Cleveland, do. Constant Lewis, do. Holden Henman, do. Jeremiah Ingham, do. Joseph Ignasius Prass, do. Cyrus M. Hussey, cooper, Rowland Coffin, do. George Comstock, seaman, and William Lay, do.

On the 15th day of December, we sailed from Edgarton, on a whaling voyage, to the Pacific Ocean, but in working out, having carried away the cross-jack-yard, we returned to port, and after having refitted and sent aloft another, we sailed again on the 19th, and on the same day anchored in Holmes' Hole. On the following day a favourable opportunity offering to proceed to sea, we got under way, and after having cleared the land, discharged the pilot, made sail, and performed the necessary duties of stowing the anchors, unbending and coiling away the cables, &c.

On the 1st of January 1823, we experienced a heavy gale from N. W. which was but the first in the catalogue of difficulties we were fated to encounter.—As this was our first trial of a seaman's life, the scene presented to our view, "mid the howling storm," was one of terrific grandeur, as well as of real danger. But as the ship scudded well, and the wind was fair, she was kept before it, under a close reefed main-top-sail and foresail, although during the gale, which lasted forty-eight hours, the sea frequently threatened to board us, which was prevented by the skillful management of the helm. On the 9th of January we made the Cape Verd Islands, bearing S.W. twenty-five miles distant, and on the 17th, crossed the Equator. On the 29th of the same month we saw sperm whales, lowered our boats, and succeeded in taking one; the blubber of which, when boiled out, yielded us seventy-five barrels of oil. Pursuing our voyage, on the twenty-third of February we passed the Falkland Islands, and about the 5th of March, doubled the great promontory of South America, Cape Horn, and stood to the Northward.

We saw whales once only before we reached the Sandwich Islands, which we made on the first of May early in the morning. When drawing in with the Island of Hawaii about four in the afternoon, the man at the mast head gave notice that he saw a shoal of black fish on the lee bow; which we soon found to be canoes on their way to meet us. It falling calm at this time prevented their getting along side until night fall, which they did, at a distance of more than three leagues from the land. We received from them a very welcome supply of potatoes, sugar cane, yams, cocoanuts, bananas, fish, &c. for which we gave them in return, pieces of iron hoop, nails, and similar articles. We stood off and on during the next day, and after obtaining a sufficient supply of vegetables and fruit, we shaped our course for Oahu, at which place we arrived on the following day, and after lying there twenty hours, sailed for the coast of Japan, in company with the whaling ships *Palladium* of Boston, and *Pocahontas* of Falmouth; from which ships we parted company when two days out.—After cruising in the Japan seas several months, and obtaining five hundred and fifty barrels of oil, we again shaped our course for the Sandwich Islands, to obtain a supply of vegetables, &c.

While lying at Oahu, six of the men deserted in the night; two of them having been re-taken were put in irons, but one of them having found means to divest himself of his irons, set the other at liberty, and both escaped.

To supply their places, we shipped the following persons, viz: Silas Payne, John Oliver, Anthony Hanson, a native of Oahu, Wm. Humphries, a black man, and steward, and Thomas Lilliston.—Having accommodated ourselves with as many vegetables and much fruit as could be preserved, we again put to sea, fondly anticipating a successful cruise, and a speedy and happy meeting with our friends. After leaving Oahu we ran to the south of the Equator, and after cruising a short time for whales without much success, we steered for Fannings Island, which lies in lat. 3° 49' N. and long. 158° 29' W. While cruising off this Island an event occurred which, whether we consider the want of motives, or the cold blooded and obstinate cruelty with which it was perpetrated, has not often been equalled.

We speak of the want of motives, because, although some occurrences which we shall mention, had given the crew some ground for dissatisfaction, there had been no abuse or severity which could in the least degree excuse or palliate so barbarous a mode of redress and revenge. During our cruise to Japan the season before, many complaints were uttered by the crew among themselves, with respect to the manner and quantity in which they received their meat, the quantity sometimes being more than sufficient for the number of men, and at others not enough to supply the ship's company; and it is fair to presume, that the most dissatisfied, deserted the ship at Oahu.

But the reader will no doubt consider it superfluous for us to attempt an unrequired vindication of the conduct of the officers of the *Globe* whose aim was to maintain a correct discipline, which should result in the furtherance of the voyage and be a benefit to all concerned, more especially when he is informed, that part of the men shipped at Oahu, in the room of the deserters, were abandoned wretches, who frequently were the cause of severe reprimands from the officers, and in one instance one of them received a severe flogging.

The reader will also please to bear in mind, that Samuel B. Comstock, the ringleader of the mutiny, was an officer, (being a boat-steerer,) and as is customary, ate in the cabin. The conduct and deportment of the Captain towards this individual, was always decorous and gentlemanly, a proof of intentions long premeditated to destroy the ship. Some of the crew were determined to leave the ship provided she touched at Fannings Island, and we believe had concerted a plan of escape, but of which the perpetration of a deed chilling to humanity, precluded the necessity. We were at this time in company with the ship *Lyra*, of New-Bedford, the Captain of which, had been on board the *Globe* during the most of the day, but had returned in the evening to his own ship. An agreement had been made by him with the Captain of the *Globe*, to set a light at midnight as a signal for tacking.

It may not be amiss to acquaint the reader of the manner in which whalemen keep watch during the night. They generally carry three boats, though some carry four, five, and sometimes six, the *Globe*, however, being of the class carrying three. The Captain, mate, and second mate stand no watch except there is blubber to be boiled; the boat-steerers taking charge of the watch and managing the ship with their respective boats crews, and in this instance dividing the night into three parts, each taking a third. It so happened that Smith after keeping the first watch, was relieved by Comstock, (whom we shall call by his surname in contradistinction to his brother George) and the waist boat's crew, and the former watch retired below to their births and hammocks. George Comstock took the helm, and during his trick, received orders from his brother to "keep the ship a good full," swearing that the ship was too nigh the wind.

When his time at the helm had expired he took the rattle (an instrument used by whalemen, to announce the expiration of the hour, the watch, &c.) and began to shake it, when Comstock came to him, and in the most peremptory manner, ordered him to desist, saying "if you make the least damn bit of noise I'll send you to hell!" He then lighted a lamp and went into the steerage. George becoming alarmed at this conduct of his unnatural brother, again took the rattle for the purpose of alarming some one; Comstock arrived in time to prevent him, and with threatenings dark and diabolical, so congealed the blood of his trembling brother, that even had he possessed the power of alarming the unconscious and fated victims below, his life would have been the forfeit of his temerity!

Comstock, now laid something heavy upon a small work bench near the cabin gangway, which was afterwards found to be a boarding knife. It is an instrument used by whalers to cut the blubber when hoisting it in, is about four feet in length, two or three inches wide, and necessarily kept very sharp, and for greater convenience when in use, is two edged.

In giving a detail of this chilling transaction, we shall be guided by the description given of it by the younger Comstock, who, as has been observed, was upon deck at the time, and afterwards learned several particulars from his brother, to whom alone they could have been known. Comstock went down into the cabin, accompanied by Silas Payne or Paine, of Sag-Harbour, John Oliver, of Shields, Eng., William Humphries, (the steward) of Philadelphia, and Thomas Lilliston; the latter, however, went no farther than the cabin gangway, and then ran forward and turned in. According to his own story he did not think they would attempt to put their designs in execution, until he saw them actually descending into the cabin, having gone so far, to use his own expression, to show himself as brave as any of them. But we believe he had not the smallest idea of assisting the villains. Comstock entered the cabin so silently as not to be perceived by the man at the helm, who was first apprised of his having begun the work of death, by the sound of a heavy blow with an axe, which he distinctly heard.

The Captain was asleep in a hammock, suspended in the cabin, his state room being uncomfortably warm; Comstock approaching him with the axe, struck him a blow upon the head, which was nearly severed in two by the first stroke! After repeating the blow, he ran to Payne, who it seems was stationed with the before mentioned boarding knife, to attack the mate, as soon as the Captain was killed. At this instant, Payne making a thrust at the mate, he awoke, and terrified, exclaimed, "what! what! what!" "Is this——Oh! Payne! Oh! Comstock!" "Don't kill me, don't;" "have I not always——"

Here Comstock interrupted him, saying, "Yes! you have always been a d—d rascal; you tell lies of me out of the ship will you? It's a d—d good time to beg now, but you're too late," here the mate sprang, and grasped

73

him by the throat. In the scuffle, the light which Comstock held in his hand was knocked out, and the axe fell from his hand; but the grasp of Mr. Beetle upon his throat, did not prevent him from making Payne understand that his weapon was lost, who felt about until he found it, and having given it to Comstock, he managed to strike him a blow upon the head, which fractured his skull; when he fell into the pantry where he lay groaning until despatched by Comstock! The steward held a light at this time, while Oliver put in a blow as often as possible!

The second and third mates, fastened in their state rooms, lay in their births listening, fearing to speak, and being ignorant of the numerical strength of the mutineers, and unarmed, thought it best to wait the dreadful issue, hoping that their lives might yet be spared.

Comstock leaving a watch at the second mate's door, went upon deck to light another lamp at the binnacle, it having been again accidentally extinguished. He was there asked by his terrified brother, whose agony of mind we will not attempt to portray, if he intended to hurt Smith, the other boat-steerer. He replied that he did; and inquired where he was. George fearing that Smith would be immediately pursued, said he had not seen him.—Comstock then perceiving his brother to be shedding tears, asked sternly, "What are you crying about?" "I am afraid," replied George, "that they will hurt me!" "I will hurt you," said he, "if you talk in that manner!"

But the work of death was not yet finished. Comstock, took his light into the cabin, and made preparations for attacking the second and third mates, Mr. Fisher, and Mr. Lumbert. After loading two muskets, he fired one through the door, in the direction as near as he could judge of the officers, and then inquired if either was shot! Fisher replied, "yes, I am shot in the mouth!" Previous to his shooting Fisher, Lumbert asked if he was going to kill him? To which he answered with apparent unconcern, "Oh no, I guess not."

They now opened the door, and Comstock making a pass at Mr. Lumbert, missed him, and fell into the state room. Mr. Lumbert collared him, but he escaped from his hands. Mr. Fisher had got the gun, and actually presented the bayonet to the monster's heart! But Comstock assuring him that his life should be spared if he gave it up, he did so;

when Comstock immediately ran Mr. Lumbert through the body several times!!

He then turned to Mr. Fisher, and told him there was no hope for him!!—"You have got to die," said he, "remember the scrape you got me into, when in company with the *Enterprise* of Nantucket." The "scrape" alluded to, was as follows. Comstock came up to Mr. Fisher to wrestle with him.—Fisher being the most athletick of the two, handled him with so much ease, that Comstock in a fit of passion struck him. At this Fisher seized him, and laid him upon deck several times in a pretty rough manner.

Comstock then made some violent threats, which Fisher paid no attention to, but which now fell upon his soul with all the horrors of reality. Finding his cruel enemy deaf to his remonstrances, and entreaties, he said, "If there is no hope, I will at least die like a man!" and having by order of Comstock, turned back too, said in a firm voice, "I am ready!!"

Comstock then put the muzzle of the gun to his head, and fired, which instantly put an end to his existence!—Mr. Lumbert, during this time, was begging for life, although no doubt mortally wounded. Comstock, turned to him and said, "I am a bloody man! I have a bloody hand and will be avenged!" and again run him through the body with a bayonet! He then begged for a little water; "I'll give you water," said he, and once more plunging the weapon in his body, left him for dead!

Thus it appears that this more than demon, murdered with his own hand, the whole! Gladly would we wash from "memory's waste" all remembrance of that bloody night. The compassionate reader, however, whose heart sickens within him, at the perusal, as does ours at the recital, of this tale of woe, will not, we hope, disapprove our publishing these melancholy facts to the world. As, through the boundless mercy of Providence, we have been restored, to the bosom of our families and homes, we deemed it a duty we owe to the world, to record our "unvarnished tale."

Smith, the other boat-steerer, who had been marked as one of the victims, on hearing the noise in the cabin, went aft, apprehending an altercation between the Captain and some of the other officers, little dreaming that innocent blood was flowing in torrents. But what was his astonishment, when he beheld Comstock, brandishing the boarding knife, and heard him exclaim, "I am the bloody man, and will have revenge!" Horror struck, he hurried forward, and asked the crew in the forecastle, what he should do.

Some urged him to secrete himself in the hold, others to go aloft until Comstock's rage should be abated; but alas! the reflection that the ship afforded no secure hiding place, determined him to confront the ringleader, and if he could not save his life by fair means, to sell it dearly! He was soon called for by Comstock, who upon meeting him, threw his bloody arms around his neck, and embracing him, said, "you are going to be with us, are you not?" The reader will discover the good policy of Smith when he unhesitatingly answered, "Oh, yes, I will do any thing you require."

All hands were now called to make sail, and a light at the same time was set as a signal for the *Lyra* to tack;—while the *Globe* was kept upon the same tack, which very soon caused a separation of the two ships. All the reefs were turned out, top-gallant-sails set, and all sail made on the ship, the wind being quite light.

The mutineers then threw the body of the Captain overboard, after wantonly piercing his bowels with a boarding knife, which was driven with an axe, until the point protruded from his throat!! In Mr. Beetle, the mate, the lamp of life had not entirely gone out, but he was committed to the deep.

Orders were next given to have the bodies of Mr. Fisher, and Mr. Lumbert brought up. A rope was fastened to Fisher's neck, by which he was hauled upon deck. A rope was made fast to Mr. Lumbert's feet, and in this way was he got upon deck, but when in the act of being thrown from the ship, he caught the plank-shear; and appealed to Comstock, reminding him of his promise to save him, but in vain; for the monster forced him from his hold, and he fell into the sea! As he appeared to be yet capable of swimming, a boat was ordered to be lowered, to pursue and finish him, fearing he might be picked up by the *Lyra*; which order was as soon countermanded as given, fearing, no doubt, a desertion of his murderous companions.

We will now present the reader, with a journal of our passage to the Mulgrave Islands, for which group we shaped our course.

1824, Jan. 26th. At 2 A. M. from being nearly calm a light breeze sprung up, which increased to a fresh breeze by 4 A. M. This day cleaned out the cabin, which was a scene of blood and destruction of which the recollection at this day chills the blood in our veins.—Every thing bearing marks of the murder, was brought on deck and washed.

Lat. 5° 50' N. Long. 159° 13' W.

Jan. 27th. These twenty-four hours commenced with moderate breezes from the eastward. Middle and latter part calm. Employed in cleaning the small arms which were fifteen in number, and making cartridge boxes.

Lat. 3° 45' N. Long. 160° 45' W.

Jan. 28. This day experienced fine weather, and light breezes from N. by W. The black steward was hung for the following crime.

George Comstock who was appointed steward after the mutiny, and business calling him into the cabin, he saw the former steward, now called the purser, engaged in loading a pistol. He asked him what he was doing that for. His reply was, "I have heard something very strange, and I'm going to be ready for it." This information was immediately carried to Comstock, who called to Payne, now mate, and bid him follow him.

On entering the cabin they saw Humphreys, still standing with the pistol in his hand. On being demanded what he was going to do with it, he said he had heard something which made him afraid of his life!

Comstock told him if he had heard any thing, that he ought to have come to him, and let him know, before he began loading pistols. He then demanded to know, what he had heard. Humphreys answered at first in a very suspicious and ambiguous manner, but at length said, that Gilbert Smith, the boat-steerer who was saved, and Peter Kidder, were going to re-take the ship. This appeared highly improbable, but they were summoned to attend a council at which Comstock presided, and asked if they had entertained any such intentions.

They positively denied ever having had conversation upon the subject. All this took place in the evening. The next morning the parties were summoned, and a jury of two men called. Humphreys under a guard of six men, armed with muskets, was arraigned, and Smith and Kidder, seated upon a chest near him. The prisoner was asked a few questions touching his intentions, which he answered but low and indistinctly. The trial, if it may be so called, had progressed thus far, when Comstock made a speech in the following words. "It appears that William Humphreys has been accused guilty, of a treacherous and base act, in loading a pistol for the purpose of shooting Mr. Payne and myself. Having been tried the jury will now give in their verdict, whether Guilty or Not Guilty. If guilty he shall be hanged to a studding-sail boom, rigged out eight feet upon the fore-yard, but if found not guilty, Smith and Kidder, shall be hung upon the aforementioned gallows!"

But the doom of Humphreys had been sealed the night before, and kept secret except from the jury, who returned a verdict of Guilty.— Preparations were immediately made for his execution! His watch was taken from him, and he was then taken forward and seated upon the rail, with a cap drawn over his face, and the rope placed round his neck.

Every man was ordered to take hold of the execution rope, to be ready to run him up when Comstock should give the signal, by ringing the ship's bell!

He was now asked if he had any thing to say, as he had but fourteen seconds to live! He began by saying, "little did I think I was born to come to this———;" the bell struck! and he was immediately swung to the yard-arm! He died without a struggle; and after he had hung a few minutes, the rope was cut, to let him fall overboard, but getting entangled aloft, the body was towed some distance along side, when a runner hook, was attached to it, to sink it, when the rope was again cut and the body disappeared. His chest was now overhauled, and sixteen dollars in specie found, which he had taken from the Captain's trunk. Thus ended the life of one of the mutineers, while the blood of innocent victims was scarcely washed from his hands, much less the guilty stain from his soul.

Feb. 7th. These twenty-four hours commenced with thick squally weather. Middle part clear and fine weather.—Hove to at 2 A.M., and at 6 made sail, and steered W. by S. At ½ past 8 made an Island ahead, one of the Kingsmill group. Stood in with the land and received a number of canoes along side, the natives in them however having nothing to sell us but a few beads of their own manufacture. We saw some cocoanut, and other trees upon the shore, and discovered many of the natives upon the beach, and some dogs. The principal food of these Islanders is, a kind of bread fruit, which they pound very fine and mix it with fish.

Feb. 8. Commences squally with fresh breezes from the northward.— Took a departure from Kingsmill Island; one of the groupe of that name, in Lat. 1° 27' N. and Long. 175° 14' E. In the morning passed through the channel between Marshall's and Gilbert's Islands; luffed to and despatched a boat to Marshall's Island, but did not land, as the natives appeared hostile, and those who swam off to the boat, endeavoured to steal from her. When about to leave, a volley of musketry was discharged at them, which probably killed or wounded some of them. The boat then gave chase to a canoe, paddled by two of the natives, which were fired upon when within gunshot, when they immediately ceased paddling; and on the boat approaching them, discovered that one of the natives was wounded. In the most supplicating manner they held up a jacket, man-ufactured from a kind of flag, and some beads, being all they possessed, giving their inhuman pursuers to understand, that all should be theirs if they would spare their lives! The wounded native laid down in the bottom of the boat, and from his convulsed frame and trembling lip, no doubt remained but that the wound was mortal. The boat then returned on board and we made sail for the Mulgrave Islands. Here was another sac-rifice; an innocent child of nature shot down, merely to gratify the most wanton and unprovoked cruelty, which could possibly possess the heart of man. The unpolished savage, a stranger to the more tender sympathies of the human heart, which are cultivated and enjoyed by civilized nations, nurtures in his bosom a flame of revenge, which only the blood of those who have injured him, can damp; and when years have rolled away, this act of cruelty will be remembered by these Islanders, and made the pre-text to slaughter every white man who may fall into their hands.

Feb. 11th. Commenced with strong breezes from the Northward. At ½ past meridian made the land bearing E. N. E. four leagues distant. Stood in and received a number of canoes along side. Sent a boat on shore; and brought off a number of women, a large quantity of cocoanuts, and some fish.—Stood off shore most of the night, and

Feb. 12th, in the morning stood in shore again and landed the women.—We then stood along shore looking out for an anchorage, and reconnoitering the country, in the hope of finding some spot suitable for cultivation; but in this we were disappointed, or more properly speaking, they, the mutineers; for we had no will of our own, while our bosoms were torn with the most conflicting passions, in which Hope and Despair alternately gained the ascendency.

Feb. 13th. After having stood off all night, we in the morning stood in, and after coasting the shores of several small Islands, we came to one, low and narrow, where it was determined the Ship should be anchored. When nearly ready to let go, a man was sent into the chains to sound, who pronounced twelve fathoms; but at the next cast, could not get bottom. We continued to stand in, until we got regular sounding, and anchored within five rods of the shore, on a coral rock bottom, in seven fathoms water. The ship was then moored with a kedge astern, sails furled, and all hands retired to rest, except an anchor watch.

Feb. 14th, was spent in looking for a landing place. In the morning a boat was sent to the Eastward, but returned with the information that no good landing place could be found, the shore being very rocky. At 2 P. M. she was sent in an opposite direction, but returned at night without having met with better success; when it was determined to land at the place where we lay; notwithstanding it was very rocky.—Nothing of consequence was done, until

Sunday, 15th Feb. 1824, when all hands were set to work to construct a raft out of the spare spars, upon which to convey the provisions, &c. on shore.

The laws by which we were now governed had been made by Comstock, soon after the mutiny, and read as follows:

"That if any one saw a sail and did not report it immediately, he should be put to death! If any one refused to fight a ship he should be put to death; and the manner of their death, this—They shall be bound hand and foot and boiled in the try pots, of boiling oil!" Every man was made to seal and sign this instrument, the seals of the mutineers being black, and the remainder, blue and white. The raft or stage being completed, it was anchored, so that one end rested upon the rocks, the other being kept sea-ward by the anchor.

During the first day many articles were brought from the ship in boats, to the raft, and from thence conveyed on shore. Another raft, however, was made, by laying spars upon two boats, and boards again upon them, which at high water would float well up on the shore. The following, as near as can be recollected, were the articles landed from the ship; (and the intention was, when all should have been got on shore, to haul the ship on shore, or as near it as possible and burn her.) One mainsail, one foresail, one mizen-topsail, one spanker, one driver, one maintop gallantsail, two lower studdingsails, two royals, two topmast-studdingsails, two top-gallant-studdingsails, one mizen-staysail, two mizen-top-gallantsails, one fly-gib, (thrown overboard, being a little torn,) three boat's sails (new,) three or four casks of bread, eight or ten barrels of flour, forty barrels of beef and pork, three or more 60 gal. casks of molasses, one and a half barrels of sugar, one barrel dried apples, one cask vinegar, two casks of rum, one or two barrels domestic coffee, one keg W. I. coffee, one and a half chests of tea, one barrel of pickles, one do. cranberries, one box chocolate, one cask of tow-lines, three or more coils of cordage, one coil rattling, one do. lance warp, ten or fifteen balls spunyarn, one do. worming, one stream cable, one larboard bower anchor, all the spare spars, every chest of clothing, most of the ship's tools, &c. &c. The ship by this time was considerably unrigged.

On the following day, Monday 16th February, Payne the second in the mutiny, who was on board the ship attending to the discharge of articles from her, sent word to Comstock, who with Gilbert Smith and a number of the crew were on shore, attending to the landing of the raft; "That if he did not act differently with regard to the plunder, such as making presents to the natives of the officers' fine clothing, &c. he would do no more, but quit the ship and come on shore."

Comstock had been very liberal to the natives in this way, and his object was, no doubt, to attach them as much as possible to his person, as it must have been suggested to his guilty mind, that however he himself might have become a misanthrope, yet there were those around him, whose souls shuddered at the idea of being forever exiled from their country and friends, whose hands were yet unstained by blood, but who might yet imbrue them, for the purpose of escape from lonely exile, and cruel tyranny.

When the foregoing message was received from Payne, Comstock commanded his presence immediately on shore, and interrogated him, as to what he meant by sending such a message. After considerable altercation, which took place in the tent, Comstock was heard to say, "I helped to take the ship, and have navigated her to this place.—I have also done all I could to get the sails and rigging on shore, and now you may do what you please with her; but if any man wants any thing of me, I'll take a musket with him!"

"That is what I want," replied Payne, "and am ready!" This was a check upon the murderer, who had now the offer of becoming a duellist; and he only answered by saying, "I will go on board once more, and then you may do as you please."

He then went on board, and after destroying the paper upon which were recorded the "Laws," returned, went into the tent with Payne, and putting a sword into a scabbard, exclaimed, "this shall stand by me as long as I live."

We ought not to omit to mention that during the time he was on board the ship, he challenged the persons there, to fight him, and as he was leaving, exclaimed "I am going to leave you; Look out for yourselves!"

After obtaining from Payne permission to carry with him a cutlass, a knife, and some hooks and lines, he took his departure, and as was afterwards ascertained, immediately joined a gang of natives, and endeavoured to excite them to slay Payne and his companions! At dusk of this day he passed the tent, accompanied by about 50 of the natives, in a direction of their village, upwards of a league distant. Payne came on board, and after expressing apprehensions that Comstock would persuade the natives to kill us all, picked out a number of the crew to go on shore for the night, and stationed sentinels around the tent, with orders to shoot any one, who should attempt to approach without giving the countersign. The night, however, passed, without any one's appearing; but early on the morning of the

17th Feb.; Comstock was discovered at some distance coming towards the tent. It had been before proposed to Smith by Payne, to shoot him; but poor Smith like ourselves, dare do no other than remain upon the side of neutrality.

Oliver, whom the reader will recollect as one of the wretches concerned in the mutiny, hurried on shore, and with Payne and others, made preparations to put him to death. After loading a number of muskets they stationed themselves in front of the tent, and waited his approach—a bushy spot of ground intervening, he did not make his appearance until within a short distance of the tent, which, as soon as he saw, drew his sword and walked quick towards it, in a menacing manner; but as soon as he saw a number of the muskets levelled at him, he waved his hand, and cried out, "don't shoot me, don't shoot me! I will not hurt you!" At this moment they fired, and he fell!—Payne fearing he might pretend to be shot, ran to him with an axe, and nearly severed his head from his body! There were four muskets fired at him, but only two balls took effect, one entered his right breast, and passed out near the back bone, the other through his head.

Thus ended the life, of perhaps as cruel, blood-thirsty, and vindictive a being as ever bore the form of humanity.

All hands were now called to attend his burial, which was conducted in the same inconsistent manner which had marked the proceedings of the actors in this tragedy. While some were engaged in sewing the body in a piece of canvas, others were employed in digging a grave in the sand, adjacent to the place of his decease, which, by order of Payne, was made five feet deep. Every article attached to him, including his cutlass, was buried with him, except his watch; and the ceremonies consisted in reading a chapter from the bible over him, and firing a musket!

Only twenty-two days had elapsed after the perpetration of the massacre on board the ship, when with all his sins upon his head, he was hurried into eternity!

No duty was done during the remainder of the day, except the selection by Payne, of six men, to go on board the ship and take charge of her, under the command of Smith; who had communicated his intentions to a number of running away with the ship. We think we cannot do better than to give an account of their escape in the words of Smith himself. It may be well to remark, that Payne had ordered the two binacle compasses to be brought on shore, they being the only ones remaining on board, except a hanging compass suspended in the cabin. Secreting one of the binacle compasses, he took the hanging compass on shore, and the exchange was not discovered.

"At 7 P. M. we began to make preparations for our escape with the ship.—I went below to prepare some weapons for our defence should we be attacked by Payne, while the others, as silently as possible, were employed in clearing the running rigging, for every thing was in the utmost confusion. Having found one musket, three bayonets, and some whale lances, they were laid handy, to prevent the ship being boarded. A handsaw well greased was laid upon the windlass to saw off the cable, and the only remaining hatchet on board, was placed by the mizen mast, to cut the stern moorings when the ship should have sufficiently swung off. Taking one man with me, we went upon the fore-top-sail-yard, loosed the sail and turned out the reefs, while two others were loosing the main-top-sail and main sail. I will not insult the reader's good sense, by assuring him, that this was a duty, upon the success of which seemed to hang our very existence. By this time the moon was rising, which rendered it dangerous to delay, for those who had formed a resolution to swim on board, and accompany us. The bunts of the sails being yet confined aloft, by their respective gaskets, I sent a man on the fore-yard and another upon the fore-top-sail-yard, with orders to let fall, when I should give the word; one man being at the helm, and two others at the fore tack.

"It was now half past nine o'clock, when I took the handsaw, and in less than two minutes the cable was off!—The ship payed off very quick, and when her head was off the land, there being a breeze from that quarter, the hawser was cut and all the sail we could make upon the ship immediately set, a fine fair wind blowing. A raft of iron hoops, which was towing along side, was cut adrift, and we congratulated each other upon our fortunate escape; for even with a vast extent of ocean to traverse, hope excited in our bosoms a belief that we should again embrace our friends, and our joy was heightened by the reflection, that we might be the means of rescuing the innocents left behind, and having the guilty punished."

After a long and boisterous passage the ship arrived at Valparaiso, when she was taken possession of by the American Consul, Michael Hogan, Esq. and the persons on board were put in irons on board a French frigate, there being no American man-of-war in port. Their names were, Gilbert Smith, George Comstock, Stephen Kidder, Joseph Thomas, Peter C. Kidder, and Anthony Henson.

SIX

The Savage Sea

Dougal Robertson

THE BLACK VOLCANIC MOUNTAIN OF FERNANDINA, THE MOST WESTERLY of the Galapagos Islands, towered high above the tall masts of the schooner *Lucette* as she lay at anchor, rolling gently in the remnants of the long Pacific swell which surged round the rocky headland of Cape Espinosa, and sent searching fingers of white surf curling into the sheltered waters of the anchorage.

We were on the eve of our departure for the Marquesas Islands, three thousand miles to the west, and now, as the wind swung to the east under a grey mantle of rain cloud, I felt anxious to be gone, for if we left now we would be out from under the lee of the island by morning. Lyn protested vehemently at the thought of starting our journey on June the thirteenth, even when I pointed out that the most superstitious of seafarers didn't mind so long as it wasn't a Friday as well, but Douglas and Robin both now joined with my feelings of anxiety to be gone, and after a short spell of intense activity, we stowed and lashed the dinghy and secured all movables on deck and below.

By five o'clock in the afternoon we were ready for sea, and with mainsail and jibs set we heaved the anchor home, reached past the headland into the strait, then altering course to the west ran free towards the Pacific, a thousand square feet of sail billowing above *Lucette* as she moved easily along the ragged black coastline of Fernandina towards the largest stretch of ocean in the world.

By the morning of the fourteenth, the Galapagos Islands were receding into the distance astern, merging with the clouds of the overcast sky above as *Lucette*, now rolling and pitching in the heavy swell and rough seas of the Pacific trades, made steady progress west by south towards the Marquesas Islands.

In spite of the fact that we had been sailing for over a year, our stomachs still took a little time to adjust from the quietness of sheltered waters to the lively movement of the yacht in the open sea and so throughout the day those of us not actively engaged in steering and sailing *Lucette* rested as best we could in the bunks below, supplied at intervals with hot soup or coffee from Lyn's indomitable labours at the stove. Unused to the sea, Robin had been sick most of the way from Panama to the Galapagos, but he now seemed better adjusted to the physical discomfort of the constant heave of the hull. He was able to steer a fairly accurate course by compass, and although the principles of sailing were still something of a closed book to him, he could help Douglas and me with the night watches whilst Lyn and the twins helped with the watches during the day.

The wind moderated a little during the following night and breaks in the cloud enabled us to catch glimpses of stars in the predawn sky; on the morning of the fifteenth we had our first glimpse of the sun since leaving the Galapagos and with the slackening of wind and speed *Lucette* settled to a more comfortable movement in the diminishing seas.

The morning sun shone fitfully from the thinning cloud, and as I balanced myself against the surge of *Lucette*'s deck, sextant glued to my eye, I watched for the right moment when the image of the sun's rim would tip the true horizon, no easy combination when both deck and horizon are in constant motion. Douglas and Sandy were in the cockpit, one steering and the other tending the fishing line, while Robin, finding it difficult to sleep in his own bunk on the port side of the main cabin, had nipped quietly into Sandy's bunk on the starboard side of the fo'c'stle to rest after his spell on the four to eight morning watch. Neil was reading a book in his own bunk on the port side of the fo'c'stle, and Lyn had just started to clean up the usual chaos which results from a rough stretch of sailing. At last the sun, the horizon and the deck cooperated to give me a fairly accurate reading, and noting the local time by my watch at 09h 54m 45s, I collected my logarithm tables and Nautical Almanac from the chart table and retired below to the relative comfort of the after cabin to work out our longitude; it was my first position sight since leaving the islands.

With my sextant carefully replaced in its box I had turned to my books to work up a reasonably accurate dead-reckoning position when sledgehammer blows of incredible force struck the hull beneath my feet, hurling me against the bunk, the noise of the impact almost deafening my ears to the roar of inrushing water. I heard Lyn call out, and almost at the same time heard the cry of 'Whales!' from the cockpit. My senses still reeled as I dropped to my knees and tore up the floorboards to gaze in horror at the blue Pacific through the large splintered hole punched up through the hull planking between two of the grown oak frames. Water was pouring up through the hole with torrential force and although Lyn called out that it was no use, that the water was pouring in from another hole under the WC flooring as well, I jammed my foot on the broken strakes and shouted to her to give me large cloths, anything to stem the flood. She threw me a pillow and I jammed it down on top of the broken planking, rammed the floorboard on top and stood on it; the roar of the incoming water scarcely diminished, it was already above the level of the floorboards as I heard Douglas cry from the deck 'Are we sinking, Dad?' 'Yes! Abandon ship!'; my voice felt remote as numbly I watched the water rise rapidly up the engine casing; it was lapping my knees as I turned to follow Lyn, already urging Neil and Robin on deck. Wading past the galley stove, my eye glimpsed the sharp vegetable knife, and grabbing it in passing I leapt for the companionway; the water, now up to my thighs, was already lapping the top of the batteries in the engine room; it was my last glimpse of *Lucette*'s interior, our home for nearly eighteen months. Lyn was tying the twins' lifejackets on with rapid efficiency as I slashed at the lashings holding the bow of the dinghy to the mainmast; Douglas struggled to free the self-inflatable raft from under the dinghy and I ran forward to cut the remaining lashings holding the stern of the dinghy to the foremast, lifting the dinghy and freeing the raft at the same time. Lyn shouted for the knife to free the water containers and I threw it towards her; Douglas again shouted to me if he should throw the raft over, disbelieving that we were really sinking. 'Yes, get on with it!' I yelled, indicating to Robin, who now had his lifejacket on, to help him. Grasping the handles at the stern of the dinghy, I twisted it over from its inverted stowed position and slid it towards the rail, noting that the water was now nearly level with *Lucette*'s deck as she wallowed sluggishly in the seaway.

Douglas ran from the after deck with the oars and thrust them under the thwarts as I slid the dinghy seawards across the coach roof, then he took hold of the stern from me and slid the dinghy the rest of the way into the sea, Robin holding on to the painter to keep it from floating away. The raft, to our relief, our great and lasting relief, had gone off with a bang and was already half inflated, and Lyn, having severed the lashings on the water containers and flares, was carrying them to the dinghy. I caught up the knife and again shouted 'Abandon ship!' for I feared *Lucette*'s rigging might catch one of us as she went down, then cut the lashings on a bag of onions, which I gave to Sandy, instructing him to make for the raft, a bag of oranges which I threw into the dinghy and a small bag of lemons to follow. It was now too dangerous to stay aboard, and noting that Douglas, Robin and Sandy had already gone and that Neil was still sitting in the dinghy which was three-quarters full of water, I shouted that he also should make for the raft. He jumped back on *Lucette*, clutching his teddy bears, then plunged into the sea, swimming strongly for the raft. Lyn struggled through the rails into the water, still without a lifejacket, and I walked into the sea, first throwing the knife into the dinghy, the waters closing over *Lucette*'s scuppers as we left her.

I feared that the whales would now attack us and urged everyone into the raft, which was fully inflated and exhausting surplus gas noisily. After helping Lyn into the raft I swam back to the dinghy, now completely swamped, with oranges floating around it from the bag which had burst, and standing inside it to protect myself from attack, threw all the oranges and lemons within reach into the raft. The water containers had already floated away or had sunk, as had the box of flares, and since the dinghy was now three feet under the water, having only enough flotation to support itself, I made my way back to the raft again, grabbing a floating tin of petrol as I went. On leaving the dinghy I caught a last glimpse of

Lucette, the water level with her spreaders and only the tops of her sails showing. Slowly she curtsied below the waves, a lady to the last; she was gone when I looked again.

I climbed wearily into the yellow inflatable, a sense of unreality flooding through me, feeling sure that soon I would waken and find the dream gone. I looked at my watch; it was one minute to ten. 'Killer whales,' said Douglas. 'All sizes, about twenty of them. Sandy saw one with a big V in its head. I think three of them hit us at once.' My mind refused to take in the implications of the attack; I gazed at the huge genoa sail lying on the raft floor where Lyn was sitting with the twins. 'How the hell did that get there?' I asked stupidly. Douglas grinned. 'I saw the fishing line spool floating on the surface unwinding itself,' he said, 'so I grabbed it and pulled it in, the sail was hooked in the other end!'

Three killer whales; I remembered the one in captivity in Miami Seaquarium weighed three tons and that they swam at about thirty knots into an attack; no wonder the holes in *Lucette*! The others had probably eaten the injured one with the V in its head, which must have split its skull when it hit *Lucette*'s three-ton lead keel. She had served us well to the very end, and now she was gone.

Lyn gazed numbly at me, quietly reassuring the twins who had started crying, and, apart from the noise of the sea round us, we gazed in silent disbelief at our strange surroundings.

CASTAWAYS

We sat on the salvaged pieces of flotsam lying on the raft floor, our faces a pale bilious colour under the bright yellow canopy, and stared at each other, the shock of the last few minutes gradually seeping through to our consciousness. Neil, his teddy bears gone, sobbed in accompaniment to Sandy's hiccup cry, while Lyn repeated the Lord's Prayer, then, comforting them, sang the hymn 'For those in peril on the Sea.' Douglas and Robin watched at the doors of the canopy to retrieve any useful pieces of debris which might float within reach and gazed with dumb longing at the distant five-gallon water container, bobbing its polystyrene lightness ever farther away from us in the steady trade wind.

The dinghy *Ednamair* wallowed, swamped, nearby with a line attached to it from the raft and our eyes travelled over and beyond to the heaving undulations of the horizon, already searching for a rescue ship even while knowing there would not be one. Our eyes travelled fruitlessly across the limitless waste of sea and sky, then once more ranged over the scattering debris. Of the killer whales which had so recently shattered our very existence, there was no sign. Lyn's sewing basket floated close and it was brought aboard followed by a couple of empty boxes, the canvas raft cover, and a plastic cup. I leaned across to Neil and put my arm round him, 'It's alright now, son, we're safe and the whales have gone.' He looked at me reproachfully. 'We're not crying 'cos we're frightened,' he sobbed, 'we're crying 'cos Lucy's gone.' Lyn gazed at me over their heads, her eyes filling with tears. 'Me too,' she said, and after a moment added, 'I suppose we'd better find out how we stand.'

I looked at Douglas, he had grown to manhood in our eighteen months at sea together; the twins, previously shy, introspective farm lads, had become interested in the different peoples we had met and their various ways of life, and were now keen to learn more; I tried to ease my conscience with the thought that they had derived much benefit from their voyage and that our sinking was as unforeseeable as an earthquake, or an aeroplane crash, or anything to ease my conscience.

We cleared a space on the floor and opened the survival kit, which was part of the raft's equipment, and was contained in a three-foot-long polythene cylinder; slowly we took stock: Vitamin-fortified bread and glucose for ten men for two days. Eighteen pints of water, eight flares. One bailer, two large fishhooks, two small, one spinner and trace and a twenty-five-pound breaking strain fishing line. A patent knife which would not puncture the raft (or anything else for that matter), a signal mirror, torch [flashlight], first-aid box, two sea anchors, instruction book, bellows, and three paddles.

In addition to this there was the bag of a dozen onions which I had given to Sandy, to which Lyn had added a one-pound tin of biscuits and a bottle containing about half a pound of glucose sweets, ten oranges and six lemons. How long would this have to last us? As I looked round our meagre stores my heart sank and it must have shown on my face for Lyn put her hand on mine; 'We must get these boys to land,' she said quietly. 'If we do nothing else with our lives, we must get them to land!' I looked at her and nodded, 'Of course, we'll make it!' The answer came from my heart but my head was telling me a different story.

We were over two hundred miles down wind and current from the Galapagos Islands. To try to row the small dinghy into two hundred miles of rough ocean weather was an impossible journey even if it was tried by only two of us in an attempt to seek help for the others left behind in the raft. The fact that the current was against us as well only put the seal of hopelessness on the idea. There was no way back.

The Marquesas Islands lay two thousand eight hundred miles to the west but we had no compass or means of finding our position; if, by some miraculous feat of endurance, one of us made the distance the chances of striking an island were remote. The coast of Central America, more than a thousand miles to the northeast, lay on the other side of the windless Doldrums, that dread area of calms and squalls which had inspired Coleridge's

> Water, water, everywhere,
> And all the boards did shrink;
> Water, water, everywhere,
> Nor any drop to drink.

I was a Master Mariner, I thought ruefully, not an ancient one, and could count on no ghostly crew to get me out of this dilemma!

What were our chances if we followed the textbook answer, 'Stay put and wait for rescue'? In the first place we wouldn't be missed for at least five weeks and if a search was made, where would they start looking in three thousand miles of ocean? In the second place the chance of seeing a passing vessel in this area was extremely remote and could be discounted completely, for of the two possible shipping routes from Panama to Tahiti and New Zealand, one lay four hundred miles to the south and the other three hundred miles to the north.

Looking at the food, I estimated that six of us might live for ten days and since we could expect no rain in this area for at least six months, apart from an odd shower, our chances of survival beyond ten days were doubtful indeed.

My struggle to reach a decision, gloomy whichever way I looked at it, showed on my face, and Lyn leaned forward. 'Tell us how we stand,' she said, looking round, 'we want to know the truth.' They all nodded, 'What chance have we?' I could not tell them I thought they were going to die so I slowly spelled out the alternatives, and then suddenly I knew there was only one course open to us; we must sail with the trade winds to the Doldrums four hundred miles to the north. We stood a thin chance of reaching land but the only possible shipping route lay in that direction, our only possible chance of rainwater in any quantity lay in that direction even if it was four hundred miles away, and our only possible chance of reaching land lay in that direction, however small that chance might be.

We would work and fight for our lives at least; better than dying in idleness! 'We must get these boys to land,' Lyn had said. I felt the reality of the decision lifting the hopelessness from my shoulders and looked around; five pairs of eyes watched me as I spoke, Lyn once again with her arms round the twins, Douglas and Robin each at their lookout posts watching for any useful debris that might come within reach. 'We have no alternative,' I said, 'we'll stay here for twenty-four hours to see if any other wreckage appears, then we must head north and hope to find rain in the Doldrums.'

* * *

I peeped round the canopy of the raft at the dinghy; the *Ednamair* lay disconsolately awash at the end of her painter, her white gunwale just visible above the surface of the water. She was helping the sea anchor, I supposed, but we'd have to bail her out first thing in the morning, for the wooden thwarts, which contained the polystyrene flotation reserve, would loosen and come adrift if they became waterlogged.

The water exploded as a thirty-pound dorado leapt high in the air after a flying fish, landing with a slap on its side in a shower of luminescence. I glanced down to where several large fish swam under the raft, constantly rising to skim the underside of the raft's edge, sometimes hitting it a heavy blow with their high jutting foreheads.

I looked across at Lyn, rubbing the cramp out of the twins' legs. 'We'll see to the *Ednamair* after breakfast'; I looked hopefully at the water jar, but it was nearly empty. We had emptied the glucose sweets out of their glass jar so that it could be used to hold drinking water as it was decanted from the tin, for although we had discussed the issue of equal rations of water (there wasn't enough to do that) we had decided simply to pass the jar round, each person limiting him or herself to the minimum needed to carry on; at the same time, the visible water level in the jar enabled everyone to see there was no cheating. Breakfast consisted of one quarter ounce biscuit, a piece of onion and a sip of water, except for Robin and Neil who could not eat and were with difficulty persuaded to take some extra water with a seasick pill. We had used two pints of water in one day between six, hardly a maintenance ration under a tropic sun, which I remembered had been placed as high as two pints per person per day! We ate slowly, savouring each taste of onion and biscuit with a new appreciation and, although we hardly felt as if we had breakfasted on bacon and eggs, we were still sufficiently shocked at our altered circumstances not to feel hunger.

Breakfast over, Lyn, with Sandy helping, sorted out the various pieces of sail which were to be used for bedding, chatting quietly all the while to Neil and Robin. Douglas and I went to the door of the raft and, pulling the dinghy alongside, first attempted to bail it out as it lay swamped, but the waves filled it as fast as we bailed. We turned its stern towards us and, lifting slowly, allowed the bow to submerge, then when we could lift it no higher, I called 'Let go!' The dinghy flopped back in the water with three inches of freeboard, we bailed desperately with small bailers, then Douglas took one of the wooden boxes and with massive scoops bailed enough water out to allow him to board the dinghy and bail it dry. We were all cheered by the sight of little *Ednamair* afloat again, and with a cry of delight Douglas held up his Timex watch; it had been lying in the bottom of the dinghy all this time and was still going! He also found what was to prove our most valuable possession, the stainless steel kitchen knife which I had thrown in after the fruit.

After a segment of orange each for elevenses we loaded the oars, a paddle, the empty boxes, the petrol can, the hundred-foot raft painter, and the piece of the genoa designated for the dinghy sail, then climbing into the dinghy started work on the jury rig that was to turn the *Edna-mair* into a tugboat for our first stage of the journey north. Douglas, in the meantime, helped Lyn to reorganise the inside of the raft now that there was much more room, and topped up the flotation chambers with air.

I rigged one oar in the mast step with appropriate fore and back stays, then cutting notches in the raft paddle, bent the head of the sail onto it to form a square sail. The paddle was made fast to the top of the oar, and the sail foot secured to the two ends of the other oar, placed athwartships across the rowlock sockets. A violent jerk sent me sprawling into the bottom of the boat and I realised that we were operational.

I climbed back aboard the raft for a lunch of a small piece of fortified bread, of which there was about a pound and a half in the emergency rations, along with eight ounces of glucose and a mouthful of water; I felt very thirsty after my exertions in the raft. *Ednamair* was now straining at the leash so I called to Douglas to trip the sea anchor and haul it aboard; the time was two o'clock in the afternoon and we had started our voyage to the Doldrums, and, I shuddered at the thought of the alternative, rain. I estimated our position at Latitude 1° South and Longitude 94°40' West or, more accurately, two hundred miles west of Cape Espinosa.

The white plastic-covered luff wire was now snapping taut with considerable force as *Ednamair* yawed at the end of her towrope, so having little use for the petrol I lashed the can to the centre of the towing wire to act as a tension buffer which it did quite effectively. We now turned our attention to the flotation chambers of the raft to see if we could find any leaks. The double canopy alone was worth a gallon of water a day to us in keeping out the heat of the sun, and its emergency rations were available to us now only because they were already stowed inside the raft.

We examined the raft's flotation chambers as well as we could, pouring water over all the exposed surface areas, but could find no leaks, although there were one or two repair patches, and finally put down the loss of air to seepage through the treated fabric of the raft. We arranged a regular routine of topping up on each watch to keep the raft as rigid as possible, for the continuous flexing of the softened chambers by the waves was bound to cause wear.

I lay down to think in the long hours of the night of how long it would take us to reach the Doldrums and of our chances of finding rain there; an exercise that was to occupy my nights with increasing urgency as our meagre store of water cans gradually dwindled. Robin had puffed rather ineffectually at the inflating tube before he went off watch, but the raft was still pretty soft, so I stuck the end in my mouth and gave it a good blow at both ends; Robin would get better at it as he got used to the idea.

DAY THREE

My watch, in the dawn hours of the morning, started with a clear sky, but, as the sun tinted the clouds, the wind freshened again from the south and the tall flowery cumulus, pink peaked with grey bases, seemed heavy enough to give rain. As soon it was light I pulled in the dinghy and climbed aboard to inspect the sail fastenings and stays, one of which had worked loose in the night. While I was securing the stay I caught sight of a small black shape under the wooden box by the thwart; I stooped and lifted our first contribution from the sea, a flying fish of about eight inches. I gutted and descaled it, then passed it over to Lyn, now awake, for her to marinate it in a squeeze of lemon juice, which acted as a cooking agent. We breakfasted at seven, an hour later, each savouring our tiny piece of fish done to a turn in the lemon juice, followed by a crunchy piece of onion and a mouthful of water. The raft had begun pitching heavily again, surging on the crests of the breaking sea and dropping steeply into the troughs. To our disappointment, both Neil and Robin started being seasick again and though we offered them seasick pills they decided to do without and try to get used to the motion of the raft instead.

The waves began to break over the stern of the raft, and with swells of up to twenty feet high, it looked as if we were in for a bad day. *Ednamair* yawed violently as the wind gusted in her sail and she pulled hard on the towrope, lifting it clear of the water at times. I decided to take a reef in the sail to ease the strain on the towing straps of the raft, so Douglas hauled the dinghy alongside the raft and held her while I balanced precariously on the seat. To reef her, I simply tied a rope around the belly of the sail, giving it an hourglass effect and reducing its effective pulling power by half. I had just completed the operation and was standing up again to return when a large breaker surged round the raft and caught the dinghy broadside. As she tilted, I lost balance and fell, grabbing at the mast to prevent myself falling into the sea; *Ednamair* tilted sharply with the increased leverage and the sea rushed in over the gunwale in a wave. Before I could let go the mast and drop to the floor of the dinghy, it was swamped. Luckily we retained about three inches of freeboard and before the next wave could complete the damage, I dived through the door of the canopy into the raft, and the dinghy, relieved of my weight, floated a little higher. We bailed desperately for several minutes from the raft and then, gaining on the influx of water slopping over the gunwale, we finally got enough freeboard to allow me to return to the *Ednamair* and bail it dry again. In the night, I had thought of the possibility of us taking to the dinghy altogether and leaving the raft, but this incident served to highlight the difficulty of any such move; the subject of trim with a very small freeboard would be of paramount importance and now I doubted if the dinghy could take the six of us and remain afloat in the open sea.

After our exhausting morning, we rested awhile, lunching on a mouthful of water and a few 'crumbs' of a type of fortified bread which, although made up in tablet form, disintegrated at the first touch and made the conveyance of the crumbs from container to mouth an operation that required great care to avoid spilling and usually resulted in some waste, even when we licked the stray crumbs off our clothes. This was followed by a piece of orange.

The clouds thickened as the day advanced and the high cumulus began to drop rain in isolated showers. The wind freshened still further and with the surf of breaking waves slopping through the canopy door at the rear of the raft, we closed the drawstrings on the flaps as much as was possible without cutting off all ventilation. With the large blanket pin I punched bigger holes in the empty water cans and made plugs to fit them in case a shower should cross us and give us water, while Douglas blew lustily into the pipe to make the raft as rigid as possible in the heavy seas. *Ednamair* bounced around at the end of her towrope like a pup on a leash and I was considering taking the sail down altogether when the patter of raindrops on the canopy warned us that we were about to get rain. A pipe led down from the centre of the rain catchment area on the roof and, pulling this to form a depression in the roof, we prepared to gather our first rainwater. With fascinated eyes we gazed at the mouth of the pipe, at the liquid that dribbled from the end, bright yellow, and saltier than the sea. As soon as the salt had been washed off the roof, we managed to collect half a pint of yellowish rubbery-tasting liquid before the shower passed over. I looked at the jar of fluid (one could hardly call it water) sadly; we would need to do a lot better than that if we were to survive.

The raft, now pitching heavily, required blowing up every hour to keep it rigid, and the undulations and jerks did nothing to ease the spasms of seasickness which Neil and Robin were suffering; they both looked drawn and pale, refusing even water in spite of Lyn's pleading. As the raft slid up the twenty-foot swells to the breaking combers at the top, Lyn prayed desperately for calm weather and for rain, urging that the rest of us should join her in prayer with such insistence that I had to remind her that freedom of thought and religion was a matter of individual choice and no one should be coerced.

I passed the water jar around for 'sippers' before our meagre ration of biscuit, reminding everyone that our supplies were now very low and that only minimal amounts should be taken. 'We must try to drink less than two pints per day between us,' I said. 'We have only twelve tins left and we still have over three hundred miles to go.' A quart of water each for the next three hundred miles, it didn't sound much.

As darkness closed in and the first watchkeeper settled to his two-hour vigil, I could feel the bump and bite of the dorado fish through the bottom of the raft and resolved to try to catch one in the morning. Neil and Sandy were sleeping soundly after helping to blow up the raft and mop up the water which was now coming through the floor at a greater rate than before. They looked so vulnerable that my heart turned over at the prospect of what lay ahead for them; death by thirst, or starvation, or just a slow deterioration into exhaustion. I heard Lyn's voice many times that night, in my mind: 'We must get these boys to land,' and sleep would not come to ease the burden of my conscience.

DAY TEN

As soon as daylight had faded the stars from the clearing skies, we tripped and housed the sea anchor, shook the reef out of the sail and continued on our way to the Doldrums. We had paid lip service to the standard practices of rescue by remaining in the shipping lane for as long as we could, but I felt that our present circumstances called for more than standard practice and was anxious that no more time should be wasted, for we were still some distance from the rain area and our stocks of water were dwindling once more.

As soon as we were moving again I dumped the offal and bailed the blood out of the dinghy; dozens of scavenger fish appeared from nowhere, the sea swirling as they fought to devour the scraps of coagulated turtle blood. In a few minutes, the now familiar fins of four sharks were seen as they cruised around looking for the source of the blood. The sea boiled as one of them attacked a dorado, the shark leaping its full ten-foot length clear of the water in a tremendous strike. Although they were our constant reminders of what lay in store for us if we failed, we could not help admiring the beautiful streamlined shape of these white-tipped sharks as they cruised in smooth unhurried serenity with their attendant bevies of pilot fish close to the raft. Our admiration did not deter me from thumping one of them with a paddle when it came too close (it beat a hasty retreat) and as if they had taken the hint we weren't troubled by any of the others, but from then onwards we were never without at least one shark in attendance.

At 3°30' North and 250 miles west of Cape Espinosa, our noon position confirmed that the Doldrums, a mere ninety miles now, were well within striking distance and that our first leg of the journey was nearly over. High cirrus clouds moved contrary to the trade winds, their unsubstantial vapours conveying little to the searcher for weather signs, and I turned my attention to the dinghy, scraping out the turtle shell and collecting all the pieces of bone from the flippers. The half-cured meat had turned a deep brown colour under the heat of the sun and I took a little of it back to the raft, to spin out our luncheon of flipper bones and eggs.

During the afternoon the plug in the bottom of the raft was dislodged and water flooded into the forward compartment through a now much enlarged hole. We plugged it eventually by ramming an aircraft dinghy instruction book, made of waterproof material, into the hole, a creditable use for it, and while Robin bailed the compartment dry again, I wondered how long it would be before the raft became untenable altogether and we became dependent on *Ednamair* for our lives. There was no doubt in my mind that we should have to do this eventually, but the prospect of the six of us fitting into, and living in, the confined limits of the nine-foot-six-inch boat along with our food and water supplies and other items of equipment appalled me, for the slightest imbalance would bring the sea flooding in over the small freeboard.

The life belts, which were filled with kapok, had been used as pillows, and for keeping our bodies from lying in the pools of water which collected in the raft during the night, but now they had become so saturated that I took them over to the dinghy and placed them between the thwarts to dry out. In the meanwhile, we again searched for leaks, for there was one, as yet unlocated, in the after section which was causing us much bodily discomfort. I decided we would have to rip the side screens out of the raft to find the leak which was coming from under them and set about doing this before darkness fell, using the blunt-nosed raft knife for the purpose to avoid cutting into any of the flotation chambers.

The continuous contact with the salt water had aggravated our skin eruptions and we all suffered from an increasing number of saltwater boils on our arms and legs, shoulders and buttocks; they were extremely painful when brought in contact with the terylene sail and other rough objects, and would soon present an additional health hazard unless we could keep out of the seawater and stop the eruptions spreading.

We were still examining the raft inch by inch when daylight faded and we settled down to another comfortless night, the constant plying of the bailing cup broken only when the watchkeeper stopped to blow up the flotation chambers.

DAY EIGHTEEN

A new arrival in the way of bird life came on this, our first morning in the dinghy; a blue-footed booby circled us curiously and landed in the sea not far away. It preened its feathers and surveyed us with the rather comical expression peculiar to these birds. I caught my breath, then shouted as I saw a shark nosing upwards towards the bird; the booby looked at me curiously, then sensing the presence of danger, stuck its head under the water. The shark, now only a few feet away, moved swiftly towards it, but to my surprise the booby, instead of taking off, pecked at that shark's nose three or four times, then as the shark turned away, spread its wings and flew off. The shark was young and perhaps just curious, but I wondered how the booby would have fared if it had been an older and hungrier shark.

It had been cold in the night without the shelter from the canopy and we were grateful for the warming sun. After sorting out the meat, discarding the slimy pieces (even the scavenger fish were not interested in them!), we pulled the sea anchor aboard and set the sail, sheeted to the bow. The light southerly breeze allowed us to steer northeast, using the steering oar to hold the dinghy on course; we were on our way again, and with six hundred miles to go, we were nearly halfway to the coast!

Douglas and I had changed places with Lyn and Robin, a precarious business involving much bad language on my part and fearful reaction on theirs, the tiny dinghy tipping dangerously as frantic yells of 'Trim!' rent the Pacific air. The change was necessary to allow Douglas and me to steer, for neither Lyn nor Robin could use the steering oar or find the direction in which to steer, and although Douglas could scull expertly this was the first time he had used the oar as a rudder.

As we settled down again, the dinghy only making half a knot in the slight breeze, we talked of the North Staffordshire countryside where Lyn and the children had been born, of rolling hills and valleys in the Peak district. It was at this time that we started talking of the thing that was eventually to become our main topic of conversation: a kitchen-type restaurant in the North Staffordshire town of Leek, to be called Dougal's Kitchen. It was a wonderful opportunity to talk about food.

Our estimated noon position was 5°30' North, 245 miles west of Cape Espinosa; we had made our first easting since *Lucette* had sunk and I felt that we were now far enough north to allow some set and drift for the countercurrent which runs east through the Doldrums; we really were on our way home! The sores and boils on our limbs had already begun to dry and while they were still badly inflamed and septic, the surrounding skin felt much better and there was no further extension of the infected areas. Our clothes had begun to disintegrate rapidly now, and our principal concern was to avoid being sunburned on hitherto unexposed parts of our bodies (my contortions to avoid putting pressure on my blistered posterior were sufficient warning to the others); it was the warmth these clothes afforded us at night that concerned us, far more than any moral aspect. Indeed our absence of clothing was never discussed in terms of morality and while the capes that had been cut from the doors of the raft saved us many a night of misery by containing a little of our body warmth, we never wore them during the day unless it rained, our singlets or shirts affording adequate cover from the sun while we exposed the various parts of our distressed anatomy to the dry fresh air. We steered a steady northeasterly course all day and then towards evening the wind freshened a little, building the waves big enough to slop in over the square stern of the dinghy, so with much manoeuvring to maintain an even keel, the steering oar was lashed across the stern, the sail brought aft and sheeted to the two ends of the oar.

This move allowed the dinghy to ride bow onto the waves again and we proceeded more slowly, stern first, but the danger of being swamped by a wave was much lessened. Steering in this position was done by means of pulling the sail down on the side the stern was required to move towards, and we were able to angle the dinghy across the wind by as much as forty-five degrees, if the sea was not too rough, by this method. The fore and aft trim was of much importance now, for if the bow was too light it tended to fall away from the wind, bringing the dinghy broadside to the waves, a most vulnerable position; so I streamed the sea anchor from the bow and left it half-tripped so that it would not hinder our progress too much while keeping the bow pointed to the waves. We also moved the two persons from the back seat into the bottom of the dinghy to give it more forward trim.

With the sea anchor streamed we found we could lash the sail in position, making *Ednamair* self-steering and allowing us to continue watches as before, but now Lyn insisted that I be spared the necessity of taking a watch at all, for I was liable to be called out at all times and the heavy work of tending rigging and turtle dressing was most onerous in my exhausted condition. (Douglas was quite eager to take his share in dressing turtles but he is heavy handed and I dared not risk breaking the knife again.)

The night closed in on *Ednamair*, a lonely speck in the vast reaches of the ocean, and as we arranged and rearranged our comfortless limbs we felt that we had conquered a major obstacle to our survival. We could manage to live in the dinghy.

Day Thirty

The gentle breeze fell calm during the night and at dawn the promise of another dry day was reflected in the sunrise. The limpid blue of the sea flashed as the dorado sped under and around *Ednamair*, then the cry of 'Turtle!' from Sandy made us move hastily to our positions, clearing the dinghy for action.

A large stag turtle nosed curiously at a trailing rope, and with a swift grab we secured first one then both back flippers. A wild struggle ensued, for this was a tough one and with painful lacerations to our hands, we finally landed him, lashing out wildly with clawed flippers in the bottom of the dinghy. We secured him, Douglas holding one flipper and snapping beak, Neil and Robin a back flipper each, and myself a front flipper under my knee to have both hands free for the coup de grâce. The tough hide made difficult work of it and we all sustained bruises and cuts to our legs before the deed was done and the turtle lay quiet.

It was well past noon by the time the meat was hanging and the shells and offal dumped. It had been tough work, but the meat was a good deep red and tastier than usual. Neil had helped to collect the fat and Douglas had done his stuff on the flipper bones. Robin had finally been persuaded to help Neil collect the fat but he didn't seem to have much liking for the job. We nursed our wounds and cut the meat into small pieces for drying in the hot sunshine. The shark was still occupying the rigging, so since there was no wind, the sail was taken down and the small pieces of meat spread out across the stern seat and the centre thwart while we all crouched in the bottom of the dinghy, limbs overlapping in the cramped space.

We lunched well on shark and fresh turtle meat, nibbling at turtle fat afterwards and crunching the bones to extract the rich marrow from the centre. We were all blessed with fairly strong teeth and although the rest of our anatomy suffered in many degrees from the privations we had undergone, our teeth remained clean and unfurred without any external assistance from brushes. The diet obviously suited them!

The sun shone all day, but we suffered it gladly for the drying meat and fish needed every minute of it. The quicker it dried the better it cured so we poured cups of salt water over each other to keep cool and turned the meat over at regular intervals.

It was only when I was making up the log for the day and was about to enter up the small change in our position since noon the day before, that Neil leaned across to me and whispered, 'Hey, Dad, put this in your log. On the thirtieth day Neil had a shit.' I looked to see if he was serious; he grinned an impish smile and said, 'It's right,' so I put it in. It was, after all, a fairly remarkable incident and that's what logs are for, as well as the routine remarks.

While our skin problems were generally improving in a slow sort of way, my hands had become a mass of hacks and cuts. Every time we caught a turtle I usually collected one or two cuts to mark the occasion and this, aggravated with sticking fishhooks into myself, brought the combination of cuts and old boil scars to a pitch where I looked like the victim of some ancient torture. Yet after the initial hurt of these cuts they gave me very little pain and I wondered if the salt water anaesthetised them in some way.

Evening threw quiet shadows over the sea as we packed the drying food under cover for the night. The small pieces of turtle were placed in one section of Lyn's bamboo sewing basket, while the shark strips, now smelling pretty strongly, were placed in a separate piece of sail. The sea was almost mirror calm and loud splashes broke the unaccustomed silence as dorado leapt after flying fish. A louder splash made the sea foam, as a larger predator, probably a shark, attacked a dorado which leapt desperately to escape. The fins of the larger sharks were never far away but we ignored them now as long as they left us alone.

DAY THIRTY-SIX

Slowly the wind rose from the south. At first it was a fine gentle breeze, then blew with increasing force until the breaking tips of the waves gleamed in the darkness.

As *Ednamair* pitched and yawed, shipping more and more water over the midships section, I set Douglas steering her into the waves while I opened the sea anchor out and adjusted the trim of the dinghy to keep a high, weather side. The squalls strengthened and Douglas and I stood watch on watch, helping the tiny boat through the violence of the rising seas. Lyn and Robin were still unable to steer so that they took over the bailing when necessary. I felt uncomfortable without the assurance of the flotation collar and prepared a strangle cord on the water sleeve to enable me to make it into an airtight float very swiftly if an emergency arose.

The squalls brought rain, intermittent and of moderate precipitation, to make the night cold and uncomfortable. We bailed and sang songs to keep warm, the memory of drought too recent for us to feel churlish with the weather. Collecting rainwater became difficult in the strong wind but we managed to gather enough to rinse the salt out of the sleeve and put a half gallon of good fresh water into it before the rain finally tailed off into a drizzle. The wind eased with the rain, and dawn found us shivering and huddled together, eating dried turtle and shark to comfort our sodden skins. The turtle of yesterday was forgotten in the discomfort of the new day.

Each day had now acquired a built-in objective in that we had to try to gain as much as possible over our reserves of stores and water until there would be enough in stock to get us to the coast. I looked upon each turtle as the last, each fish as the one before I lost the hook, by an error in strike. It only needed a six-inch mistake to make the difference between a dynamic pull of about eighty pounds and one of a hundred and eighty with the consequent breaking of the unevenly tensioned lines, and I knew that sooner or later it had to happen.

Lyn washed and mended our clothes, which now had the appearance of some aboriginal garb. Douglas had only his shirt left (Lyn was trying to sew his shredded undershorts together in some attempt to make him presentable when we reached land); Lyn's housecoat, now in ribbons, was more ornament than use, and my tattered underpants and vest were stiff with turtle blood and fat. Robin and the twins were in rather better garb, for their labours made less demands on their clothing. I suppose we would have been thought a most indecent lot in civilised society. (On second thoughts, I've seen some weird products of modern society whose appearance was rather similar so that perhaps we would merely have been thought a little avant-garde.) Robin and I had beards with unkempt moustaches which hung over our upper lips; saltwater boils and scars covered our arms, legs and buttocks and were scattered on other parts of our anatomy, intermingled with clawmarks from turtles, as well as cuts and scratches from other sources. The adults were not desperately thin but the twins, Neil in particular, had become very emaciated.

Knee cramps troubled us from time to time, but generally speaking, apart from Sandy who had a slight bronchial cough which Lyn's expert ear had detected the day previously (for she had a constant fear of a static pneumonia developing in our cramped situation), we were in better physical condition than when we had abandoned the raft. Many of our sores had healed and our bodies were functioning again. We were eating and drinking more, and our ability to gnaw bones and suck nutrition from them increased with our knowledge of the easiest ways to attack them.

We were no longer just surviving, but were improving in our physical condition. As I looked around at our little company, only Neil gave me cause for worry, for his thin physique made it difficult to determine whether he was improving or not, and though he was a most imaginative child, he seldom complained unless in real physical pain. Lyn was careful to see that his supplementary diet was kept as high as possible, and I scraped bone marrow to add to the twins' turtle 'soup' (a mixture of pieces of dried turtle, meat juice, water, eggs when available, and fresh or dried fish).

Our thirty-sixth day ended much as it had started; wet, cold and windy, seas slopping into *Ednamair* as she bounced in the steep short waves, the bailer's familiar scrape and splash, and the helmsman hunched on the stern and peering at each wave to determine its potential danger to our craft. Robin, trying to snatch forty winks in his 'off' time, suddenly sat up with a cry of distress. 'There's no meat on my bone!' he shouted.

Then looking at his thumb (which he had been sucking) with a puzzled expression on his face, he lay down to sleep again.

The twins chortled in the bows for an hour afterwards. Late that evening Sandy said he thought he must have 'done it' accidentally for there was diarrhoea all over his clothes. I passed Sandy over to Lyn while I cleaned up the sheets, moving Neil around to get the muck cleaned off the dinghy, when Lyn said 'You'd better send Neil along when I've finished with this one, Sandy hasn't done anything at all!' Neil's voice full of injured innocence came from the bow, 'Well, how was I to know?' We chortled for half an hour over that one!

DAY THIRTY-EIGHT

After breakfast of some raw steak and the flesh of a scavenger fish (which I speared on the end of the knife) marinated in the meat juice collected overnight, we felt more able to see through the day. It hadn't rained much, and I had a good-sized lump on my head where the shark had left its mark.

A small shower, followed by some drizzle, had increased our water reserves by a pint and the overcast sky gave little prospect of a good drying day, but we hung out the meat in small strips to make the most of it. A large white-tipped shark cruised nearby, reminding me of my lump, and the escort of eight pilot fish in perfect formation across its back lent it the appearance of an underwater aeroplane.

I prepared the gaff while Lyn and Robin sorted out the turtle meat for drying and the twins readjusted the canopy and handed out some strips of dried dorado which needed airing for an hour. We now checked over our considerable amount of dry stores every morning to ensure that it kept in good condition. The fish strips quickly went damp and soggy in the humid atmosphere and the small pieces of turtle meat, if they were allowed to become compacted, warmed up as if affected by spontaneous combustion.

The dorado were reluctant to come near the *Ednamair* with the shark still cruising around, but after we had made one or two swipes at it with the paddle, it went away. I planned to land another two dorado that morning, one for eating immediately, to save the turtle steak for drying, and the other to increase our already good stocks of dried fish. I angled the gaff towards two likely bull dorado of rather a large size, then a large female shot close above the hook; I struck swiftly and missed, but at that instant a small bull of about fifteen pounds followed the female's track and my hook sank into it in a perfect strike! The fish flew into the dinghy with unerring precision and it was secured and killed in the space of seconds.

Feeling very pleased with ourselves, we admired the high forehead of the bull while I made some adjustments to the nylon lines which weren't taking the strain evenly, then I told Douglas to gut it and keep the offal. I had noticed that although the dorado didn't eat the offal, they gathered round curiously as the scavenger fish fought over it. I had the idea a good fish could be taken unawares at this time, so I had Robin throw some offal over just ahead of the gaff. The scavenger fish rushed in, a boil of foam as they fought over the scraps, while the dorado swooped close by. I chose a twenty-five-pound female dorado and struck.

The hook gave, then with a ripping sound the lines snapped one after the other, and the gaff went light. I looked swiftly at Douglas but he was pulling in the reserve line slowly. 'Didn't feel a thing,' he said. My initial reaction was one of extreme dejection; that fish had gone with our last big hook, no more fresh dorado. The nylon must have been cracked and I failed to notice; the tensions of the lines had been different too or they would have broken together; the disturbed water had probably distorted my aim, but it was no use being wise now, there wasn't another hook to be wise with. My spirits picked up a little as I realised that our stocks of dorado exceeded those of turtle meat and we had enough of both now to get us to the coast, even if we caught no more fresh turtle to supplement our rations. I still had another small hook to use for inshore fishing if that should be necessary, and if we felt like a taste of fresh fish I could always try a stab at another scavenger fish; we had been fattening them up for a while now, with our regular dumpings of turtle and fish offal.

Noon position 8°21′ North and 85 miles west of Espinosa, twelve miles nearer land, was not a great boost to our morale but I pointed out that throughout all the time we had been adrift we had either been becalmed or the wind had been favourable. There hadn't been a day yet when I had had to record an adverse run. The calming seas also indicated that we might soon be able to row although the heavy cross swell would have to diminish a little too before that would be possible.

Lyn bathed the twins that afternoon and after their daily exercises and a half hour apiece on the centre thwart to move around a bit, they retreated under the canopy again as a heavy shower threatened. The dorado, caught in the morning, now hung in wet strips from the forestay while the drying turtle meat festooned the stays and cross lines which had been rigged to carry the extra load of meat from two turtles. We worked a little on the thole pins, binding canvas on them to save wear on the rope, then realising that we were neglecting the most important job of making a flotation piece, took the unused piece of sleeve and started to bind one end with fishing line. The clouds grew thicker as the afternoon advanced; it was going to be a wet night again and perhaps we would be able to fill the water sleeve. Seven gallons of water seemed like wealth beyond measure in our altered sense of values.

I chopped up some dried turtle meat for tea, and Lyn put it with a little wet fish to soak in meat juice. She spread the dry sheets for the twins under the canopy, then prepared their little supper as we started to talk of Dougal's Kitchen and if it should have a wine license. As we pondered the delights of Gaelic coffee, my eye, looking past the sail, caught sight of something that wasn't sea. I stopped talking and stared; the others all looked at me. 'A ship,' I said. 'There's a ship and it's coming towards us!' I could hardly believe it but it seemed solid enough. 'Keep still now!' In the sudden surge of excitement, everyone wanted to see. 'Trim her! We mustn't capsize now!' All sank back to their places.

I felt my voice tremble as I told them that I was going to stand on the thwart and hold a flare above the sail. They trimmed the dinghy as I stood on the thwart. 'Right, hand me a flare, and remember what happened with the last ship we saw!' They suddenly fell silent in memory of that terrible despondency when our signals had been unnoticed. 'Oh God!' prayed Lyn. 'Please let them see us.'

I could see the ship quite clearly now, a Japanese tunny fisher. Her grey and white paint stood out clearly against the dark cross swell. 'Like a great white bird,' Lyn said to the twins, and she would pass within about a mile of us at her nearest approach. I relayed the information as they listened excitedly, the tension of not knowing, of imminent rescue, building like a tangible, touchable, unbearable unreality around me. My eye caught the outlines of two large sharks, a hundred yards to starboard. 'Watch the trim,' I warned. 'We have two man-eating sharks waiting if we capsize!' Then, 'I'm going to light the flare now, have the torch ready in case it doesn't work.'

I ripped the caps off, pulled out the striker and struck the primer. The flare smoked then sparked into life, the red glare illuminating *Ednamair* and the sea around us in the twilight. I could feel my index finger roasting under the heat of the flare and waved it to and fro to escape the searing heat radiating outwards in the calm air, then unable to bear the heat any longer, I dropped my arm, nearly scorching Lyn's face, and threw the flare high in the air. It curved in a brilliant arc and dropped into the sea. 'Hand me another, I think she's altered course!' My voice was hoarse with pain and excitement and I felt sick with apprehension that it might only be the ship corkscrewing in the swell, for she had made no signal that she had seen us. The second flare didn't work. I cursed it in frustrated anguish as the priming substance chipped off instead of lighting. 'The torch!' I shouted, but it wasn't needed, she had seen us, and was coming towards us.

I flopped down on the thwart. 'Our ordeal is over,' I said quietly. Lyn and the twins were crying with happiness; Douglas, with tears of joy in his eyes, hugged his mother. Robin laughed and cried at the same time, slapped me on the back and shouted, 'Wonderful! We've done it. Oh! Wonderful!' I put my arms about Lyn, feeling the tears stinging my own eyes: 'We'll get these boys to land after all.' As we shared our happiness and watched the fishing boat close with us, death could have taken me quite easily just then, for I knew that I would never experience another such pinnacle of contentment.

Seven

The Shetland Bus

David Howarth

During the German occupation of Norway, from 1940 to 1945, every Norwegian knew that small boats were constantly sailing from the Shetland Isles to Norway to land weapons and supplies and to rescue refugees. The Norwegians who stayed in Norway and struggled there against the invaders were fortified by this knowledge, and gave the small boats the familiar name. "To take the Shetland bus" became a synonym in Norway for escape when danger was overwhelming. This record of the adventures of the Norwegian sailors who manned the boats is offered as a tribute from an English colleague to Norwegian seamanship, and as a humble memorial to those who lost their lives.

* * *

On 17 March Larsen left Scalloway to carry out the last trip of the season to Traena. By the end of the month, when he was due back, it would be too light up there to visit the district again till September.

Bergholm was faster than the smaller boats, and they made good time to Traena in fine weather, arriving off the islands in three and a half days. Larsen sighted the coast and fixed his position in daylight, and when darkness fell he closed in and felt his way in among the skerries. He reached one of the sounds between the islands, where the rocks rose steeply from the water, and laid the *Bergholm* alongside them. One of the passengers jumped ashore and climbed over the steep hills to some houses on the other side of the island.

When he came back he brought with him a man who had volunteered to come with them to another very small island, where a single family lived who he thought could take charge of the passengers and cargo until the small local boats were able to ferry them across to the mainland. Larsen took *Bergholm* through the sounds to this little island and moored her to the quay there. The owner of the island had a boathouse on the quay, in which were two dinghies and a lot of nets. They woke him up, and found he was quite willing to keep the cargo in the boathouse and to take care of the passengers. Most of the rest of the night was spent in taking out the boats and nets, packing in the cargo, and arranging the boats and nets on top of it.

By the time this was finished it was too late to put to sea again that morning, and as it was a good place to lie, sheltered from observation on all sides but one, Larsen stayed there till the following evening. All day, from the island, they watched a German patrol boat steaming up and down its beat nearby, and there were several alarms when it seemed to be approaching a point from which its crew could have seen the *Bergholm*. Had it done so they would have had to fight their way out, so they cleared away the guns and started the engine. It was only an armed Norwegian Arctic whaler, and they could probably have sunk it, but the aftereffects of a fight on the passengers who had been landed and the local people who had helped them would have been disastrous, and Traena would have been finished as a landing place. So it was lucky that its beat seemed to stop just short of the point from which the quay would have been visible, and that each time discovery seemed imminent it turned back on its tracks. As darkness approached it steamed off towards the mainland, and at eight in the evening *Bergholm* left for home. It was still very fine and clear.

At two o'clock the next afternoon they were steaming on their homeward course, parallel to the coast and about seventy-five miles off it, when a twin-engined plane approached them from astern and flew round them, very low and about three hundred yards away. As they expected the plane to attack at a moment's notice, and as they were much farther offshore than an innocent fishing boat had any right to be, they dropped their camouflage and manned the guns. But it did not attack; it flew off towards the coast.

The crew of the plane had certainly seen their guns, and it seemed sure that when it reached the coast and their position was reported, a real attack would be made. Larsen altered course to the westward; but after a bit he reflected that in such perfectly clear weather, at eight knots, he had no chance whatever of evading a search, so he returned to the course he had set for Shetland. The crew tested all their weapons and brought all the ammunition on deck. They had a single .5 Colt machine gun mounted forward and a twin one aft, two twin Lewis guns amidships, and two unmounted Brens.

About six o'clock the attack came. Two twin-engined seaplanes approached the boat from the port beam and circled it at a height of two hundred feet. Then, diving to mast height, they flew across her bow, firing with cannon. *Bergholm* returned the fire with all her guns. Not much damage was done to the boat, but for a few seconds the decks were swept with cannon shell splinters, and Klausen, on the port Lewis mounting, received so many wounds that Larsen sent him below.

The planes stood off and circled for about five minutes. Perhaps the fire put up by *Bergholm* was more than they had expected, and they were discussing it on their shortwave radio. After a time they swooped again, both attacking from the starboard side. As they approached, Larsen at the wheel tried to turn the boat to bring all the guns to bear. Another storm of shells and splinters hit her. The Colt and Lewis tracers were seen hitting the planes. Enoksen, at the twin Colt, staggered away from his guns with his face and hands hidden with blood. When they went to help him they found he was riddled with shell splinters from head to foot, and he could not see, so he also had to be sent below. Kalve, at the bow Colt, was hit in one hand and one foot. As the planes roared by he swung his gun round and aimed it with his remaining hand, then jammed his other elbow onto the trigger. Faeroy and Vika, the two engineers, were firing the two Bren guns. Hansen had gone below to try to send a radio signal to us, but the aerial was shot away. By then the boat was badly damaged, but she was still underway, and they knew they had damaged the planes.

Suddenly as they watched for the next attack, one plane broke away and flew off low towards the coast.

The other one went on circling round, then dived again. Faeroy and Noreiger had taken over the Colts. Enoksen was trying to get up the ladder again from the cabin, but he was hit again and fell back down the hatch. Faeroy was also wounded, but he was able to stay at the gun.

On its next attack the plane dropped a stick of six bombs. None of them fell near the boat, but its cannon fire was still accurate, and Faeroy was wounded again and could not do any more.

Then there was nobody left to man the Lewis guns or Brens. Noreiger was still at one Colt, and Vika took over the other. Larsen was still at the wheel in what remained of the wheelhouse, manoeuvring the boat to meet each attack.

In the next run another stick of bombs came down, and the last of them fell a few feet from the stern. It shook the boat badly. Noreiger and Vika both shot accurately, and Larsen saw strikes on the plane. But as it receded once more Vika fell, and when Larsen ran to help him he found his foot was shot off above the ankle. Five of the eight men aboard were out of action. Hansen had come up from below and reported that the radio was dead and the boat was leaking. Larsen, wondering how to dispose his remaining men to meet the next attack, looked up at the plane. It was disappearing to the eastward, smoking.

The whole fight had lasted just over half an hour. This short time had wrought a terrible difference on *Bergholm* and her crew; but dusk was falling, and they could be sure that the night would give them respite. Larsen, Noreiger and Hansen, who were not wounded, first went to attend to the other five men. Vika was the most seriously hurt. Someone had already put a tourniquet on his leg, but they knew he was dying, and they thought he knew it too. He was conscious, and sometimes smiled, but he did not speak or complain. Faeroy, Enoksen, and Klausen were in great pain from the number of steel splinters in their hands and heads and bodies, and they could not move. Enoksen, however, was not blinded, as they had thought at first, it was only blood which had run into his eyes, and the shock of a shell which exploded in front of his face, which had made him unable to see. Kalve, who only had one leg and one hand out of action, was able to move and to give some help with the work that had to be done. They disinfected the men's wounds and bandaged them, then turned their attention to the boat.

The engine was still running, and with the wheel lashed she was holding nearly to her course; but the water in the bilges was rising, and the two pumps on the engine could not hold it in check. Kalve and Noreiger manned the hand pump, but still the water rose, and in spite of all they could do, at about eight o'clock it reached the air intakes of the engine, and the engine stopped.

In the meantime Larsen had inspected the rest of the boat. The decks were full of holes and covered with blood and empty cartridge cases. The masts were still standing, but a lot of the rigging was shot away, and wire and rope were swinging from side to side as the boat rolled. The wheelhouse, in which Larsen had stood unscratched through the whole engagement, was literally shot to pieces. The windows were all gone and the inside was littered with broken glass. All the doors were shot away, and the wooden walls and roof were smashed by exploding shells, so that nothing but the broken framework remained.

Most important of all was the lifeboat, which was stowed on top of a deckhouse on the port side of the wheelhouse. Most of its gunwale was split off, and it had seven shell holes in its bottom. Larsen and Noreiger set to work to patch the holes with canvas and sheets cut from bully beef tins. Hansen collected food and water and navigating instruments, and the lifeboat's mast and sail and oars. By midnight they had made the boat tight enough to be kept afloat, and they launched it and stowed the essential stores aboard. Larsen tore up his marked charts and ciphers and threw them overboard. Then came a grievous struggle to get the wounded men up the steep companionway from the cabin and into the boat without hurting them too much. At last they got Vika laid in the bows, on the floor of the boat, and Enoksen and Faeroy amidships. Klausen and Kalve had to sit up in the stern, as the boat was only sixteen feet long and there was no room for them to lie down. The three who were not wounded arranged to take turns at rowing, two at a time, each rowing for four hours and resting for two. At one in the morning they abandoned the *Bergholm*. It was dead calm.

The first thing they had to do was to get as far away as possible before dawn, when the Germans would very likely send out a plane to see what had happened to the wreck. She might still be floating; a wooden ship will often float with gunwales awash. If so, there was a chance that the Germans, seeing no life aboard, would assume that they had all been killed; but it was more likely that they would see that the lifeboat was gone, and would make a search for it.

But apart from getting away from the scene of the fight, Larsen had to decide where to make for. They were seventy-five miles from the nearest point on the Norwegian coast, and three hundred fifty miles from Shetland. After thinking it over, he decided that it was very unlikely that they could reach Shetland in the lifeboat. It was heavily laden, and with most of its top plank on each side shot away it had very little freeboard, so that a very moderate sea would have swamped it. Besides, with the best of luck it would take them, say, ten days to get there, and none of the wounded men could be expected to survive so long in an open boat. On the other hand, he did not like to take the shortest route to Norway, partly because he thought it was what the Germans would expect him to do, and partly because it led to a part of the coast, near Trondheim, where he had no friends he could rely on, and he thought that even if they got there safely it would be difficult to get away again.

So he made up his mind to steer for Ålesund, a hundred and fifty miles away, twice as far as the Trondheim coast. Larsen was a seaman by nature, and the prospect of rowing so far in a leaky boat did not worry him, provided that it gave some small hope of getting the wounded crew alive to Shetland.

The two men rowing took one oar each, sharing the midship thwart. The third unwounded man, taking his two hours off from the oars, could not lie down because there was no room, and was occupied with helping the wounded men to shift their positions and to take food and water. Kalve was able to bail with his undamaged hand, and he did so continuously. At four o'clock on the first morning Vika asked for water and aspirins. They had no aspirins, but Larsen gave him the water and he seemed satisfied. When he went to him an hour later he had died. They wanted to bury his body in Norway, but later on their journey they wrapped it in a blanket and lifted it overboard. They remained stubbornly sure that they would reach safety in the end.

At dawn on the first morning, when they had been rowing for six hours, it was still calm and crystal clear. They saw a plane searching the place where the *Bergholm* had been. It flew in increasing circles, and they realised that its crew must have found the wreck, seen that the lifeboat was gone, and started to look for them. Planes remained in sight for the whole of the day, quartering the ocean in which a boat seems dreadfully conspicuous and vulnerable. Whenever the planes approached, the rowers shipped their oars, in case the flash of sunlight on the wet blades should show them up. Often the planes came so close that the men in the boat were certain they had been seen, and nerved themselves to a fresh attack like those of the day before; but each time the plane sheered off, and after a day of suspense at last the darkness fell and covered them. They rowed all night.

On the second day a light breeze came, and as there was no plane in sight they hoisted the sail, and for a time they made good progress. But at dusk it fell calm again, and they rowed for the third night.

The third day was sunny and calm again, and they rowed all day without seeing anything.

During the fourth night they saw a light ahead. They made towards it, and saw it was a fishing boat. They hailed her and drew alongside. Larsen climbed stiffly aboard. He told the fishermen that he and his men had been torpedoed in a merchant ship, and he asked them if they had enough fuel to get to Shetland. The fishermen were friendly and sympathetic, but they said they were only allowed to carry enough fuel to get to the fishing grounds and back, in order to stop them sailing across to the other side. They came from Kristiansund, and were willing to take Larsen and his crew back there with them and to help them to escape; but Larsen was doubtful whether it would be possible to escape from there, so he thanked them but refused the offer. He also refused food, saying they had plenty in the lifeboat. The fishermen gave him his exact position, which was thirty-five miles offshore, and he returned to the boat and started to row again.

At dawn on the next day, their fourth in the boat, they could see land, but it was still a long way off. As they struggled on towards it they saw a lot of fishing boats coming out towards them. This was an unwelcome sight, for it meant that by the evening, when the boats got back to port, the approach of a shipwrecked crew would be common knowledge. But it could not be helped, and when the boats reached them they hailed the first, and Larsen asked again if they had enough fuel to get to Shetland. This time the skipper tried to persuade them against going to Shetland, and after some hedging he came out into the open and said that if they would come back with him he would use his influence to help them; he thought it would only mean a month or two in prison, and then they could join the merchant service for the Germans. Luckily, before the conversation had reached this stage, his crew had given the men in the boat some cooked fish and coffee. It was the first hot food they had had for four days, and they felt much better for it; and they thanked the skipper politely, and went on their way. It turned out later that they had hit on the only local quisling [traitor]. When he got home that evening he reported them to the Germans, and was seen the next day walking in Ålesund with a German officer.

After leaving the fishing boat they rowed at the best speed they could make. They were sure he meant to report them, but they did not think he would lose a day's fishing to do it; so their only hope was to get into hiding on the coast before he could reach home. As they neared the coast a tidal stream against them slowed their progress, and they had a hard struggle to make any headway against it; but about three in the afternoon, very thankfully, they reached the first of the islands, and ran the boat in among some rocks, and waited for darkness.

There were still about ten miles to go, among the islands, to the place where Björnöy lived. They set off at eight, and got there soon after midnight. They were very tired, and some of the wounded men were very ill.

Larsen went ashore and knocked at the door of a man called Nils Sorviknes, who had helped him when he had been there before. It was some time before he could get an answer, and he leaned against the doorpost in the last stages of exhaustion. But at last the door opened, and Sorviknes, astonished to see him again, took him inside. Larsen told him what had happened, and told him he had six men waiting in the lifeboat. He asked what chance there was of getting a boat to go to Shetland. Sorviknes said he would talk to Björnöy, but that it was too late to do anything that night; and when Larsen told him the name of the fishing boat they had spoken to, Sorviknes knew it at once as belonging to a quisling, and was sure that by then the Germans would have been told that they were in the district. So the first essential was to hide them before dawn. He thought for a while, then advised Larsen to go to a man called Lars Torholmen, who lived with his wife and two sisters in the only house on a very small island a couple of miles away. He gave him a letter to Torholmen, and told him to lie low there till he got further instructions.

Back in the lifeboat, Larsen took to the oars for the last time and rowed to the small island. He went ashore again, and woke Torholmen, who took him in without any hesitation. As soon as he gave him the letter from Sorviknes and explained about the wounded men in the boat, Torholmen and his two sisters came down to the shore and between them they carried the wounded men to the house. The mother and the sisters put them to bed and fed them and washed their wounds. It was beginning to get light, and the boat had to be disposed of before daybreak. There was no time to take it out and sink it, so Larsen and Torholmen rowed it round to the other side of the island and hid it in a boathouse. It was a compromising thing to keep on the island, but there was nothing better they could do. As the sun rose they got back to the house, and Larsen was also put to bed. They all slept for the whole of the day.

They stayed with Torholmen for a week, living in two rooms at the top of the house, and being well looked after by the three ladies. With good food and rest their strength began to return, but some of the wounded men were still in great pain from the shell splinters in their bodies.

Björnöy, on whom they were relying to find them a boat to get away, was skipper of a local ferry, and Nils Sorviknes was one of his crew. So they knew that Sorviknes would be able to tell Björnöy about them during the ferry's morning run to Ålesund on the day they arrived. But the story Larsen had been able to tell Sorviknes was incomplete, so one of Torholmen's sisters invented some errands in Ålesund, and went as a passenger on the boat to give him more details. She also asked him whether he knew of a doctor who could be trusted to come to Torholmen to see the wounded men, as she was worried by signs of sepsis and gangrene.

On their second day with Torholmen, which was a Sunday, Björnöy came to see Larsen and to tell him what he was doing. The problem of getting them away had been made much more difficult and dangerous by their meeting with the quisling skipper. As they had expected, he had reported their arrival to the Germans and a tremendous search was going on. Within an hour of their arrival at Torholmen's house the Germans had started an air search of the whole of that part of Norway, and had sent two armed trawlers to the part of the coast the quisling had thought they were making for. They had also dispatched a ferry steamer to land parties of soldiers on each of the string of large islands which ran north from Ålesund; and they had evidently seen or photographed the registration number which was painted on the *Bergholm* at the time, for they sent a detachment of thirty men to the village which the number denoted.

Advertisements offering rewards for their capture were printed in the papers, and everybody in the district was talking about them. The little island where they were hiding was in the very middle of the area which was being searched, and Björnöy thought that until the excitement had died down it would be foolish to risk the slightest move which might draw attention to the place. He did not think it was safe even to bring a doctor out to the island. He offered to take the wounded men to the doctor, in a rowing boat by night, but it would be a dangerous journey, partly over land, where the men would have to be carried and the whole party would be at the mercy of anyone who happened to see them. Larsen and Björnöy agreed that in any case the doctor could not do much unless the men went to a hospital to have the splinters extracted, and they decided to give up the idea unless any of them got much worse.

Björnöy was against using a local boat to escape to Shetland if it could be avoided. The Germans were obviously very anxious to capture the *Bergholm*'s crew. If a local boat disappeared they would be sure to guess that the crew had escaped in it, and their punishment of the owners would be severe and might even lead to discovery of the whole organisation to which Björnöy belonged. Although this might be risked as a last resort, he had first got in touch with Karl-Johan, and had a radio message sent to England to ask us to send a boat over to fetch them.

This signal, of course, was the first news we had had in Shetland of what had happened to the *Bergholm*. Knowing that she had been attacked confirmed us all in our opinion that the use of fishing boats was getting too dangerous to be worthwhile. By the time we received it, it was the beginning of April. The nights were already very short, and to send a fishing boat into a district already so thoroughly on the alert would be very risky. A naval M.T.B., on the other hand, would not only have a much better chance of doing the job, but also of fighting its way out if it was spotted. The Navy was very willing to send one; but unluckily the weather by then was very bad, and until it moderated it was impossible for these fast light craft to leave harbour. Every one of the fishing-boat crews which survived at the base was ready and eager to go, and the fishing boats could easily have weathered the gale. But Rogers would not let them. It was natural that everyone's first reaction was to set off at once on a rescue expedition; but when he weighed it up he concluded that it was wrong to take so great a risk of losing a second crew, and that so far as we could tell it was probably safer for the *Bergholm* crew themselves to stay in hiding till an M.T.B. could go, than to embark on a fishing boat which might so very easily be lost on the voyage home. It was hard for him to decide to leave the crew in their dangerous position, but he was certainly right; and the decision was transmitted to Ålesund, and ultimately to the men on the island.

Meanwhile the search continued around and over their hiding place. From the windows of the house they could see aircraft quartering the district, and every day Torholmen brought them news of what the Germans were doing. The search lasted for nearly a week. Then a rumour spread around in Ålesund that a fishing boat had been stolen. Larsen heard it and was pleased, because even though the Germans would not be able to trace it to its source and prove that it was true, they would certainly not be able to prove that it was not, and as time went on they would probably be more inclined to believe it. It would at least offer them a plausible explanation of the disappearance of the boatload of men. He never knew whether it was a spontaneous rumour, or whether it had been started by some friend as a means of bringing the search to an end. At all events it seemed to help.

The intensity of the search gradually died down, and after about ten days the Germans seemed to have given it up as a bad job. Why they missed the little island where the men were hiding remains a mystery. Perhaps it was because it was so small, or because it was so close to the headquarters town of Ålesund; or perhaps they hardly expected that men who had rowed a hundred and fifty miles would row a farther ten among the islands.

As soon as it began to seem that the Germans were convinced the crew had escaped from the district, Björnöy proceeded with his plans for getting them away. He had asked through Karl-Johan that the rescue boat should be sent to the island of Skorpen, a dozen miles farther south, where we had already landed and picked up many agents. Johann Skorpen, the fisherman who lived there, was well used to hiding people and would look after the crew somewhere on his island till the boat could come.

Moving them down there would be the most difficult job. There was a control point on the way at which papers had to be produced, but boats were not usually searched there; and if Björnöy could find a boat with a plausible reason for going through this control, he thought it would be better to risk this than to bring one of our boats too close to Ålesund.

The movement was deputed to Sverre Roald, another member of the organisation, who lived in the island of Vigra, close at hand, and was a neighbour and relation of our foreman shipwright Sevrin Roald. After the men had been a week with Torholmen, Roald came one night to tell them what he had arranged. The next night he came again, and they embarked in his little motorboat. But by then the calm spell which had lasted throughout their journey in the lifeboat had given place to strong southerly winds, and after trying to stem the short seas between the islands they had to give it up and go back to Torholmen.

The next night the wind had dropped a little, and they tried again. This time they reached the island of Vigra in safety, where Roald transferred them to another boat: the small decked fishing boat which was to take them through the control to Skorpen.

There is an important radio station in Vigra, which was guarded by German sentries. The fishing boat was lying within a hundred yards of the station buildings, within sight of the sentries. The trip to Skorpen had to be postponed through bad weather, and the men stayed on board the boat for five days. It was an inconvenient position to be in, because they could not go on deck in daylight but had to stay confined in the little cabin of the boat. But on the other hand it was reasonably safe, because the Germans would not expect the seven men they were looking for to be hiding in a boat under constant watch. Sverre Roald had to be careful in his visits, but he managed to see them every night, bringing them food and the latest news of how things were going. At last they were able to get away. They passed the control quite easily and safely, and reached Skorpen, where Roald handed them over to Johann Skorpen and went back to Vigra. Skorpen installed them in a cowshed on the opposite side of the island to his house, and brought them a primus stove and some food and coffee.

Skorpen was not worried at having them there, because his house and island had been searched only a week before and he did not think the Germans would bother him again for some time. Some M.T.B.s manned by the Norwegian Navy had attacked a convoy in that district, and it seemed that the Germans had not seen the boats leaving the coast and believed they had been sunk and that the crews were hiding. They had been searching for them round Skorpen, just as they had been searching for the *Bergholm* crew round Ålesund, and a party of soldiers had been landed in each island. Luckily Skorpen had seen them coming, and had retreated to the hills, taking with him his radio set, since it was forbidden to possess such a thing, and leaving his wife, who would be less suspect than himself, to deal with the search party. The officer in charge of the party had opened her door and said, 'Where are you hiding them?' It must have alarmed the lady, whose house had harboured so many different agents and refugees; but she pretended not to be able to understand the German officer's Norwegian. This universal means of avoiding difficult questions must have been very annoying for Germans who were perhaps not very sure whether they were really able to make themselves understood or not. It was very effective. The party searched the house and found an old pair of headphones from a disused crystal set, which they took away with apparent satisfaction.

As this search had been made so recently, and as the *Bergholm* crew were now out of the immediate area where the Germans had supposed them to be, they settled down with a feeling of comparative security to wait for the boat from Shetland.

Our headquarters had arranged with Karl-Johan that they should send a code message in the B.B.C. Norwegian news bulletin on the night before the rescue boat was due to arrive at Skorpen. Karl-Johan had warned Skorpen to expect this message, and Skorpen listened every evening on the radio which he had retrieved from the hilltop. When Roald got home and sent a message to Karl-Johan that the men had safely arrived at the rendezvous, a signal was sent to our headquarters and passed on to us in Shetland. The weather was still bad, and although the naval M.T.B.s were ready to sail we had to wait another week before they could leave the harbour. As soon as the wind subsided the code message was broadcast by the B.B.C. and an M.T.B. left Shetland. Skorpen heard the message, and when the boat entered the sound of Skorpen the next evening the seven men were waiting on the shore. Afterwards Larsen said, 'We are glad to be on our way.'

EIGHT

On the Grand Banks

Rudyard Kipling

To the end of his days, Harvey will never forget that sight. The sun was just clear of the horizon they had not seen for nearly a week, and his low red light struck into the riding-sails of three fleets of anchored schooners—one to the north, one to the westward, and one to the south. There must have been nearly a hundred of them, of every possible make and build, with, far away, a square-rigged Frenchman, all bowing and courtesying one to the other. From every boat dories were dropping away like bees from a crowded hive; and the clamour of voices, the rattling of ropes and blocks, and the splash of the oars carried for miles across the heaving water. The sails turned all colours, black, pearly-grey, and white, as the sun mounted; and more boats swung up through the mists to the southward.

The dories gathered in clusters, separated, reformed, and broke again, all heading one way; while men hailed and whistled and cat-called and sang, and the water was speckled with rubbish thrown overboard.

"It's a town," said Harvey. "Disko was right. It is a town!"

"I've seen smaller," said Disko. "There's about a thousand men here; an' yonder's the Virgin." He pointed to a vacant space of greenish sea, where there were no dories.

The "We're Here" skirted round the northern squadron, Disko waving his hand to friend after friend, and anchored as neatly as a racing yacht at the end of the season. The Bank fleet pass good seamanship in silence; but a bungler is jeered all along the line.

"Jest in time fer the caplin," cried the *Mary Chilton*.

"'Salt 'most wet?" asked the *King Philip*.

"Hey, Tom Platt! Come t' supper to-night?" said the *Henry Clay*; and so questions and answers flew back and forth. Men had met one another before, dory-fishing in the fog, and there is no place for gossip like the Bank fleet. They all seemed to know about Harvey's rescue, and asked if he were worth his salt yet. The young bloods jested with Dan, who had a lively tongue of his own, and inquired after their health by the town—nicknames they least liked. Manuel's countrymen jabbered at him in their own language; and even the silent cook was seen riding the jib-boom and shouting Gaelic to a friend as black as himself. After they had buoyed the cable—all around the Virgin is rocky bottom, and carelessness means chafed ground-tackle and danger from drifting—after they had buoyed the cable, their dories went forth to join the mob of boats anchored about a mile away. The schooners rocked and dipped at a safe distance, like mother ducks watching their brood, while the dories behaved like mannerless ducklings.

As they drove into the confusion, boat banging boat, Harvey's ears tingled at the comments on his rowing. Every dialect from Labrador to Long Island, with Portuguese, Neapolitan, Lingua Franca, French, and Gaelic, with songs and shoutings and new oaths, rattled round him, and he seemed to be the butt of it all. For the first time in his life he felt shy—perhaps that came from living so long with only the "We're Heres"—among the scores of wild faces that rose and fell with the reeling small craft. A gentle, breathing swell, three furlongs from trough to barrel, would quietly shoulder up a string of variously painted dories. They hung for an instant, a wonderful frieze against the sky-line, and their men pointed and hailed. Next moment the open mouths, waving arms, and bare chests disappeared, while on another swell came up an entirely new line of characters like paper figures in a toy theatre. So Harvey stared. "Watch out!" said Dan, flourishing a dip-net. "When I tell you dip, you dip. The caplin'll school any time from naow on. Where'll we lay, Tom Platt?"

Pushing, shoving, and hauling, greeting old friends here and warning old enemies there, Commodore Tom Platt led his little fleet well to leeward of the general crowd, and immediately three or four men began to haul on their anchors with intent to lee-bow the "We're Heres". But a yell of laughter went up as a dory shot from her station with exceeding speed, its occupant pulling madly on the roding.

"Give her slack!" roared twenty voices. "Let him shake it out."

"What's the matter?" said Harvey, as the boat flashed away to the southward. "He's anchored, isn't he?"

"Anchored, sure enough, but his graound-tackle's kinder shifty," said Dan, laughing. "Whale's fouled it. . . . Dip, Harve! Here they come!"

The sea round them clouded and darkened, and then frizzed up in showers of tiny silver fish, and over a space of five or six acres the cod began to leap like trout in May; while behind the cod three or four broad grey-black backs broke the water into boils.

Then everybody shouted and tried to haul up his anchor to get among the school, and fouled his neighbour's line and said what was in his heart, and dipped furiously with his dip-net, and shrieked cautions and advice to his companions, while the deep fizzed like freshly opened soda-water, and cod, men, and whales together flung in upon the luckless bait. Harvey was nearly knocked overboard by the handle of Dan's net. But in all the wild tumult he noticed, and never forgot, the wicked, set little eye—something like a circus elephant's eye—of a whale that drove along almost level with the water, and, so he said, winked at him. Three boats found their rodings fouled by these reckless mid-sea hunters, and were towed half a mile ere their horses shook the line free.

Then the caplin moved off and five minutes later there was no sound except the splash of the sinkers overside, the flapping of the cod, and the whack of the muckles as the men stunned them. It was wonderful fishing. Harvey could see the glimmering cod below, swimming slowly in droves, biting as steadily as they swam. Bank law strictly forbids more than one hook on one line when the dories are on the Virgin or the Eastern Shoals; but so close lay the boats that even single hooks snarled, and Harvey found himself in hot argument with a gentle, hairy Newfoundlander on one side and a howling Portuguese on the other.

Worse than any tangle of fishing-lines was the confusion of the dory-rodings below water. Each man had anchored where it seemed good to him, drifting and rowing round his fixed point. As the fish struck on less quickly, each man wanted to haul up and get to better ground; but every third man found himself intimately connected with some four or five neighbours. To cut another's roding is crime unspeakable on the Banks; yet it was done, and done without detection, three or four times that day. Tom Platt caught a Maine man in the black act and knocked him over the gunwale with an oar, and Manuel served a fellow-countryman in the same way. But Harvey's anchor-line was cut, and so was Penn's, and they were turned into relief-boats to carry fish to the "We're Here" as the dories filled. The caplin schooled once more at twilight, when the mad clamour was repeated; and at dusk they rowed back to dress down by the light of kerosene-lamps on the edge of the pen.

It was a huge pile, and they went to sleep while they were dressing. Next day several boats fished right above the cap of the Virgin; and Harvey, with them, looked down on the very weed of that lonely rock, which rises to within twenty feet of the surface. The cod were there in legions, marching solemnly over the leathery kelp. When they bit, they bit all together; and so when they stopped. There was a slack time at noon, and the dories began to search for amusement. It was Dan who sighted the *Hope of Prague* just coming up, and as her boats joined the company they were greeted with the question: "Who's the meanest man in the Fleet?"

Three hundred voices answered cheerily:

"Nick Bra-ady." It sounded an organ chant.

"Who stole the lamp-wicks?" That was Dan's contribution.

"Nick Bra-ady," sang the boats.

"Who biled the salt bait fer soup?" This was an unknown backbiter a quarter of a mile away.

Again the joyful chorus. Now, Brady was not especially mean, but he had that reputation, and the Fleet made the most of it. Then they discovered a man from a Truro boat who, six years before, had been convicted of using a tackle with five or six hooks—a "scrowger," they call it—on the Shoals. Naturally, he had been christened "Scrowger Jim"; and though he had hidden himself on the Georges ever since, he found his honours waiting for him full blown. They took it up in a sort of fire-cracker cho-rus: "Jim! O Jim! Jim! O Jim! Ssscrowger Jim!" That pleased everybody. And when a poetical Beverly man—he had been making it up all day, and talked about it for weeks—sang, "The *Carrie Pitman's* anchor doesn't hold her for a cent!" the dories felt that they were indeed fortunate. Then they had to ask that Beverly man how he was off for beans, because even poets must not have things all their own way. Every schooner and nearly every man got it in turn. Was there a careless or dirty cook anywhere? The dories sang about him and his food. Was a schooner badly found? The Fleet was told at full length. Had a man hooked tobacco from a messmate? He was named in meeting; the name tossed from roller to roller. Disko's infallible judgments, Long Jack's market-boat that he had sold years ago, Dan's sweetheart (oh, but Dan was an angry boy!), Penn's bad luck with dory-anchors, Salters's views on manure, Manuel's little slips from virtue ashore, and Harvey's ladylike handling of the oar—all were laid before the public; and as the fog fell around them in silvery sheets beneath the sun, the voices sounded like a bench of invisible judges pronouncing sentence.

The dories roved and fished and squabbled till a swell underran the sea. Then they drew more apart to save their sides, and some one called that if the swell continued the Virgin would break. A reckless Galway man with his nephew denied this, hauled up anchor, and rowed over the very rock itself. Many voices called them to come away, while oth-ers dared them to hold on. As the smooth-backed rollers passed to the southward, they hove the dory high and high into the mist, and dropped her in ugly, sucking, dimpled water, where she spun round her anchor, within a foot or two of the hidden rock. It was playing with death for mere bravado; and the boats looked on in uneasy silence till Long Jack rowed up behind his countrymen and quietly cut their roding.

"Can't ye hear ut knockin'?" he cried. "Pull for your miserable lives! Pull!"

The men swore and tried to argue as the boat drifted; but the next swell checked a little, like a man tripping on a carpet. There was a deep sob and a gathering roar, and the Virgin flung up a couple of acres of foaming water, white, furious, and ghastly over the shoal sea. Then all the boats greatly applauded Long Jack, and the Galway men held their tongue.

"Ain't it elegant?" said Dan, bobbing like a young seal at home. "She'll break about once every ha'af hour now, 'less the swell piles up good. What's her reg'lar time when she's at work, Tom Platt?"

"Once ivry fifteen minutes, to the tick. Harve, you've seen the greatest thing on the Banks; an' but for Long Jack you'd seen some dead men too."

There came a sound of merriment where the fog lay thicker and the schooners were ringing their bells. A big bark nosed cautiously out of the mist, and was received with shouts and cries of, "Come along, darlin'," from the Irishry.

"Another Frenchman?" said Harvey.

"Hain't you eyes? She's a Baltimore boat; goin' in fear an' tremblin'," said Dan. "We'll guy the very sticks out of her. 'Guess it's the fust time her skipper ever met up with the Fleet this way."

She was a black, buxom, eight-hundred-ton craft. Her mainsail was looped up, and her topsail flapped undecidedly in what little wind was moving. Now a bark is feminine beyond all other daughters of the sea, and this tall, hesitating creature, with her white and gilt figurehead, looked just like a bewildered woman half lifting her skirts to cross a muddy street under the jeers of bad little boys. That was very much her situation. She knew she was somewhere in the neighbourhood of the Virgin, had caught the roar of it, and was, therefore, asking her way. This is a small part of what she heard from the dancing dories:

"The Virgin? Fwhat are you talkin' of? This is Le Have on a Sunday mornin'. Go home an' sober up."

"Go home, ye tarrapin! Go home an' tell 'em we're comin'."

Half a dozen voices together, in a most tuneful chorus, as her stern went down with a roll and a bubble into the troughs: "Thay-aah—she—strikes!"

"Hard up! Hard up fer your life! You're on top of her now."

"Daown! Hard daown! Let go everything!"

"All hands to the pumps!"

"Daown jib an' pole her!"

Here the skipper lost his temper and said things. Instantly fishing was suspended to answer him, and he heard many curious facts about his boat and her next port of call. They asked him if he were insured; and whence he had stolen his anchor, because, they said, it belonged to the *Carrie Pitman*; they called his boat a mud-scow, and accused him of dumping garbage to frighten the fish; they offered to tow him and charge it to his wife; and one audacious youth slipped almost under the counter, smacked it with his open palm, and yelled: "Gid up, Buck!"

The cook emptied a pan of ashes on him, and he replied with cod-heads. The bark's crew fired small coal from the galley, and the dories threatened to come aboard and "razee" her. They would have warned her at once had she been in real peril; but, seeing her well clear of the Virgin, they made the most of their chances. The fun was spoilt when the rock spoke again, a half-mile to windward, and the tormented bark set everything that would draw and went her ways; but the dories felt that the honours lay with them.

All that night the Virgin roared hoarsely and next morning, over an angry, white-headed sea, Harvey saw the Fleet with flickering masts waiting for a lead. Not a dory was hove out till ten o'clock, when the two Jeraulds of the Day's Eye, imagining a lull which did not exist, set the example. In a minute half the boats were out and bobbing in the cockly swells, but Troop kept the "We're Heres" at work dressing-down. He saw no sense in "dares"; and as the storm grew that evening they had the pleasure of receiving wet strangers only too glad to make any refuge in the gale. The boys stood by the dory-tackles with lanterns, the men ready to haul, one eye cocked for the sweeping wave that would make them drop everything and hold on for the dear life. Out of the dark would come a yell of "Dory, dory!" They would hook up and haul in a drenched man and a half-sunk boat, till their decks were littered down with nests of dories and the bunks were full. Five times in their watch did Harvey, with Dan, jump at the foregaff where it lay lashed on the boom, and cling with arms, legs, and teeth to rope and spar and sodden canvas as a big wave filled the decks. One dory was smashed to pieces, and the sea pitched the man head first on to the decks, cutting his forehead open; and about dawn, when the racing seas glimmered white all along their cold edges, another man, blue and ghastly, crawled in with a broken hand, asking news of his brother. Seven extra mouths sat down to breakfast: a Swede; a Chatham skipper; a boy from Hancock, Maine; one Duxbury, and three Provincetown men.

There was a general sorting out among the Fleet next day; and though no one said anything, all ate with better appetites when boat after boat reported full crews aboard. Only a couple of Portuguese and an old man from Gloucester were drowned, but many were cut or bruised; and two schooners had parted their tackle and been blown to the southward, three days' sail. A man died on a Frenchman—it was the same bark that had traded tobacco with the "We're Heres." She slipped away quite quietly one wet, white morning, moved to a patch of deep water, her sails all hanging anyhow, and Harvey saw the funeral through Disko's spy-glass. It was only an oblong bundle slid overside. They did not seem to have any form of service, but in the night, at anchor, Harvey heard them across the star-powdered black water, singing something that sounded like a hymn. It went to a very slow tune.

La brigantine Qui va tourner, Roule et s'incline Pour m'entrainer. Oh, Vierge Marie, Pour moi priez Dieu! Adieu, patrie; Québec, adieu!

Tom Platt visited her, because, he said, the dead man was his brother as a Freemason. It came out that a wave had doubled the poor fellow over the heel of the bowsprit and broken his back. The news spread like a flash, for, contrary to general custom, the Frenchman held an auction of the dead man's kit,—he had no friends at St. Malo or Miquelon,—and everything was spread out on the top of the house, from his red knitted cap to the leather belt with the sheath-knife at the back. Dan and Harvey were out on twenty-fathom water in the Hattie S., and naturally rowed over to join the crowd. It was a long pull, and they stayed some little time while Dan bought the knife, which had a curious brass handle. When they dropped overside and pushed off into a drizzle of rain and a lop of sea, it occurred to them that they might get into trouble for neglecting the lines. "Guess 'twon't hurt us any to be warmed up," said Dan, shivering under his oilskins, and they rowed on into the heart of a white fog, which, as usual, dropped on them without warning.

"There's too much blame tide hereabouts to trust to your instinks," he said. "Heave over the anchor, Harve, and we'll fish a piece till the thing lifts. Bend on your biggest lead. Three pound ain't any too much in this water. See how she's tightened on her rodin' already."

There was quite a little bubble at the bows, where some irresponsible Bank current held the dory full stretch on her rope; but they could not see a boat's length in any direction. Harvey turned up his collar and bunched himself over his reel with the air of a wearied navigator. Fog had no special terrors for him now. They fished awhile in silence, and found the cod struck on well. Then Dan drew the sheath-knife and tested the edge of it on the gunwale.

"That's a daisy," said Harvey. "How did you get it so cheap?"

"On account o' their blame Cath'lic superstitions," said Dan, jabbing with the bright blade. "They don't fancy takin' iron frum off of a dead man, so to speak. 'See them Arichat Frenchmen step back when I bid?"

"But an auction ain't taking anything off a dead man. It's business."

"We know it ain't, but there's no goin' in the teeth o' superstition. That's one o' the advantages o' livin' in a progressive country." And Dan began whistling:

"Oh, Double Thatcher, how are you? Now Eastern Point comes inter view. The girls an' boys we soon shall see, At anchor off Cape Ann!"

"Why didn't that Eastport man bid, then? He bought his boots. Ain't Maine progressive?"

"Maine? Pshaw! They don't know enough, or they hain't got money enough, to paint their haouses in Maine. I've seen 'em. The Eastport man he told me that the knife had been used—so the French captain told him—used up on the French coast last year."

"Cut a man? Heave's the muckle." Harvey hauled in his fish, rebaited, and threw over.

"Killed him! 'Course, when I heard that I was keener 'n ever to get it."

"Christmas! I didn't know it," said Harvey, turning round. "I'll give you a dollar for it when I—get my wages. Say, I'll give you two dollars."

"Honest? D'you like it as much as all that?" said Dan, flushing. "Well, to tell the truth, I kinder got it for you—to give; but I didn't let on till I saw how you'd take it. It's yours and welcome, Harve, because we're dory-mates, and so on and so forth, an' so followin'. Catch a-holt!"

He held it out, belt and all.

"But look at here. Dan, I don't see—"

"Take it. 'Tain't no use to me. I wish you to hev it."

The temptation was irresistible. "Dan, you're a white man," said Harvey. "I'll keep it as long as I live."

"That's good hearin'," said Dan, with a pleasant laugh; and then, anxious to change the subject: "Look's if your line was fast to somethin'."

"Fouled, I guess," said Harve, tugging. Before he pulled up he fastened the belt round him, and with deep delight heard the tip of the sheath click on the thwart. "Concern the thing!" he cried. "She acts as though she were on strawberry-bottom. It's all sand here, ain't it'?"

Dan reached over and gave a judgmatic tweak. "Holibut'll act that way 'f he's sulky. Thet's no strawberry-bottom. Yank her once or twice. She gives, sure. 'Guess we'd better haul up an' make certain."

They pulled together, making fast at each turn on the cleats, and the hidden weight rose sluggishly.

"Prize, oh! Haul!" shouted Dan, but the shout ended in a shrill, double shriek of horror, for out of the sea came—the body of the dead Frenchman buried two days before! The hook had caught him under the right armpit, and he swayed, erect and horrible, head and shoulders above water. His arms were tied to his side, and—he had no face. The boys fell over each other in a heap at the bottom of the dory, and there they lay while the thing bobbed alongside, held on the shortened line.

"The tide—the tide brought him!" said Harvey, with quivering lips, as he fumbled at the clasp of the belt.

"Oh, Lord! Oh, Harve!" groaned Dan, "be quick. He's come for it. Let him have it. Take it off."

"I don't want it! I don't want it!" cried Harvey. "I can't find the bu-buckle."

"Quick, Harve! He's on your line!"

Harvey sat up to unfasten the belt, facing the head that had no face under its streaming hair. "He's fast still," he whispered to Dan, who slipped out his knife and cut the line, as Harvey flung the belt far over-side. The body shot down with a plop, and Dan cautiously rose to his knees, whiter than the fog.

"He come for it. He come for it. I've seen a stale one hauled up on a trawl and I didn't much care, but he come to us special."

"I wish—I wish I hadn't taken the knife. Then he'd have come on your line."

"Dunno as thet would ha' made any differ. We're both scared out o' ten years' growth. Oh, Harve, did ye see his head?"

"Did I? I'll never forget it. But look at here, Dan; it couldn't have been meant. It was only the tide."

"Tide! He come for it, Harve. Why, they sunk him six mile to south'ard o' the Fleet, an' we're two miles from where she's lyin' now. They told me he was weighted with a fathom an' a half o' chain-cable."

"Wonder what he did with the knife—up on the French coast?"

"Something bad. 'Guess he's bound to take it with him to the Judgment, an' so—What are you doin' with the fish?"

"Heaving 'em overboard," said Harvey.

"What for? We sha'n't eat 'em."

"I don't care. I had to look at his face while I was takin' the belt off. You can keep your catch if you like. I've no use for mine."

Dan said nothing, but threw his fish over again.

"'Guess it's best to be on the safe side," he murmured at last. "I'd give a month's pay if this fog 'u'd lift. Things go abaout in a fog that ye don't see in clear weather—yo-hoes an' hollerers and such like. I'm sorter relieved he come the way he did instid o' walkin'. He might ha' walked."

"Do-on't, Dan! We're right on top of him now. 'Wish I was safe aboard, bein' pounded by Uncle Salters."

"They'll be lookin' fer us in a little. Gimme the tooter." Dan took the tin dinner-horn, but paused before he blew.

"Go on," said Harvey. "I don't want to stay here all night."

"Question is, haow he'd take it. There was a man frum down the coast told me once he was in a schooner where they darsen't ever blow a horn to the dories, becaze the skipper—not the man he was with, but a captain that had run her five years before—he'd drownded a boy alongside in a drunk fit; an' ever after, that boy he'd row alongside too and shout, 'Dory! dory!' with the rest."

"Dory! dory!" a muffled voice cried through the fog. They cowered again, and the horn dropped from Dan's hand.

"Hold on!" cried Harvey; "it's the cook."

"Dunno what made me think o' thet fool tale, either," said Dan. "It's the doctor, sure enough."

"Dan! Danny! Oooh, Dan! Harve! Harvey! Oooh, Haarveee!"

"We're here," sung both boys together. They heard oars, but could see nothing till the cook, shining and dripping, rowed into them.

"What iss happened?" said he. "You will be beaten at home."

"Thet's what we want. Thet's what we're sufferin' for," said Dan. "Anything homey's good enough fer us. We've had kinder depressin' company." As the cook passed them a line, Dan told him the tale.

"Yess! He come for hiss knife," was all he said at the end.

Never had the little rocking "We're Here" looked so deliciously home—like as when the cook, born and bred in fogs, rowed them back to her. There was a warm glow of light from the cabin and a satisfying smell of food forward, and it was heavenly to hear Disko and the others, all quite alive and solid, leaning over the rail and promising them a first-class pounding. But the cook was a black master of strategy. He did not get the dories aboard till he had given the more striking points of the tale, explaining as he backed and bumped round the counter how Harvey was the mascot to destroy any possible bad luck. So the boys came overside as rather uncanny heroes, and every one asked them questions instead of pounding them for making trouble. Little Penn delivered quite a speech on the folly of superstitions; but public opinion was against him and in favour of Long Jack, who told the most excruciating ghost-stories to nearly midnight. Under that influence no one except Salters and Penn said anything about "idolatry" when the cook put a lighted candle, a cake of flour and water, and a pinch of salt on a shingle, and floated them out astern to keep the Frenchman quiet in case he was still restless. Dan lit the candle because he had bought the belt, and the cook grunted and muttered charms as long as he could see the ducking point of flame.

Said Harvey to Dan, as they turned in after watch: "How about progress and Catholic superstitions?"

"Huh! I guess I'm as enlightened and progressive as the next man, but when it comes to a dead St. Malo deck-hand scarin' a couple o' pore boys stiff fer the sake of a thirty-cent knife, why, then, the cook can take hold fer all o' me. I mistrust furriners, livin' or dead."

Next morning all, except the cook, were rather ashamed of the ceremonies, and went to work double tides, speaking gruffly to one another.

The "We're Here" was racing neck and neck for her last few loads against the "Parry Norman"; and so close was the struggle that the Fleet took sides and betted tobacco. All hands worked at the lines or dressing-down till they fell asleep where they stood—beginning before dawn and ending when it was too dark to see. They even used the cook as pitcher, and turned Harvey into the hold to pass salt, while Dan helped to dress down. Luckily a "Parry Norman" man sprained his ankle falling down the fo'c'sle, and the "We're Heres" gained. Harvey could not see how one more fish could be crammed into her, but Disko and Tom Platt stowed and stowed, and planked the mass down with big stones from the ballast, and there was always "jest another day's work." Disko did not tell them when all the salt was wetted. He rolled to the lazarette aft the cabin and began hauling out the big mainsail. This was at ten in the morning. The riding-sail was down and the main- and topsail were up by noon, and dories came alongside with letters for home, envying their good fortune. At last she cleared decks, hoisted her flag,—as is the right of the first boat off the Banks,—up-anchored, and began to move. Disko pretended that he wished to accommodate folk who had not sent in their mail, and so worked her gracefully in and out among the schooners. In reality, that was his little triumphant procession, and for the fifth year running it showed what kind of mariner he was. Dan's accordion and Tom Platt's fiddle supplied the music of the magic verse you must not sing till all the salt is wet:

Hih! Yih! Yoho!
Send your letters raound!

All our salt is wetted, an' the anchor's off the graound!
Bend, oh, bend your mains'l!, we're back to Yankeeland—
With fifteen hunder' quintal,
An' fifteen hunder' quintal,
"Teen hunder" toppin' quintal,
"Twix" old "Queereau an" Grand.

The last letters pitched on deck wrapped round pieces of coal, and the Gloucester men shouted messages to their wives and womenfolk and owners, while the "We're Here" finished the musical ride through the Fleet, her head-sails quivering like a man's hand when he raises it to say good-bye.

Harvey very soon discovered that the "We're Here", with her riding-sail, strolling from berth to berth, and the "We're Here" headed west by south under home canvas, were two very different boats. There was a bite and kick to the wheel even in "boy's" weather; he could feel the dead weight in the hold flung forward mightily across the surges, and the streaming line of bubbles overside made his eyes dizzy.

Disko kept them busy fiddling with the sails; and when those were flattened like a racing yacht's, Dan had to wait on the big topsail, which was put over by hand every time she went about. In spare moments they pumped, for the packed fish dripped brine, which does not improve a cargo. But since there was no fishing, Harvey had time to look at the sea from another point of view. The low-sided schooner was naturally on most intimate terms with her surroundings. They saw little of the horizon save when she topped a swell; and usually she was elbowing, fidgeting, and coaxing her steadfast way through grey, grey-blue, or black hollows laced across and across with streaks of shivering foam; or rubbing herself caressingly along the flank of some bigger water-hill. It was as if she said: "You wouldn't hurt me, surely? I'm only the little 'We're Here'." Then she would slide away chuckling softly to herself till she was brought up by some fresh obstacle. The dullest of folk cannot see this kind of thing hour after hour through long days without noticing it; and Harvey, being anything but dull, began to comprehend and enjoy the dry chorus of wave-tops turning over with a sound of incessant tearing; the hurry of the winds working across open spaces and herding the purple-blue cloud-shadows; the splendid upheaval of the red sunrise; the folding and packing away of the morning mists, wall after wall withdrawn across the white floors; the salty glare and blaze of noon; the kiss of rain falling over thousands of dead, flat square miles; the chilly blackening of everything at the day's end; and the million wrinkles of the sea under the moonlight, when the jib-boom solemnly poked at the low stars, and Harvey went down to get a doughnut from the cook.

But the best fun was when the boys were put on the wheel together, Tom Platt within hail, and she cuddled her lee-rail down to the crashing blue, and kept a little home-made rainbow arching unbroken over her windlass. Then the jaws of the booms whined against the masts, and the sheets creaked, and the sails filled with roaring; and when she slid into a hollow she trampled like a woman tripped in her own silk dress, and came out, her jib wet half-way up, yearning and peering for the tall twin-lights of Thatcher's Island.

They left the cold grey of the Bank sea, saw the lumber-ships making for Quebec by the Straits of St. Lawrence, with the Jersey salt-brigs from Spain and Sicily; found a friendly northeaster off Artimon Bank that drove them within view of the East light of Sable Island,—a sight Disko did not linger over,—and stayed with them past Western and Le Have, to the northern fringe of George's. From there they picked up the deeper water, and let her go merrily.

"Hattie's pulling on the string," Dan confided to Harvey. "Hattie an' ma. Next Sunday you'll be hirin' a boy to throw water on the windows to make ye go to sleep. 'Guess you'll keep with us till your folks come. Do you know the best of gettin' ashore again?"

"Hot bath?" said Harvey. His eyebrows were all white with dried spray.

"That's good, but a night-shirt's better. I've been dreamin' o' night-shirts ever since we bent our mainsail. Ye can wiggle your toes then. Ma'll hev a new one fer me, all washed soft. It's home, Harve. It's home! Ye can sense it in the air. We're runnin' into the aidge of a hot wave naow, an' I can smell the bayberries. Wonder if we'll get in fer supper. Port a trifle."

The hesitating sails flapped and lurched in the close air as the deep smoothed out, blue and oily, round them. When they whistled for a wind only the rain came in spiky rods, bubbling and drumming, and behind the rain the thunder and the lightning of mid-August. They lay on the deck with bare feet and arms, telling one another what they would order at their first meal ashore; for now the land was in plain sight. A Glouces-ter swordfish-boat drifted alongside, a man in the little pulpit on the bowsprit flourishing his harpoon, his bare head plastered down with the wet. "And all's well!" he sang cheerily, as though he were watch on a big liner. "Wouverman's waiting fer you, Disko. What's the news o' the Fleet?"

Disko shouted it and passed on, while the wild summer storm pounded overhead and the lightning flickered along the capes from four different quarters at once. It gave the low circle of hills round Gloucester Harbour, Ten Pound Island, the fish-sheds, with the broken line of house-roofs, and each spar and buoy on the water, in blinding photographs that came and went a dozen times to the minute as the "We're Here" crawled in on half-flood, and the whistling-buoy moaned and mourned behind her. Then the storm died out in long, separated, vicious dags of blue-white flame, followed by a single roar like the roar of a mortar-battery, and the shaken air tingled under the stars as it got back to silence.

"The flag, the flag!" said Disko, suddenly, pointing upward.

"What is ut?" said Long Jack.

"Otto! Ha'af mast. They can see us frum shore now."

"I'd clean forgot. He's no folk to Gloucester, has he?"

"Girl he was goin' to be married to this fall."

"Mary pity her!" said Long Jack, and lowered the little flag half-mast for the sake of Otto, swept overboard in a gale off Le Have three months before.

Disko wiped the wet from his eyes and led the "We're Here" to Wouverman's wharf, giving his orders in whispers, while she swung round moored tugs and night-watchmen hailed her from the ends of inky-black piers. Over and above the darkness and the mystery of the procession, Harvey could feel the land close round him once more, with all its thou-sands of people asleep, and the smell of earth after rain, and the familiar noise of a switching-engine coughing to herself in a freight-yard; and all those things made his heart beat and his throat dry up as he stood by the foresheet. They heard the anchor-watch snoring on a lighthouse-tug, nosed into a pocket of darkness where a lantern glimmered on either side; somebody waked with a grunt, threw them a rope, and they made fast to a silent wharf flanked with great iron-roofed sheds full of warm emptiness, and lay there without a sound.

Then Harvey sat down by the wheel, and sobbed and sobbed as though his heart would break, and a tall woman who had been sitting on a weigh-scale dropped down into the schooner and kissed Dan once on the cheek; for she was his mother, and she had seen the "We're Here" by the lightning-flashes. She took no notice of Harvey till he had recovered himself a little and Disko had told her his story. Then they went to Disko's house together as the dawn was breaking; and until the telegraph office was open and he could wire to his folk, Harvey Cheyne was perhaps the loneliest boy in all America. But the curious thing was that Disko and Dan seemed to think none the worse of him for crying.

Wouverman was not ready for Disko's prices till Disko, sure that the "We're Here" was at least a week ahead of any other Gloucester boat, had given him a few days to swallow them; so all hands played about the streets, and Long Jack stopped the Rocky Neck trolley, on principle, as he said, till the conductor let him ride free. But Dan went about with his freckled nose in the air, bungful of mystery and most haughty to his family.

"Dan, I'll hev to lay inter you ef you act this way," said Troop, pensively. "Sence we've come ashore this time you've bin a heap too fresh."

"I'd lay into him naow ef he was mine," said Uncle Salters, sourly. He and Penn boarded with the Troops.

"Oho!" said Dan, shuffling with the accordion round the back-yard, ready to leap the fence if the enemy advanced. "Dad, you're welcome to your own jedgment, but remember I've warned ye. Your own flesh an' blood ha' warned ye! 'Tain't any o' my fault ef you're mistook, but I'll be on deck to watch ye. An' ez fer yeou, Uncle Salters, Pharaoh's chief butler ain't in it 'longside o' you! You watch aout an' wait. You'll be ploughed under like your own blamed clover; but me—Dan Troop—I'll flourish like a green bay-tree because I warn't stuck on my own opinion."

Disko was smoking in all his shore dignity and a pair of beautiful carpet-slippers. "You're gettin' ez crazy as poor Harve. You two go araound gigglin' an' squinchin' an' kickin' each other under the table till there's no peace in the haouse," said he.

"There's goin' to be a heap less—fer some folks," Dan replied. "You wait an' see."

He and Harvey went out on the trolley to East Gloucester, where they tramped through the bayberry-bushes to the lighthouse, and lay down on the big red boulders and laughed themselves hungry. Harvey had shown Dan a telegram, and the two swore to keep silence till the shell burst.

"Harve's folk?" said Dan, with an unruffled face after supper. "Well, I guess they don't amount to much of anything, or we'd ha' heard frum 'em by naow. His pop keeps a kind o' store out West. Maybe he'll give you's much as five dollars, dad."

"What did I tell ye?" said Salters. "Don't sputter over your vittles, Dan."

NINE

Loss of the Whaleship *Essex*

Owen Chase

THE SHIP *ESSEX*, COMMANDED BY CAPTAIN GEORGE POLLARD, JUNIOR, was fitted out at Nantucket, and sailed on the 12th day of August, 1819, for the Pacific Ocean, on a whaling voyage. Of this ship I was first mate. She had lately undergone a thorough repair in her upper works, and was at that time, in all respects, a sound, substantial vessel: she had a crew of twenty-one men, and was victualled and provided for two years and a half.

* * *

I have not been able to recur to the scenes which are now to become the subject of description, although a considerable time has elapsed, without feeling a mingled emotion of horror and astonishment at the almost incredible destiny that has preserved me and my surviving companions from a terrible death.

Frequently, in my reflections on the subject, even after this lapse of time, I find myself shedding tears of gratitude for our deliverance, and blessing God, by whose divine aid and protection we were conducted through a series of unparalleled suffering and distress, and restored to the bosoms of our families and friends.

On the 20th of November (cruising in latitude 0° 40' S., longitude 119° 0' W.) a shoal of whales was discovered off the lee-bow. The weather at this time was extremely fine and clear, and it was about eight o'clock in the morning that the man at the masthead gave the usual cry of "there she blows." The ship was immediately put away, and we ran down in the direction for them. When we had got within half a mile of the place where they were observed, all our boats were lowered down, manned, and we started in pursuit of them.

The captain and the second mate, in the other two boats, kept up the pursuit, and soon struck another whale. They being at this time a considerable distance to leeward, I went forward, braced around the mainyard, and put the ship off in a direction for them.

I observed a very large spermaceti whale, as well as I could judge about eighty-five feet in length; he broke water about twenty rods off our weather-bow, and was lying quietly, with his head in a direction for the ship. He spouted two or three times, and then disappeared. In less than two or three seconds he came up again, about the length of the ship off, and made directly for us, at the rate of about three knots. The ship was then going with about the same velocity. His appearance and attitude gave us at first no alarm; but while I stood watching his movements, and observing him but a ship's length off, coming down for us with great celerity, I involuntarily ordered the boy at the helm to put it hard up; intending to sheer off and avoid him. The words were scarcely out of my mouth, before he came down upon us with full speed, and struck the ship with his head, just forward of the fore-chains; he gave us such an appalling and tremendous jar, as nearly threw us all on our faces.

The ship brought up as suddenly and violently as if she had struck a rock, and trembled for a few seconds like a leaf. We looked at each other with perfect amazement, deprived almost of the power of speech. Many minutes elapsed before we were able to realize the dreadful accident; during which time he passed under the ship, grazing her keel as he went along, came up alongside of her to leeward, and lay on the top of the water (apparently stunned with the violence of the blow) for the space of a minute; he then suddenly started off, in a direction to leeward. After a few moments' reflection, and recovering, in some measure, from the sudden consternation that had seized us, I of course concluded that he had stove a hole in the ship, and that it would be necessary to set the pumps going.

Accordingly they were rigged, but had not been in operation more than one minute before I perceived the head of the ship to be gradually settling down in the water; I then ordered the signal to be set for the other boats, which, scarcely had I despatched, before I again discovered the whale, apparently in convulsions, on the top of the water, about one hundred rods to leeward. He was enveloped in the foam of the sea, that his continual and violent thrashing about in the water had created around him, and I could distinctly see him smite his jaws together, as if distracted with rage and fury. He remained a short time in this situation, and then started off with great velocity, across the bows of the ship, to windward. By this time the ship had settled down a considerable distance in the water, and I gave her up for lost. I, however, ordered the pumps to be kept constantly going, and endeavoured to collect my thoughts for the occasion. I turned to the boats, two of which we then had with the ship, with an intention of clearing them away, and getting all things ready to embark in them, if there should be no other resource left; and while my attention was thus engaged for a moment, I was aroused with the cry of a man at the hatch way, "here he is—he is making for us again."

I turned around, and saw him about one hundred rods directly ahead of us, coming down apparently with twice his ordinary speed, and to me at that moment, it appeared with tenfold fury and vengeance in his aspect. The surf flew in all directions about him, and his course towards us was marked by a white foam of a rod in width, which he made with the continual violent thrashing of his tail; his head was about half out of water, and in that way he came upon, and again struck the ship. I was in hopes when I descried him making for us, that by a dexterous movement of putting the ship away immediately, I should be able to cross the line of his approach, before he could get up to us, and thus avoid what I knew, if he should strike us again, would prove our inevitable destruction.

I bawled out to the helmsman, "hard up!" but she had not fallen off more than a point, before we took the second shock. I should judge the speed of the ship to have been at this time about three knots, and that of the whale about six. He struck her to windward, directly under the cat-head, and completely stove in her bows. He passed under the ship again, went off to leeward, and we saw no more of him.

Not a moment, however, was to be lost in endeavouring to provide for the extremity to which it was now certain we were reduced. We were more than a thousand miles from the nearest land, and with nothing but a light open boat, as the resource of safety for myself and companions. I ordered the men to cease pumping, and everyone to provide for himself; seizing a hatchet at the same time, I cut away the lashings of the spare boat, which lay bottom-up across two spars directly over the quarter deck, and cried out to those near me to take her as she came down.

From the time we were first attacked by the whale, to the period of the fall of the ship, and of our leaving her in the boat, more than ten minutes could not certainly have elapsed! God only knows in what way, or by what means, we were enabled to accomplish in that short time what we did; the cutting away and transporting the boat from where she was deposited would of itself, in ordinary circumstances, have consumed as much time as that, if the whole ship's crew had been employed in it. My companions had not saved a single article but what they had on their backs; but to me it was a source of infinite satisfaction, if any such could be gathered from the horrors of our gloomy situation, that we had been fortunate enough to have preserved our compasses, navigators, and quadrants.

After the first shock of my feelings was over, I enthusiastically contemplated them as the probable instruments of our salvation; without them all would have been dark and hopeless. Gracious God! what a picture of distress and suffering now presented itself to my imagination. The crew of the ship were saved, consisting of twenty human souls. All that remained to conduct these twenty beings through the stormy terrors of the ocean, perhaps many thousand miles, were three open light boats. The prospect of obtaining any provisions or water from the ship, to subsist upon during the time, was at least now doubtful. How many long and watchful nights, thought I, are to be passed?

We lay at this time in our boat, about two ship lengths off from the wreck, in perfect silence, calmly contemplating her situation, and absorbed in our own melancholy reflections, when the other boats were discovered rowing up to us. They had but shortly before discovered that some accident had befallen us, but of the nature of which they were entirely ignorant. The sudden and mysterious disappearance of the ship was first discovered by the boat-steerer in the captain's boat, and with a horrorstruck countenance and voice, he suddenly exclaimed, "Oh, my God! where is the ship?" Their operations upon this were instantly suspended, and a general cry of horror and despair burst from the lips of every man, as their looks were directed for her, in vain, over every part of the ocean.

They immediately made all haste towards us. The captain's boat was the first that reached us. He stopped about a boat's length off, but had no power to utter a single syllable: he was so completely overpowered with the spectacle before him that he sat down in his boat, pale and speechless. I could scarcely recognise his countenance, he appeared to be so much altered, awed, and overcome with the oppression of his feelings, and the dreadful reality that lay before him. He was in a short time however enabled to address the inquiry to me, "My God, Mr. Chase, what is the matter?"

I answered, "We have been stove by a whale."

After a few moments' reflection he observed that we must cut away her masts, and endeavour to get something out of her to eat. Our thoughts were now all accordingly bent on endeavours to save from the wreck whatever we might possibly want, and for this purpose we rowed up and got onto her. Search was made for every means of gaining access to her hold; and for this purpose the lanyards were cut loose, and with our hatchets we commenced to cut away the masts, that she might right up again, and enable us to scuttle her decks. In doing which we were occupied about three quarters of an hour, owing to our having no axes, nor indeed any other instruments, but the small hatchets belonging to the boats.

After her masts were gone she came up about two-thirds of the way upon an even keel. While we were employed about the masts the captain took his quadrant, shoved off from the ship, and got an observation. We now commenced to cut a hole through the planks, directly above two large casks of bread, which most fortunately were between decks, in the waist of the ship, and which being in the upper side, when she upset, we had strong hopes was not wet. It turned out according to our wishes, and from these casks we obtained six hundred pounds of hard bread. Other parts of the deck were then scuttled, and we got without difficulty as much fresh water as we dared to take in the boats, so that each was supplied with about sixty-five gallons; we got also from one of the lockers a musket, a small canister of powder, a couple of files, two rasps, about two pounds of boat nails, and a few turtle.

In the afternoon the wind came on to blow a strong breeze; and having obtained everything that occurred to us could then be got out, we began to make arrangements for our safety during the night. A boat's line was made fast to the ship, and to the other end of it one of the boats was moored, at about fifty fathoms to leeward; another boat was then attached to the first one, about eight fathoms astern; and the third boat, the like distance astern of her.

Night came on just as we had finished our operations; and such a night as it was to us! so full of feverish and distracting inquietude, that we were deprived entirely of rest. The wreck was constantly before my eyes. I could not, by any effort, chase away the horrors of the preceding day from my mind.

NOVEMBER 21ST

The morning dawned upon our wretched company. The weather was fine, but the wind blew a strong breeze from the SE. and the sea was very rugged. Watches had been kept up during the night, in our respective boats, to see that none of the spars or other articles (which continued to float out of the wreck) should be thrown by the surf against, and injure the boats. At sunrise, we began to think of doing something; what, we did not know: we cast loose our boats, and visited the wreck, to see if anything more of consequence could be preserved, but everything looked cheerless and desolate, and we made a long and vain search for any useful article; nothing could be found but a few turtle; of these we had enough already; or at least, as many as could be safely stowed in the boats, and we wandered around in every part of the ship in a sort of vacant idleness for the greater part of the morning.

We were presently aroused to a perfect sense of our destitute and forlorn condition, by thoughts of the means which we had for our subsistence, the necessity of not wasting our time, and of endeavouring to seek some relief wherever God might direct us. Our thoughts, indeed, hung about the ship, wrecked and sunken as she was, and we could scarcely discard from our minds the idea of her continuing protection.

Some great efforts in our situation were necessary, and a great deal of calculation important, as it concerned the means by which our existence was to be supported during, perhaps, a very long period, and a provision for our eventual deliverance. Accordingly, by agreement, all set to work in stripping off the light sails of the ship, for sails to our boats; and the day was consumed in making them up and fitting them. We furnished ourselves with masts and other light spars that were necessary, from the wreck. Each boat was rigged with two masts, to carry a flying-jib and two sprit-sails; the sprit-sails were made so that two reefs could be taken in them, in case of heavy blows.

We continued to watch the wreck for any serviceable articles that might float from her, and kept one man during the day, on the stump of her foremast, on the lookout for vessels. Our work was very much impeded by the increase of the wind and sea, and the surf breaking almost continually into the boats gave us many fears that we should not be able to prevent our provisions from getting wet; and above all served to increase the constant apprehensions that we had of the insufficiency of the boats themselves during the rough weather that we should necessarily experience.

In order to provide as much as possible against this, and withal to strengthen the slight materials of which the boats were constructed, we procured from the wreck some light cedar boards (intended to repair boats in cases of accidents) with which we built up additional sides, about six inches above the gunwale; these, we afterwards found, were of infinite service for the purpose for which they were intended; in truth, I am satisfied we could never have been preserved without them.

I got an observation today, by which I found we were in latitude 0° 6' S., longitude 119° 30' W., having been driven by the winds a distance of forty-nine miles the last twenty-four hours; by this it would appear that there must have been a strong current, setting us to the N.W. during the whole time.

The captain, after visiting the wreck, called a council, consisting of himself and the first and second mates, who all repaired to his boat, to interchange with the variable winds, and then, endeavour to get eastward to the coast of Chile or Peru. Accordingly, preparations were made for our immediate departure; the boat which it was my fortune, or rather misfortune to have, was the worst of the three, she was old and patched up, having been stove a number of times during the cruise. At best, a whaleboat is an extremely frail thing; the most so of any other kind of boat; they are what is called clinker built, and constructed of the lightest materials, for the purpose of being rowed with the greatest possible celerity according to the necessities of the business for which they are intended.

In consideration of my having the weakest boat, six men were allotted to it; while those of the captain and second mate took seven each, and at half past twelve we left the wreck, steering our course, with nearly all sail set, S.SE. At four o'clock in the afternoon we lost sight of her entirely. Many were the lingering and sorrowful looks we cast behind us.

The wind was strong all day; and the sea ran very high, our boat taking in water from her leaks continually, so that we were as that depended, most especially, on a reasonable calculation, and on our own labours, we conceived that our provision and water, on a small allowance, would last us sixty days; that with the trade wind, on the course we were then lying, we should be able to average the distance of a degree a day, which, in twenty-six days, would enable us to attain the region of the variable winds, and then, in thirty more, at the very utmost, should there be any favor in the elements, we might reach the coast.

Our allowance of provision at first consisted of bread, one biscuit, weighing about one pound three ounces, and half a pint of water a day, for each man. This small quantity (less than one third which is required by an ordinary person), small as it was, we however took without murmuring, and, on many occasions afterwards, blest God that even this pittance was allowed to us in our misery. The darkness of another night overtook us; and after having for the first time partook of our allowance of bread and water, we laid our weary bodies down in the boat, and endeavoured to get some repose.

November 23rd

In my chest, which I was fortunate enough to preserve, I had several small articles which we found of great service to us; among the rest, some eight or ten sheets of writing paper, a lead pencil, a suit of clothes, three small fishhooks, a jackknife, a whetstone, and a cake of soap. I commenced to keep a sort of journal with the little paper and pencil which I had; and the knife, besides other useful purposes, served us as a razor. It was with much difficulty, however, that I could keep any sort of record, owing to the incessant rocking and unsteadiness of the boat, and the continual dashing of the spray of the sea over us.

The boat contained, in addition to the articles enumerated, a lantern, tinder box, and two or three candles, which belonged to her, and with which they are kept always supplied while engaged in taking whale. In addition to all which, the captain had saved a musket, two pistols, and a canister containing about two pounds of gunpowder; the latter he distributed in equal proportions between the three boats, and gave the second mate and myself each a pistol.

When morning came we found ourselves quite near together, and the wind had considerably increased since the day before; we were consequently obliged to reef our sails; and although we did not apprehend any very great danger from the then-violence of the wind, yet it grew to be very uncomfortable in the boats from the repeated dashing of the waves that kept our bodies constantly wet with the salt spray. We, however, stood along our course until twelve o'clock, when we got an observation, as well as we were able to obtain one, while the water flew all over us, and the sea kept the boat extremely unsteady. We found ourselves this day in latitude 0° 58' S. having repassed the equator.

We abandoned the idea altogether of keeping any correct longitudinal reckoning, having no glass, nor log-line. The wind moderated in the course of the afternoon a little, but at night came on to blow again almost a gale. We began now to tremble for our little barque; she was so ill calculated, in point of strength, to withstand the racking of the sea, while it required the constant labours of one man to keep her free of water. We were surrounded in the afternoon with porpoises that kept playing about us in great numbers, and continued to follow us during the night.

November 26th

Our sufferings, heaven knows, were now sufficiently increased, and we looked forward, not without an extreme dread, and anxiety, to the gloomy and disheartening prospect before us. We experienced a little abatement of wind and rough weather today, and took the opportunity of drying the bread that had been wet the day previously; to our great joy and satisfaction also, the wind hauled out to E.NE. and enabled us to hold a much more favorable course; with these exceptions, no circumstance of any considerable interest occurred in the course of this day.

November 30th

This was a remarkably fine day; the weather not exceeded by any that we had experienced since we left the wreck. At one o'clock, I proposed to our boat's crew to kill one of the turtle; two of which we had in our possession. I need not say that the proposition was hailed with the utmost enthusiasm; hunger had set its ravenous gnawings upon our stomachs, and we waited with impatience to suck the warm flowing blood of the animal. A small fire was kindled in the shell of the turtle, and after dividing the blood (of which there was about a gill) among those of us who felt disposed to drink it, we cooked the remainder, entrails and all, and enjoyed from it an unspeakably fine repast. The stomachs of two or three revolted at the sight of the blood, and refused to partake of it; not even the outrageous thirst that was upon them could induce them to taste it; for myself, I took it like a medicine, to relieve the extreme dryness of my palate, and stopped not to inquire whether it was anything else than a liquid. After this, I may say exquisite banquet, our bodies were considerably recruited, and I felt my spirits now much higher than they had been at any time before. By observation, this day we found ourselves in latitude 7° 53' S. Our distance from the wreck, as nearly as we could calculate, was then about four hundred and eighty miles.

December 3rd

With great joy we hailed the last crumb of our damaged bread, and commenced this day to take our allowance of healthy provisions. The salutary and agreeable effects of this change were felt at first in so slight a degree as to give us no great satisfaction; but gradually, as we partook of our small allowance of water, the moisture began to collect in our mouths, and the parching fever of the palate imperceptibly left it.

An accident here happened to us which gave us a great momentary spell of uneasiness. The night was dark, and the sky was completely overcast, so that we could scarcely discern each other's boats, when at about ten o'clock, that of the second mate was suddenly missing. I felt for a moment considerable alarm at her unexpected disappearance; but after a little reflection I immediately hove to, struck a light as expeditiously as possible, and hoisted it at the masthead, in a lantern. Our eyes were now directed over every part of the ocean, in search of her, when, to our great joy, we discerned an answering light, about a quarter of a mile to leeward of us; we ran down to it, and it proved to be the lost boat.

Strange as the extraordinary interest which we felt in each other's company may appear, and much as our repugnance to separation may seem to imply of weakness, it was the subject of our continual hopes and fears. It is truly remarked that misfortune more than anything else serves to endear us to our companions.

There is no question but that an immediate separation, therefore, was the most politic measure that could be adopted, and that every boat should take its own separate chance: while we remained together, should any accident happen of the nature alluded to, no other course could be adopted than that of taking the survivors into the other boats, and giving up voluntarily what we were satisfied could alone prolong our hopes and multiply the chances of our safety, or unconcernedly witness their struggles in death, perhaps beat them from our boats, with weapons, back into the ocean.

December 8th

In the afternoon of this day the wind set in E.SE. and began to blow much harder than we had yet experienced it; by twelve o'clock at night it had increased to a perfect gale, with heavy showers of rain, and we now began, from these dreadful indications, to prepare ourselves for destruction. We continued to take in sail by degrees, as the tempest gradually increased, until at last we were obliged to take down our masts. At this juncture we gave up entirely to the mercy of the waves. The sea and rain had wet us to the skin, and we sat down, silently, and with sullen resignation, awaiting our fate. We made an effort to catch some fresh water by spreading one of the sails, but after having spent a long time, and obtained but a small quantity in a bucket, it proved to be quite as salty as that from the ocean: this we attributed to its having passed through the sail which had been so often wet by the sea, and upon which, after drying so frequently in the sun, concretions of salt had been formed. It was a dreadful night—cut off from any imaginary relief—nothing remained but to await the approaching issue with firmness and resignation.

The appearance of the heavens was dark and dreary, and the blackness that was spread over the face of the waters dismal beyond description. The heavy squalls, that followed each other in quick succession, were preceded by sharp flashes of lightning, that appeared to wrap our little barge in flames. The sea rose to a fearful height, and every wave that came looked as if it must be the last that would be necessary for our destruction. To an overruling Providence alone must be attributed our salvation from the horrors of that terrible night. It can be accounted for in no other way: that a speck of substance, like that which we were, before the driving terrors of the tempest, could have been conducted safely through it. At twelve o'clock it began to abate a little in intervals of two or three minutes, during which we would venture to raise up our heads and look to windward.

Our boat was completely unmanageable; without sails, mast, or rudder, and had been driven, in the course of the afternoon and night, we knew not whither, nor how far. When the gale had in some measure subsided we made efforts to get a little sail upon her, and put her head towards the course we had been steering. My companions had not slept any during the whole night, and were dispirited and broken down to such a degree as to appear to want some more powerful stimulus than the fears of death to enable them to do their duty. By great exertions, however, towards morning we again set a double-reefed mainsail and jib upon her, and began to make tolerable progress on the voyage. An unaccountable good fortune had kept the boats together during all the troubles of the night: and the sun rose and showed the disconsolate faces of our companions once more to each other.

DECEMBER 10TH –13TH

I have omitted to notice the gradual advances which hunger and thirst for the last six days, had made upon us. As the time had lengthened since our departure from the wreck, and the allowance of provision, making the demands of the appetite daily more and more importunate, they had created in us an almost uncontrollable temptation to violate our resolution, and satisfy, for once, the hard yearnings of nature from our stock; but a little reflection served to convince us of the imprudence and unmanliness of the measure, and it was abandoned with a sort of melancholy effort of satisfaction.

I had taken into custody, by common consent, all the provisions and water belonging to the boat, and was determined that no encroachments should be made upon it with my consent; nay, I felt myself bound, by every consideration of duty, by every dictate of sense, of prudence, and discretion, without which, in my situation, all other exertions would have been folly itself, to protect them, at the hazard of my life. For this purpose I locked up in my chest the whole quantity, and never, for a single moment, closed my eyes without placing some part of my person in contact with the chest; and having loaded my pistol, kept it constantly about me. I should not certainly have put any threats in execution as long as the most distant hopes of reconciliation existed; and was determined, in case the least refractory disposition should be manifested (a thing which I contemplated not unlikely to happen, with a set of starving wretches like ourselves) that I would immediately divide our subsistence into equal proportions, and give each man's share into his own keeping.

Then, should any attempt be made upon mine, which I intended to mete out to myself according to exigencies, I was resolved to make the consequences of it fatal. There was, however, the most upright and obedient behaviour in this respect manifested by every man in the boat, and I never had the least opportunity of proving what my conduct would have been on such an occasion.

While standing on our course this day we came across a small shoal of flying fish: four of which, in their efforts to avoid us, flew against the mainsail, and dropped into the boat; one having fallen near me, I eagerly snatched up and devoured; the other three were immediately taken by the rest, and eaten alive. For the first time I, on this occasion, felt a disposition to laugh, upon witnessing the ludicrous and almost desperate efforts of my five companions, who each sought to get a fish. They were very small of the kind, and constituted but an extremely delicate mouthful, scales, wings, and all, for hungry stomachs like ours.

From the eleventh to the thirteenth of December inclusive, our progress was very slow, owing to light winds and calms; and nothing transpired of any moment, except that on the eleventh we killed the only remaining turtle, and enjoyed another luxuriant repast, that invigorated our bodies, and gave a fresh flow to our spirits. The weather was extremely hot, and we were exposed to the full force of a meridian sun, without any covering to shield us from its burning influence, or the least breath of air to cool its parching rays.

December 15th –16th

Our boat continued to take in water so fast from her leaks, and the weather proving so moderate, we concluded to search out the bad places, and endeavour to mend them as well as we should be able. After a considerable search, and, removing the ceiling near the bows, we found the principal opening was occasioned by the starting of a plank or streak in the bottom of the boat, next to the keel. To remedy this, it was now absolutely necessary to have access to the bottom. The means of doing which did not immediately occur to our minds. After a moment's reflection, however, one of the crew, Benjamin Lawrence, offered to tie a rope around his body, take a boat's hatchet in his hand, and thus go under the water, and hold the hatchet against a nail, to be driven through from the inside, for the purpose of clenching it. This was, accordingly, all effected, with some little trouble, and answered the purpose much beyond our expectations. Our latitude was this day 21° 42' S.

The oppression of the weather still continuing through the sixteenth, bore upon our health and spirits with an amazing force and severity. The most disagreeable excitements were produced by it, which, added to the disconsolate endurance of the calm, called loudly for some mitigating expedient—some sort of relief to our prolonged sufferings. By our observations today we found, in addition to our other calamities, that we had been urged back from our progress, by the heave of the sea, a distance of ten miles; and were still without any prospect of wind.

DECEMBER 20TH

This was a day of great happiness and joy. After having experienced one of the most distressing nights in the whole catalogue of our sufferings, we awoke to a morning of comparative luxury and pleasure. About seven o'clock, while we were sitting dispirited, silent, and dejected, in our boats, one of our companions suddenly and loudly called out, "There is land!"

We were all aroused in an instant, as if electrified, and casting our eyes to leeward, there indeed, was the blessed vision before us, "as plain and palpable" as could be wished for. A new and extraordinary impulse now took possession of us. We shook off the lethargy of our senses, and seemed to take another, and a fresh existence.

One or two of my companions, whose lagging spirits and worn-out frames had begun to inspire them with an utter indifference to their fate, now immediately brightened up, and manifested a surprising alacrity and earnestness to gain, without delay, the much wished for shore. It appeared at first a low, white beach, and lay like a basking paradise before our longing eyes. It was discovered nearly at the same time by the other boats, and a general burst of joy and congratulation now passed between us. It is not within the scope of human calculation, by a mere listener to the story, to divine what the feelings of our hearts were on this occasion. Alternate expectation, fear, gratitude, surprise, and exultation, each swayed our minds, and quickened our exertions.

We ran down for it, and at eleven o'clock, a.m., we were within a quarter of a mile of the shore. It was an island, to all appearance, as nearly as we could determine it, about six miles long and three broad; with a very high, rugged shore, and surrounded by rocks; the sides of the mountains were bare, but on the tops it looked fresh and green with vegetation.

Upon examining our navigators, we found it was Ducie's Island, lying in latitude 24° 40' S., longitude 124° 40' W.

A short moment sufficed for reflection, and we made immediate arrangements to land. None of us knew whether the island was inhabited or not, nor what it afforded, if anything; if inhabited, it was uncertain whether by beasts or savages; and a momentary suspense was created by the dangers which might possibly arise by proceeding without due preparation and care. Hunger and thirst, however, soon determined us, and having taken the musket and pistols, I, with three others, effected a landing upon some sunken rocks, and waded thence to the shore. Upon arriving at the beach, it was necessary to take a little breath, and we laid down for a few minutes to rest our weak bodies before we could proceed. Let the reader judge, if he can, what must have been our feelings now! Bereft of all comfortable hopes of life, for the space of thirty days of terrible suffering; our bodies wasted to mere skeletons, by hunger and thirst, and death itself staring us in the face; to be suddenly and unexpectedly conducted to a rich banquet of food and drink, which subsequently we enjoyed for a few days, to our full satisfaction; and he will have but a faint idea of the happiness that here fell to our lot. We now, after a few minutes, separated, and went different directions in search of water; the want of which had been our principal privation, and called for immediate relief.

I had not proceeded far in my excursion, before I discovered a fish, about a foot and a half in length, swimming along in the water close to the shore. I commenced an attack upon him with the breach of my gun, and struck him, I believe, once and he ran under a small rock, that lay near the shore, from whence I took him with the aid of my ramrod, and brought him up on the beach, and immediately fell to eating. My companions soon joined in the repast; and in less than ten minutes, the whole was consumed, bones, and skin, and scales, and all. With full stomachs, we imagined we could now attempt the mountains, where, if in any part of the island, we considered water would be most probably obtained.

I accordingly clambered, with excessive labour, suffering, and pain, up amongst the bushes, roots, and underwood, of one of the crags, looking in all directions in vain, for every appearance of water that might present itself. There was no indication of the least moisture to be found, within the distance to which I had ascended, although my strength did not enable me to get them, which only served to whet his appetite, and from which nothing like the least satisfaction had proceeded.

I immediately resolved in my own mind, upon this information, to advise remaining until morning, to endeavour to make a more thorough search the next day, and with our hatchets to pick away the rock which had been discovered, with the view of increasing, if possible, the run of the water. We all repaired again to our boats, and there found that the captain had the same impressions as to the propriety of our delay until morning. We therefore landed; and having hauled our boats up on the beach, laid down in them that night, free from all the anxieties of watching and labour, and amid all our sufferings, gave ourselves up to an unreserved forgetfulness and peace of mind, that seemed so well to accord with the pleasing anticipations that this day had brought forth.

It was but a short space, however, until the morning broke upon us; and sense, and feeling, and gnawing hunger, and the raging fever of thirst then redoubled my wishes and efforts to explore the island again. We had obtained, that night, a few crabs, by traversing the shore a considerable distance, and a few very small fish; but waited until the next day, for the labours of which, we considered a night of refreshing and undisturbed repose would better qualify us.

DECEMBER 21ST

We had still reserved our common allowance, but it was entirely inadequate for the purpose of supplying the raging demands of the palate; and such an excessive and cruel thirst was created, as almost to deprive us of the power of speech. The lips became cracked and swollen, and a sort of glutinous saliva collected in the mouth, disagreeable to the taste, and intolerable beyond expression. Our bodies had wasted away to almost skin and bone, and possessed so little strength as often to require each other's assistance in performing some of its weakest functions. Relief, we now felt, must come soon, or nature would sink. The most perfect discipline was still maintained in respect to our provisions; and it now became our whole object, if we should not be able to replenish our subsistence from the island, to obtain, by some means or other, a sufficient refreshment to enable us to prosecute our voyage.

Our search for water accordingly again commenced with the morning; each of us took a different direction, and prosecuted the examination of every place where there was the least indication of it; the small leaves of the shrubbery, affording a temporary alleviation, by being chewed in the mouth, and but for the peculiarly bitter taste which those of the island possessed, would have been an extremely grateful substitute. In the course of our rambles too, along the sides of the mountain, we would now and then meet with tropic birds, of a beautiful figure and plumage, occupying small holes in the sides of it, from which we plucked them without the least difficulty. Upon our approaching them they made no attempts to fly, nor did they appear to notice us at all. These birds served us for a fine repast; numbers of which were caught in the course of the day, cooked by fires which we made on the shore, and eaten with the utmost avidity. We found also a plant, in taste not unlike the peppergrass, growing in considerable abundance in the crevices of the rocks, and which proved to us a very agreeable food, by being chewed with the meat of the birds. These, with birds' nests, some of them full of young, and others of eggs, a few of which we found in the course of the day, served us for food, and supplied the place of our bread; from the use of which, during our stay here, we had restricted ourselves.

But water, the great object of all our anxieties and exertions, was nowhere to be found, and we began to despair of meeting with it on the island. Our state of extreme weakness, and many of us without shoes or any covering for the feet, prevented us from exploring any great distance; lest by some sudden faintness, or overexertion, we should not be able to return, and at night be exposed to attacks of wild beasts, which might inhabit the island, and be alike incapable of resistance, as beyond the reach of the feeble assistance that otherwise could be afforded to each. The whole day was thus consumed in picking up whatever had the least shape or quality of sustenance, and another night of misery was before us, to be passed without a drop of water to cool our parching tongues. In this state of affairs, we could not reconcile it to ourselves to remain at this place; a day, an hour, lost to us unnecessarily here, might cost us our preservation. A drop of the water that we then had in our possession might prove, in the last stages of our debility, the very cordial of life. I addressed the substance of these few reflections to the captain, who agreed with me in opinion, upon the necessity of taking some decisive steps in our water. My principal hope was founded upon my success in picking the rocks where the moisture had been discovered the day before, and thither I hastened as soon as my strength would enable me to get there.

Upon examining the place from whence we had obtained this miraculous and unexpected succour, we were equally astonished and delighted with the discovery. It was on the shore, above which the sea flowed to the depth of nearly six feet; and we could procure the water, therefore, from it only when the tide was down. The crevice from which it rose was in a flat rock, large surfaces of which were spread around, and composed the face of the beach. We filled our two kegs before the tide rose, and went back again to our boats. The remainder of this day was spent in seeking for fish, crabs, birds, and sufficiently known whether it would be advisable to make any arrangements for a more permanent abode.

DECEMBER 23RD

At eleven o'clock, a.m., we again visited our spring: the tide had fallen to about a foot below it, and we were able to procure, before it rose again, about twenty gallons of water. It was at first a little brackish, but soon became fresh, from the constant supply from the rock and the departure of the sea. Our observations this morning tended to give us every confidence in its quantity and quality, and we, therefore, rested perfectly easy in our minds on the subject, and commenced to make further discoveries about the island. Each man sought for his own daily living, on whatsoever the mountains, the shore, or the sea, could furnish him with; and every day, during our stay there, the whole time was employed in roving about for food.

There was no one thing on the island upon which we could in the least degree rely, except the peppergrass, and of that the supply was precarious, and not much relished without some other food. Our situation here, therefore, now became worse than it would have been in our boats on the ocean; because, in the latter case, we should be still making some progress towards the land, while our provisions lasted, and the chance of falling in with some vessel be considerably increased. It was certain that we ought not to remain here unless upon the strongest assurances in our own minds, of sufficient sustenance, and that, too, in regular supplies, that might be depended upon.

After much conversation amongst us on this subject, and again examining our navigators, it was finally concluded to set sail for Easter Island, which we found to be E.SE. from us in latitude 27° 9′ S., longitude 109° 35′ W. All we knew of this island was that it existed as laid down in the books; but of its extent, productions, or inhabitants, if any, we were entirely ignorant; at any rate, it was nearer by eight hundred and fifty miles to the coast, and could not be worse in its productions than the one we were about leaving.

DECEMBER 26TH

The day was wholly employed in preparations for our departure; our boats were hauled down to the vicinity of the spring, and our casks, and everything else that would contain it, filled with water.

There had been considerable talk between three of our companions about their remaining on this island, and taking their chance both for a living, and an escape from it; and as the time drew near at which we were to leave, they made up their minds to stay behind. The rest of us could make no objection to their plan, as it lessened the load of our boats, allowed us their share of the provisions, and the probability of their being able to sustain themselves on the island was much stronger than that of our reaching the mainland. Should we, however, ever arrive safely, it would become our duty, and we so assured them, to give information of their situation, and make every effort to procure their removal from thence; which we accordingly afterwards did.

Their names were William Wright of Barnstable, Massachusetts, Thomas Chapple of Plymouth, England, and Seth Weeks of the former place. They had begun, before we came away, to construct a sort of habitation, composed of the branches of trees, and we left with them every little article that could be spared from the boats. It was their intention to build a considerable dwelling, that would protect them from the rains, as soon as time and materials could be provided.

The captain wrote letters, to be left on the island, giving information of the fate of the ship, and that of our own; and stating that we had set out to reach Easter Island, with further particulars, intended to give notice (should our fellow sufferers die there, and the place be ever visited by any vessel) of our misfortunes. These letters were put in a tin case, enclosed in a small wooden box, and nailed to a tree, on the west side of the island, near our landing place. We had observed, some days previously, the name of a ship, *The Elizabeth*, cut out in the bark of this tree, which rendered it indubitable that one of that name had once touched here. There was, however, no date to it, or anything else, by which any further particulars could be made out.

DECEMBER 27TH

I went, before we set sail this morning, and procured for each boat a flat stone, and two armfuls of wood, with which to make a fire in our boats, should it become afterwards necessary in the further prosecution of our voyage; as we calculated we might catch a fish, or a bird, and in that case be provided with the means of cooking it; otherwise, from the intense heat of the weather, we knew they could not be preserved from spoiling. At ten o'clock, a.m., the tide having risen far enough to allow our boats to float over the rocks, we made all sail, and steered around the island, for the purpose of making a little further observation, which would not detain us any time, and might be productive of some unexpected good fortune. Before we started we missed our three companions, and found they had not come down, either to assist us to get off, nor to take any kind of leave of us. I walked up the beach towards their rude dwelling, and informed them that we were then about to set sail, and should probably never see them more.

They seemed to be very much affected, and one of them shed tears. They wished us to write to their relations, should Providence safely direct us again to our homes, and said but little else. They had every confidence in being able to procure a subsistence there as long as they remained: and, finding them ill at heart about taking any leave of us, I hastily bid them "goodbye," hoped they would do well, and came away. They followed me with their eyes until I was out of sight, and I never saw more of them.

JANUARY 10TH

Matthew P. Joy, the second mate, had suffered from debility, and the privations we had experienced, much beyond any of the rest of us, and was on the eighth removed to the captain's boat, under the impression that he would be more comfortable there, and more attention and pains be bestowed in nursing and endeavouring to comfort him. This day being calm, he manifested a desire to be taken back again; but at four o'clock in the afternoon, after having been, according to his wishes, placed in his own boat, he died very suddenly after his removal. On the eleventh, at six o'clock in the morning, we sewed him up in his clothes, tied a large stone to his feet, and, having brought all the boats to, consigned him in a solemn manner to the ocean. This man did not die of absolute starvation, although his end was no doubt very much hastened by his sufferings. He had a weak and sickly constitution, and complained of being unwell the whole voyage.

It was an incident, however, which threw a gloom over our feelings for many days. In consequence of his death, one man from the captain's boat was placed in that from which he died, to supply his place, and we stood away again on our course.

JANUARY 14TH

We had now been nineteen days from the island, and had only made a distance of about nine hundred miles: necessity began to whisper to us, that a still further reduction of our allowance must take place, or we must abandon altogether the hopes of reaching the land, and rely wholly on the chance of being taken up by a vessel. But how to reduce the daily quantity of food, with any regard to life itself, was a question of the utmost consequence. Upon our first leaving the wreck, the demands of the stomach had been circumscribed to the smallest possible compass; and subsequently before reaching the island, a diminution had taken place of nearly one half and it was now, from a reasonable calculation, become necessary even to curtail that at least one half; which must, in a short time, reduce us to mere skeletons again. We had a full allowance of water, but it only served to contribute to our debility; our bodies deriving but the scanty support which an ounce and a half of bread for each man afforded. It required a great effort to bring matters to this dreadful alternative, either to feed our bodies and our hopes a little longer, or in the agonies of hunger to seize upon and devour our provisions, and coolly await the approach of death.

I had almost determined upon this occurrence to divide our provisions, and give to each man his share of the whole stock; and should have done so in the height of my resentment had it not been for the reflection that some might, by imprudence, be tempted to go beyond the daily allowance, or consume it all at once, and bring on a premature weakness or starvation: this would of course disable them for the duties of the boat, and reduce our chances of safety and deliverance.

JANUARY 15TH

A very large shark was observed swimming about us in a most ravenous manner, making attempts every now and then upon different parts of the boat, as if he would devour the very wood with hunger; he came several times and snapped at the steering oar, and even the stern-post. We tried in vain to stab him with a lance, but we were so weak as not to be able to make any impression upon his hard skin; he was so much larger than an ordinary one, and manifested such a fearless malignity, as to make us afraid of him; and our utmost efforts, which were at first directed to kill him for prey, became in the end self-defense. Baffled however in all his hungry attempts upon us, he shortly made off.

JANUARY 20TH

Richard Peterson manifested today symptoms of a speedy dissolution; he had been lying between the seats in the boat, utterly dispirited and broken down, without being able to do the least duty, or hardly to place his hand to his head for the last three days, and had this morning made up his mind to die rather than endure further misery: he refused his allowance; said he was sensible of his approaching end, and was perfectly ready to die: in a few minutes he became speechless, the breath appeared to be leaving his body without producing the least pain, and at four o'clock he was gone.

To add to our calamities, biles began to break out upon us, and our imaginations shortly became as diseased as our bodies. I laid down at night to catch a few moments of oblivious sleep, and immediately my starving fancy was at work. I dreamt of being placed near a splendid and rich repast, where there was everything that the most dainty appetite could desire; and of contemplating the moment in which we were to commence to eat with enraptured feelings of delight; and just as I was about to partake of it, I suddenly awoke to the cold realities of my miserable situation.

Nothing could have oppressed me so much. It set such a longing frenzy for victuals in my mind, that I felt as if I could have wished the dream to continue forever, that I never might have awoke from it. I cast a sort of vacant stare about the boat, until my eyes rested upon a bit of tough cowhide, which was fastened to one of the oars; I eagerly seized and commenced to chew it, but there was no substance in it, and it only served to fatigue my weak jaws, and add to my bodily pains.

My fellow sufferers murmured very much the whole time, and continued to press me continually with questions upon the probability of our reaching land again. I kept constantly rallying my spirits to enable me to afford them comfort. I encouraged them to bear up against all evils, and if we must perish, to die in our own cause, and not weakly distrust the providence of the Almighty by giving ourselves up to despair. I reasoned with them, and told them that we would not die sooner by keeping up our hopes; that the dreadful sacrifices and privations we endured were to preserve us from death, and were not to be put in competition with the price which we set upon our lives, and their value to our families: it was, besides, unmanly to repine at what neither admitted of alleviation nor cure; and withal, that it was our solemn duty to recognise in our calamities an overruling divinity, by whose mercy we might be suddenly snatched from peril, and to rely upon him alone, "Who tempers the wind to the shorn lamb."

JANUARY 28TH

Our spirits this morning were hardly sufficient to allow of our enjoying a change of the wind, which took place to the westward. It had nearly become indifferent to us, from what quarter it blew: nothing but the slight chance of meeting a vessel remained to us now: it was this narrow comfort alone that prevented me from lying down at once to die. But fourteen days' stinted allowance of provisions remained, and it was absolutely necessary to increase the quantity to enable us to live five days longer: we therefore partook of it, as pinching necessity demanded, and gave ourselves wholly up to the guidance and disposal of our Creator.

FEBRUARY 8TH

Our sufferings were now drawing to a close; a terrible death appeared shortly to await us; hunger became violent and outrageous, and we prepared for a speedy release from our troubles; our speech and reason were both considerably impaired, and we were reduced to be at this time, certainly the most helpless and wretched of the whole human race. Isaac Cole, one of our crew, had the day before this, in a fit of despair, thrown himself down in the boat, and was determined there calmly to wait for death. It was obvious that he had no chance; all was dark he said in his mind, not a single ray of hope was left for him to dwell upon; and it was folly and madness to be struggling against what appeared so palpably to be our fixed and settled destiny. I remonstrated with him as effectually as the weakness both of my body and understanding would allow of; and what I said appeared for a moment to have a considerable effect: he made a powerful and sudden effort, half rose up, crawled forward and hoisted the jib, and firmly and loudly cried that he would not give up; that he would live as long as the rest of us—but alas! this effort was but the hectic fever of the moment, and he shortly again relapsed into a state of melancholy and despair.

This day his reason was attacked, and he became about nine o'clock in the morning a most miserable spectacle of madness: he spoke incoherently about everything, calling loudly for a napkin and water, and then lying stupidly and senselessly down in the boat again, would close his hollow eyes, as if in death. About ten o'clock, we suddenly perceived that he became speechless; we got him as well as we were able upon a board, placed on one of the seats of the boat, and covering him up with some old clothes, left him to his fate. He lay in the greatest pain and apparent misery, groaning piteously until four o'clock, when he died, in the most horrid and frightful convulsions I ever witnessed.

We kept his corpse all night, and in the morning my two companions began as a course to make preparations to dispose of it in the sea; when after reflecting on the subject all night, I addressed them on the painful subject of keeping the body for food!! Our provisions could not possibly last us beyond three days, within which time, it was not in any degree probable that we should find relief from our present sufferings, and that hunger would at last drive us to the necessity of casting lots. It was without any objection agreed to, and we set to work as fast as we were able to prepare it so as to prevent its spoiling. We separated his limbs from his body, and cut all the flesh from the bones; after which, we opened the body, took out the heart, and then closed it again—sewed it up as decently as we could, and committed it to the sea. We now first commenced to satisfy the immediate cravings of nature from the heart, which we eagerly devoured, and then ate sparingly of a few pieces of the flesh; after which we hung up the remainder, cut in thin strips about the boat, to dry in the sun: we made a fire and roasted some of it, to serve us during the next day. In this manner did we dispose of our fellow sufferer; the painful recollection of which brings to mind at this moment, some of the most disagreeable and revolting ideas that it is capable of conceiving.

We knew not then to whose lot it would fall next, either to die or be shot, and eaten like the poor wretch we had just dispatched. Humanity must shudder at the dreadful recital. I have no language to paint the anguish of our souls in this dreadful dilemma.

February 10th

We found that the flesh had become tainted, and had turned a greenish colour upon which we concluded to make a fire and cook it at once, to prevent its becoming so putrid as not to be eaten at all: we accordingly did so, and by that means preserved it for six or seven days longer; our bread during the time remained untouched; as that would not be liable to spoil, we placed it carefully aside for the last moments of our trial. About three o'clock this afternoon a strong breeze set in from the N.W. and we made very good progress, considering that we were compelled to steer the boat by management of the sails alone: this wind continued until the thirteenth, when it changed again ahead.

We contrived to keep soul and body together by sparingly partaking of our flesh, cut up in small pieces and eaten with salt water. By the fourteenth, our bodies became so far recruited, as to enable us to make a few attempts at guiding our boat again with the oar; by each taking his turn, we managed to effect it, and to make a tolerable good course. On the fifteenth, our flesh was all consumed, and we were driven to the last morsel of bread, consisting of two cakes; our limbs had for the last two days swelled very much, and now began to pain us most excessively. We were still, as near as we could judge, three hundred miles from the land, and but three days of our allowance on hand.

FEBRUARY 16TH

At night, full of the horrible reflections of our situation, and panting with weakness, I laid down to sleep, almost indifferent whether I should ever see the light again. I had not lain long, before I dreamt I saw a ship at some distance off from us, and strained every nerve to get to her, but could not. I awoke almost overpowered with the frenzy I had caught in my slumbers, and stung with the cruelties of a diseased and disappointed imagination.

FEBRUARY 17TH

In the afternoon, a heavy cloud appeared to be settling down in an E. by N. direction from us, which in my view, indicated the vicinity of some land, which I took for the island of Massafuera. I concluded it could be no other; and immediately upon this reflection, the life blood began to flow again briskly in my veins. I told my companions that I was well convinced it was land, and if so, in all probability we should reach it before two days more. My words appeared to comfort them much; and by repeated assurances of the favourable appearance of things, their spirits acquired even a degree of elasticity that was truly astonishing. The dark features of our distress began now to diminish a little, and the countenance, even amid the gloomy bodings of our hard lot, to assume a much fresher hue.

February 18th

Before daylight, Thomas Nicholson, a boy about seventeen years of age, one of my two companions who had thus far survived with me, after having bailed the boat, laid down, drew a piece of canvas over him, and cried out that he then wished to die immediately. I saw that he had given up, and I attempted to speak a few words of comfort and encouragement to him, and endeavoured to persuade him that it was a great weakness and even wickedness to abandon a reliance upon the Almighty, while the least hope, and a breath of life remained; but he felt unwilling to listen to any of the consolatory suggestions which I made to him; and, notwithstanding the extreme probability which I stated there was of our gaining the land before the end of two days more, he insisted upon lying down and giving himself up to despair.

A fixed look of settled and forsaken despondency came over his face: he lay for some time silent, sullen, and sorrowful—and I felt at once satisfied that the coldness of death was fast gathering upon him: there was a sudden and unaccountable earnestness in his manner that alarmed me, and made me fear that I myself might unexpectedly be overtaken by a like weakness, or dizziness of nature, that would bereave me at once of both reason and life; but Providence willed it otherwise.

At about seven o'clock this morning, while I was lying asleep, my companion who was steering, suddenly and loudly called out, "There's a sail!" I know not what was the first movement I made upon hearing such an unexpected cry: the earliest of my recollections are that immediately I stood up, gazing in a state of abstraction and ecstasy upon the blessed vision of a vessel about seven miles off from us; she was standing in the same direction with us, and the only sensation I felt at the moment was, that of a violent and unaccountable impulse to fly directly towards her.

I do not believe it is possible to form a just conception of the pure, strong feelings, and the unmingled emotions of joy and gratitude, that took possession of my mind on this occasion: the boy, too, took a sudden and animated start from his despondency, and stood up to witness the probable instrument of his salvation. Our only fear was now that she would not discover us, or that we might not be able to intercept her course: we, however, put our boat immediately, as well as we were able, in a direction to cut her off; and found, to our great joy, that we sailed faster than she did. Upon observing us, she shortened sail, and allowed us to come up to her.

The captain hailed us, and asked who we were. I told him we were from a wreck, and he cried out immediately for us to come alongside the ship. I made an effort to assist myself along to the side, for the purpose of getting up, but strength failed me altogether, and I found it impossible to move a step further without help. We must have formed at that moment, in the eyes of the captain and his crew, a most deplorable and affecting picture of suffering and misery. Our cadaverous countenances, sunken eyes, and bones just starting through the skin, with the ragged remnants of clothes stuck about our sunburnt bodies, must have produced an appearance to him affecting and revolting in the highest degree.

The sailors commenced to remove us from our boat, and we were taken to the cabin, and comfortably provided for in every respect. In a few minutes we were permitted to taste of a little thin food, made from tapioca, and in a few days, with prudent management, we were considerably recruited. This vessel proved to be the brig *Indian*, Captain William Crozier, of London; to whom we are indebted for every polite, friendly, and attentive disposition towards us, that can possibly characterize a man of humanity and feeling.

FEBRUARY 25TH

We arrived at Valparaiso in utter distress and poverty. Our wants were promptly relieved there.

The captain and the survivors of his boat's crew, were taken up by the American whale-ship, the *Dauphin*, Captain Zimri Coffin, of Nantucket, and arrived at Valparaiso on the seventeenth of March following.

The third boat got separated from him on the twenty-eighth of January, and has not been heard of since.

The names of all the survivors, are as follows: Captain George Pollard, Jr., Charles Ramsdale, Owen Chase, Benjamin Lawrence, and

Thomas Nicholson, all of Nantucket. There died in the captain's boat, the following: Brazilla Ray of Nantucket, Owen Coffin of the same place, who was shot, and Samuel Reed.

The captain relates, that after being separated, as herein before stated, they continued to make what progress they could towards the island of Juan Fernandez, as was agreed upon; but contrary winds and the extreme debility of the crew prevailed against their united exertions. He was with us equally surprised and concerned at the separation that took place between us; but continued on his course, almost confident of meeting with us again. On the fourteenth, the whole stock of provisions belonging to the second mate's boat was entirely exhausted, and on the twenty-fifth, Lawson Thomas died, and was eaten by his surviving companions.

On the twenty-first, the captain and his crew were in the like dreadful situation with respect to their provisions; and on the twenty-third, another man, Charles Shorter, died out of the same boat, and his body was shared for food between the crews of both boats. On the twenty-seventh, another, Isaac Shepherd, died in the third boat; and on the twenty-eighth, another man, named Samuel Reed, died out of the captain's boat. The bodies of these men constituted their only food while it lasted; and on the twenty-ninth, owing to the darkness of the night and want of sufficient power to manage their boats, those of the captain and second mate separated in latitude 35° S., longitude 100° W. On the first of February, having consumed the last morsel, the captain and the three other men that remained with him were reduced to the necessity of casting lots. It fell upon Owen Coffin to die, who with great fortitude and resignation submitted to his fate.

They drew lots to see who should shoot him: he placed himself firmly to receive his death, and was immediately shot by Charles Ramsdale, whose hard fortune it was to become his executioner. On the eleventh Brazilla Ray died; and on these two bodies the captain and Charles Ramsdale, the only two that were then left, subsisted until the morning of the twenty-third, when they fell in with the *Dauphin*.

On the eleventh of June following I arrived at Nantucket in the whale-ship the Eagle, Captain William H. Coffin. My family had received the most distressing account of our shipwreck, and had given me up for lost. My unexpected appearance was welcomed with the most grateful obligations and acknowledgements to a beneficent Creator, who had guided me through darkness, trouble, and death, once more to the bosom of my country and friends.

Ten

Yammerschooner

Joshua Slocum

The *Spray* rounded Cape Virgins and entered the Strait of Magellan. The scene was again real and gloomy; the wind, northeast, and blowing a gale, sent feather-white spume along the coast; such a sea ran as would swamp an ill-appointed ship. As the sloop neared the entrance to the strait I observed that two great tide-races made ahead, one very close to the point of the land and one farther offshore. Between the two, in a sort of channel, through combers, went the *Spray* with close-reefed sails. But a rolling sea followed her a long way in, and a fierce current swept around the cape against her; but this she stemmed, and was soon chirruping under the lee of Cape Virgins and running every minute into smoother water. However, long trailing kelp from sunken rocks waved forebodingly under her keel, and the wreck of a great steamship smashed on the beach abreast gave a gloomy aspect to the scene.

I was not to be let off easy. The Virgins would collect tribute even from the *Spray* passing their promontory. Fitful rain-squalls from the northwest followed the northeast gale. I reefed the sloop's sails, and sitting in the cabin to rest my eyes, I was so strongly impressed with what in all nature I might expect that as I dozed the very air I breathed seemed to warn me of danger. My senses heard "*Spray* ahoy!" shouted in warning. I sprang to the deck, wondering who could be there that knew the *Spray* so well as to call out her name passing in the dark; for it was now the blackest of nights all around, except away in the southwest where rose

the old familiar white arch, the terror of Cape Horn, rapidly pushed up
by a southwest gale. I had only a moment to douse sail and lash all solid
when it struck like a shot from a cannon, and for the first half hour it was
something to be remembered by way of a gale. For thirty hours it kept on
blowing hard. The sloop could carry no more than a three-reefed mainsail
and forestaysail; with these she held on stoutly and was not blown out of
the strait. In the height of the squalls in this gale she doused all sail, and
this occurred often enough.

After this gale followed only a smart breeze, and the *Spray*, passing
through the narrows without mishap, cast anchor at Sandy Point on
February 14, 1896.

Sandy Point (Punta Arenas) is a Chilean coaling station, and boasts
about two thousand inhabitants, of mixed nationality, but mostly Chil-
eans. What with sheepfarming, goldmining, and hunting, the settlers in
this dreary land seemed not the worst off in the world. But the natives,
Patagonian and Fuegian, on the other hand, were as squalid as contact
with unscrupulous traders could make them. A large percentage of the
business there was traffic in "firewater." If there was a law against selling
the poisonous stuff to the natives, it was not enforced. Fine specimens
of the Patagonian race, looking smart in the morning when they came
into town, had repented before night of ever having seen a white man,
so beastly drunk were they, to say nothing about the peltry of which they
had been robbed.

The port at that time was free, but a customhouse was in course of
construction, and when it is finished, port and tariff dues are to be collected. A soldier police guarded the place, and a sort of vigilante force
besides took down its guns now and then; but as a general thing, to my
mind, whenever an execution was made they killed the wrong man. Just
previous to my arrival the governor, himself of a jovial turn of mind, had
sent a party of young bloods to foray a Fuegian settlement and wipe out
what they could of it on account of the recent massacre of a schooner's
crew somewhere else. Altogether the place was quite newsy and supported two papers—dailies, I think. The port captain, a Chilean naval
officer, advised me to ship hands to fight Indians in the strait farther
west, and spoke of my stopping until a gunboat should be going through,

which would give me a tow. After canvassing the place, however, I found only one man willing to embark, and he on condition that I should ship another "mon and a doog." But as no one else was willing to come along, and as I drew the line at dogs, I said no more about the matter, but simply loaded my guns. At this point in my dilemma Captain Pedro Samblich, a good Austrian of large experience, coming along, gave me a bag of carpet-tacks, worth more than all the fighting men and dogs of Tierra del Fuego. I protested that I had no use for carpettacks on board. Samblich smiled at my want of experience, and maintained stoutly that I would have use for them. "You must use them with discretion," he said; "that is to say, don't step on them yourself." With this remote hint about the use of the tacks I got on all right, and saw the way to maintain clear decks at night without the care of watching.

Samblich was greatly interested in my voyage, and after giving me the tacks he put on board bags of biscuits and a large quantity of smoked venison. He declared that my bread, which was ordinary sea-biscuits and easily broken, was not nutritious as his, which was so hard that I could break it only with a stout blow from a maul. Then he gave me, from his own sloop, a compass which was certainly better than mine, and offered to unbend her mainsail for me if I would accept it. Last of all, this large-hearted man brought out a bottle of Fuegian golddust from a place where it had been cached and begged me to help myself from it, for use farther along on the voyage. But I felt sure of success without this draft on a friend, and I was right. Samblich's tacks, as it turned out, were of more value than gold.

The port captain, finding that I was resolved to go, even alone, since there was no help for it, set up no further objections, but advised me, in case the savages tried to surround me with their canoes, to shoot straight, and begin to do it in time, but to avoid killing them if possible, which I heartily agreed to do. With these simple injunctions the officer gave me my port clearance free of charge, and I sailed on the same day, February 19, 1896. It was not without thoughts of strange and stirring adventure beyond all I had yet encountered that I now sailed into the country and very core of the savage Fuegians.

A fair wind from Sandy Point brought me on the first day to St. Nicholas Bay, where, so I was told, I might expect to meet savages; but seeing no signs of life, I came to anchor in eight fathoms of water, where I lay all night under a high mountain. Here I had my first experience with the terrific squalls, called williwaws, which extended from this point on through the strait to the Pacific. They were compressed gales of wind that Boreas handed down over the hills in chunks. A full-blown williwaw will throw a ship, even without sail on, over on her beam ends; but, like other gales, they cease now and then, if only for a short time.

February 20 was my birthday, and I found myself alone, with hardly so much as a bird in sight, off Cape Froward, the southernmost point of the continent of America. By daylight in the morning I was getting my ship under way for the bout ahead.

The sloop held the wind fair while she ran thirty miles farther on her course, which brought her to Fortescue Bay, and at once among the natives' signal fires, which blazed up now on all sides. Clouds flew over the mountain from the west all day; at night my good east wind failed, and in its stead a gale from the west soon came on. I gained anchorage at twelve o'clock that night, under the lee of a little island, and then prepared myself a cup of coffee, of which I was sorely in need; for, to tell the truth, hard beating in the heavy squalls and against the current had told on my strength. Finding that the anchor held, I drank my beverage, and named the place Coffee Island. It lies to the south of Charles Island, with only a narrow channel between.

By daylight the next morning the *Spray* was again under way, beating hard; but she came to in a cove in Charles Island, two and a half miles along on her course. Here she remained undisturbed two days, with both anchors down in a bed of kelp. Indeed, she might have remained undisturbed indefinitely had not the wind moderated; for during these two days it blew so hard that no boat could venture out on the strait, and the natives being away to other hunting grounds, the island anchorage was safe. But at the end of the fierce windstorm fair weather came; then I got my anchors, and again sailed out upon the strait.

Canoes manned by savages from Fortescue now came in pursuit. The wind falling light they gained on me rapidly till coming within hail, when they ceased paddling, and a bowlegged savage stood up and called to me, "Yammerschooner! yammerschooner!" which is their begging term. I said, "No!" Now, I was not for letting on that I was alone, and so I stepped into the cabin, and, passing through the hold, came out at the fore-scuttle, changing my clothes as I went along. That made two men. Then the piece of bowsprit which I had sawed off at Buenos Aires, and which I had still on board, I arranged forward on the outlook, dressed as a seaman, attaching a line by which I could pull it into motion. That made three of us, and we didn't want to "yammerschooner"; but for all that the savages came on faster than before. I saw that besides four at the paddles in the canoe nearest to me, there were others in the bottom, and that they were shifting hands often. At eighty yards I fired a shot across the bows of the nearest canoe, at which they all stopped, but only for a moment. Seeing that they persisted in coming nearer, I fired the second shot so close to the chap who wanted to "yammerschooner" that he changed his mind quickly enough and bellowed with fear, "Bueno jo via Isla," and sitting down in his canoe, he rubbed his starboard cathead for some time. I was thinking of the good port captain's advice when I pulled the trigger, and must have aimed pretty straight; however, a miss was as good as a mile for Mr. "Black Pedro," as he it was, and no other, a leader in several bloody massacres. He made for the island now, and the others followed him. I knew by his Spanish lingo and by his full beard that he was the villain I had named, a renegade mongrel, and the worst murderer in Tierra del Fuego. The authorities had been in search of him for two years. The Fuegians are not bearded.

So much for the first day among the savages. I came to anchor at midnight in Three Island Cove, about twenty miles along from Fortescue Bay. I saw on the opposite side of the strait signal fires, and heard the barking of dogs, but where I lay it was quite deserted by natives. I have always taken it as a sign that where I found birds sitting about, or seals on the rocks, I should not find savage Indians. Seals are never plentiful in these waters, but in Three Island Cove I saw one on the rocks, and other signs of the absence of savage men.

On the next day the wind was again blowing a gale, and although she was in the lee of the land, the sloop dragged her anchors, so that I had to get her under way and beat farther into the cove, where I came to in a landlocked pool. At another time or place this would have been a rash thing to do, and it was safe now only from the fact that the gale which drove me to shelter would keep the Indians from crossing the strait. Seeing this was the case, I went ashore with gun and axe on an island, where I could not in any event be surprised, and there felled trees and split about a cord of fire-wood, which loaded my small boat several times.

While I carried the wood, though I was morally sure there were no savages near, I never once went to or from the skiff without my gun. While I had that and a clear field of over eighty yards about me I felt safe.

The trees on the island, very scattering, were a sort of beech and a stunted cedar, both of which made good fuel. Even the green limbs of the beech, which seemed to possess a resinous quality, burned readily in my great drumstove. I have described my method of wooding up in detail, that the reader who has kindly borne with me so far may see that in this, as in all other particulars of my voyage, I took great care against all kinds of surprises, whether by animals or by the elements. In the Strait of Magellan the greatest vigilance was necessary. In this instance I reasoned that I had all about me the greatest danger of the whole voyage—the treachery of cunning savages, for which I must be particularly on the alert.

The *Spray* sailed from Three Island Cove in the morning after the gale went down, but was glad to return for shelter from another sudden gale. Sailing again on the following day, she fetched Borgia Bay, a few miles on her course, where vessels had anchored from time to time and had nailed boards on the trees ashore with name and date of harbouring carved or painted. Nothing else could I see to indicate the civilized man had ever been before. I had taken a survey of the gloomy place with my spyglass, and was getting my boat out to land and take notes, when the Chilean gunboat Huemel came in, and officers, coming on board, advised me to leave the place at once, a thing that required little eloquence to persuade me to do. I accepted the captain's kind offer of a tow to the next anchorage, at the place called Notch Cove, eight miles farther along, where I should be clear of the worst of the Fuegians.

We made anchorage at the cove about dark that night, while the wind came down in fierce williwaws from the mountains. An instance of Magellan weather was afforded when the Huemel, a wellappointed gunboat of great power, after attempting on the following day to proceed on her voyage, was obliged by sheer force of the wind to return and take up anchorage again and remain till the gale abated; and lucky she was to get back!

Meeting this vessel was a little godsend. She was commanded and officered by highclass sailors and educated gentlemen. An entertainment that was gotten up on her, impromptu, at the Notch would be hard to beat anywhere. One of her midshipmen sang popular songs in French, German, and Spanish, and one (so he said) in Russian. If the audience did not know the lingo of one song from another, it was no drawback to the merriment.

I was left alone the next day, for then the Huemel put out on her voyage, the gale having abated. I spent a day taking in wood and water; by the end of that time the weather was fine. Then I sailed from the desolate place.

There is little more to be said concerning the *Spray's* first passage through the strait that would differ from what I have already recorded. She anchored and weighed many times, and beat many days against the current, with now and then a "slant" for a few miles, till finally she gained anchorage and shelter for the night at Port Tamar, with Cape Pillar in sight to the west. Here I felt the throb of the great ocean that lay before me. I knew now that I had put a world behind me, and that I was opening out another world ahead. I had passed the haunts of savages. Great piles of granite mountains of bleak and lifeless aspect were now astern; on some of them not even a speck of moss had ever grown. There was an unfinished newness all about the land. On the hill back of Port Tamar a small beacon had been thrown up showing that some man had been there. But how could one tell but that he had died of loneliness and grief? In a bleak land is not the place to enjoy solitude.

Throughout the whole of the strait west of Cape Froward I saw no animals except dogs owned by savages. These I saw often enough, and heard them yelping night and day. Birds were not plentiful. The scream of a wild fowl, which I took for a loon, sometimes startled me with its piercing cry. The steamboat duck, so called because it propels itself over the sea with its wings, and resembles a miniature sidewheel steamer in its motion, was sometimes seen scurrying on out of danger. It never flies, but, hitting the water instead of the air with its wings, it moves faster than a rowboat or a canoe. The few furseals I saw were very shy; and of fishes I saw next to none at all. I did not catch one; indeed, I seldom or never put a hook over during the whole voyage. Here in the strait I found great abundance of mussels of an excellent quality. I fared sumptuously on them. There was a sort of swan, smaller than a Muscovy duck, which might have been brought down with the gun, but in the loneliness of life about the dreary country I found myself in no mood to make one life less, except in selfdefence.

ELEVEN

The Wreck of the *Medusa*

Charlotte-Adélaïde Dard

AT NOON, ON THE 2ND OF JULY, SOUNDINGS WERE TAKEN. M. MAUDET, ensign of the watch, was convinced we were upon the edge of the Arguin Bank. The Captain said to him, as well as to everyone, that there was no cause of alarm. In the meanwhile, the wind blowing with great violence, impelled us nearer and nearer to the danger which menaced us. A species of stupor overpowered all our spirits, and everyone preserved a mournful silence, as if they were persuaded we would soon touch the bank.

The colour of the water entirely changed, a circumstance even remarked by the ladies. About three in the afternoon, being in 19° 30' north latitude, and 19° 45' west longitude, a universal cry was heard upon deck. All declared they saw sand rolling among the ripple of the sea. The Captain in an instant ordered to sound. The line gave eighteen fathoms; but on a second sounding it only gave six. He at last saw his error, and hesitated no longer on changing the route, but it was too late. A strong concussion told us the frigate had struck. Terror and consternation were instantly depicted on every face. The crew stood motionless; the passengers in utter despair.

In the midst of this general panic, cries of vengeance were heard against the principal author of our misfortunes, wishing to throw him overboard; but some generous persons interposed, and endeavoured to calm their spirits, by diverting their attention to the means of our safety. The confusion was already so great, that M. Poinsignon, commandant of a troop, struck my sister Caroline a severe blow, doubtless thinking it was one of his soldiers. At this crisis my father was buried in profound sleep, but he quickly awoke, the cries and the tumult upon deck having informed him of our misfortunes.

He poured out a thousand reproaches on those whose ignorance and boasting had been so disastrous to us. However, they set about the means of averting our danger. The officers, with an altered voice, issued their orders, expecting every moment to see the ship go in pieces. They strove to lighten her, but the sea was very rough and the current strong. Much time was lost in doing nothing; they only pursued half measures, and all of them unfortunately failed.

When it was discovered that the danger of the *Medusa* was not so great as was at first supposed, various persons proposed to transport the troops to the island of Arguin, which was conjectured to be not far from the place where we lay aground. Others advised to take us all successively to the coast of the desert of Sahara, by the means of our boats, and with provisions sufficient to form a caravan, to reach the island of Saint Louis, at Senegal. The events which afterwards ensued proved this plan to have been the best, and which would have been crowned with success; unfortunately it was not adopted. M. Schmaltz, the governor, suggested the making of a raft of a sufficient size to carry two hundred men, with provisions: which latter plan was seconded by the two officers of the frigate, and put in execution.

The fatal raft was then begun to be constructed, which would, they said, carry provisions for everyone. Masts, planks, boards, cordage, were thrown overboard. Two officers were charged with the framing of these together. Large barrels were emptied and placed at the angles of the machine, and the workmen were taught to say, that the passengers would be in greater security there, and more at their ease, than in the boats. However, as it was forgotten to erect rails, everyone supposed, and with reason, that those who had given the plan of the raft, had had no design of embarking upon it themselves.

When it was completed, the two chief officers of the frigate publicly promised, that all the boats would tow it to the shore of the Desert; and, when there, stores of provisions and firearms would be given us to form a caravan to take us all to Senegal. Why was not this plan executed? Why were these promises, sworn before the French flag, made in vain? But it is necessary to draw a veil over the past. I will only add, that if these promises had been fulfilled, everyone would have been saved, and that, in spite of the detestable egotism of certain personages, humanity would not now have had to deplore the scenes of horror consequent on the wreck of the *Medusa*!

On the 3rd of July, the efforts were renewed to disengage the frigate, but without success. We then prepared to quit her. The sea became very rough, and the wind blew with great violence. Nothing now was heard but the plaintive and confused cries of a multitude, consisting of more than four hundred persons, who, seeing death before their eyes, deplored their hard fate in bitter lamentations.

On the 4th, there was a glimpse of hope. At the hour the tide flowed, the frigate, being considerably lightened by all that had been thrown overboard, was found nearly afloat; and it is very certain, if on that day they had thrown the artillery into the water, the *Medusa* would have been saved; but M. Lachaumareys said, he could not thus sacrifice the King's cannon, as if the frigate did not belong to the King also. However, the sea ebbed, and the ship sinking into the sand deeper than ever, made them relinquish that on which depended our last ray of hope.

On the approach of night, the fury of the winds redoubled, and the sea became very rough. The frigate then received some tremendous concussions, and the water rushed into the hold in the most terrific manner, but the pumps would not work. We had now no alternative but to abandon her for the frail boats, which any single wave would overwhelm. Frightful gulfs environed us; mountains of water raised their liquid summits in the distance. How were we to escape so many dangers? Whither could we go? What hospitable land would receive us on its shores? My thoughts then reverted to our beloved country. I did not regret Paris, but I could have esteemed myself happy to have been yet in the marshes on the road to Rochefort. Then starting suddenly from my reverie, I exclaimed: "O terrible condition! that black and boundless sea resembles the eternal night which will engulf us! All those who surround me seem yet tranquil; but that fatal calm will soon be succeeded by the most frightful torments. Fools, what had we to find in Senegal, to make us trust to the most perfidious of elements! Did France not afford every necessary for our happiness? Happy! yes, thrice happy, they who never set foot on a foreign soil! Great God! succour all these unfortunate beings; save our unhappy family!"

My father perceived my distress, but how could he console me? What words could calm my fears, and place me above the apprehension of those dangers to which we were exposed? How, in a word, could I assume a serene appearance, when friends, parents, and all that was most dear to me were, in all human probability, on the very verge of destruction? Alas! my fears were but too well founded. For I soon perceived that, although we were the only ladies, besides the Misses Schmaltz, who formed a part of the governor's suit, they had the barbarity of intending our family to embark upon the raft, where were only soldiers, sailors, planters of Cape Verde, and some generous officers who had not the honour (if it could be accounted one) of being considered among the ignorant confidants of MM. Schmaltz and Lachaumareys.

My father, indignant at a proceeding so indecorous, swore we would not embark upon the raft, and that, if we were not judged worthy of a place in one of the six boats, he would himself, his wife, and children, remain on board the wrecks of the frigate. The tone in which he spoke these words, was that of a man resolute to avenge any insult that might be offered to him. The governor of Senegal, doubtless fearing the world would one day reproach him for his inhumanity, decided we should have a place in one of the boats. This having in some measure quieted our fears concerning our unfortunate situation, I was desirous of taking some repose, but the uproar among the crew was so great I could not obtain it.

Towards midnight, a passenger came to inquire at my father if we were disposed to depart; he replied, we had been forbid to go yet. However, we were soon convinced that a great part of the crew and various passengers were secretly preparing to set off in the boats. A conduct so perfidious could not fail to alarm us, especially as we perceived among those so eager to embark unknown to us, several who had promised, but a little while before, not to go without us.

M. Schmaltz, to prevent that which was going on upon deck, instantly rose to endeavour to quiet their minds; but the soldiers had already assumed a threatening attitude, and, holding cheap the words of their commander, swore they would fire upon whosoever attempted to depart in a clandestine manner. The firmness of these brave men produced the desired effect, and all was restored to order. The governor returned to his cabin; and those who were desirous of departing furtively were confused and covered with shame. The governor, however, was ill at ease; and as he had heard very distinctly certain energetic words which had been addressed to him, he judged it proper to assemble a council. All the officers and passengers being collected, M. Schmaltz there solemnly swore before them not to abandon the raft, and a second time promised, that all the boats would tow it to the shore of the Desert, where they would all be formed into a caravan. I confess this conduct of the governor greatly satisfied every member of our family; for we never dreamed he would deceive us, nor act in a manner contrary to what he had promised.

About three in the morning, some hours after the meeting of the council, a terrible noise was heard in the powder room; it was the helm which was broken. All who were sleeping were roused by it. On going on deck everyone was more and more convinced that the frigate was lost beyond all recovery. Alas! the wreck was for our family the commencement of a horrible series of misfortunes. The two chief officers then decided with one accord, that all should embark at six in the morning, and abandon the ship to the mercy of the waves. After this decision, followed a scene the most whimsical, and at the same time the most melancholy that can be well conceived. To have a more distinct idea of it, let the reader transport himself in imagination to the midst of the liquid plains of the ocean; then let him picture to himself a multitude of all classes, of every age, tossed about at the mercy of the waves upon a dismasted vessel, foundered, and half submerged; let him not forget these are thinking beings with the certain prospect before them of having reached the goal of their existence.

Separated from the rest of the world by a boundless sea, and having no place of refuge but the wrecks of a grounded vessel, the multitude addressed at first their vows to Heaven, and forgot, for a moment, all earthly concerns. Then, suddenly starting from their lethargy, they began to look after their wealth, the merchandise they had in small ventures, utterly regardless of the elements which threatened them. The miser, thinking of the gold contained in his coffers, hastening to put it in a place of safety, either by sewing it into the lining of his clothes, or by cutting out for it a place in the waistband of his trousers.

The smuggler was tearing his hair at not being able to save a chest of contraband which he had secretly got on board, and with which he had hoped to have gained two or three hundred percent. Another, selfish to excess, was throwing overboard all his hidden money, and amusing himself by burning all his effects. A generous officer was opening his portmanteau, offering caps, stockings, and shirts, to any who would take them. These had scarcely gathered together their various effects, when they learned that they could not take anything with them; those were searching the cabins and storerooms to carry away everything that was valuable.

Shipboys were discovering the delicate wines and fine liqueurs, which a wise foresight had placed in reserve. Soldiers and sailors were penetrating even into the spirit room, broaching casks, staving others, and drinking till they fell exhausted. Soon the tumult of the inebriated made us forget the roaring of the sea which threatened to engulf us. At last the uproar was at its height; the soldiers no longer listened to the voice of their captain. Some knit their brows and muttered oaths; but nothing could be done with those whom wine had rendered furious. Next, piercing cries mixed with doleful groans were heard—this was the signal of departure.

At six o'clock on the morning of the 5th, a great part of the military were embarked upon the raft, which was already covered with a large sheet of foam. The soldiers were expressly prohibited from taking their arms. A young officer of infantry, whose brain seemed to be powerfully affected, put his horse beside the barricadoes of the frigate, and then, armed with two pistols, threatened to fire upon anyone who refused to go upon the raft. Forty men had scarcely descended when it sunk to the depth of about two feet. To facilitate the embarking of a greater number, they were obliged to throw over several barrels of provisions which had been placed upon it the day before. In this manner did this furious officer get about one hundred and fifty heaped upon that floating tomb; but he did not think of adding one more to the number by descending himself, as he ought to have done, but went peaceably away, and placed himself in one of the best boats. There should have been sixty sailors upon the raft, and there were but about ten. A list had been made out on the 4th, assigning each his proper place; but this wise precaution being disregarded, everyone pursued the plan he deemed the best for his own preservation. The precipitation with which they forced one hundred and fifty unfortunate beings upon the raft was such that they forgot to give them one morsel of biscuit. However, they threw towards them twenty-five pounds in a sack, whilst they were not far from the frigate; but it fell into the sea, and was with difficulty recovered.

During this disaster, the governor of Senegal, who was busied in the care of his own dear self, effeminately descended in an armchair into the barge, where were already various large chests, all kinds of provisions, his dearest friends, his daughter and his wife. Afterwards the Captain's boat received twenty-seven persons, amongst whom were twenty-five sailors, good rowers. The shallop, commanded by M. Espiau, ensign of the ship, took forty-five passengers, and put off. The boat, called the Senegal, took twenty-five; the pinnace thirty-three; and the yawl, the smallest of all the boats, took only ten.

Almost all the officers, the passengers, the mariners and supernumeraries, were already embarked—all, but our weeping family, who still remained upon the boards of the frigate, till some charitable souls would kindly receive us into a boat. Surprised at this abandonment, I instantly felt myself roused, and, calling with all my might to the officers of the boats, besought them to take our unhappy family along with them. Soon after, the barge, in which were the governor of Senegal and all his family, approached the *Medusa*, as if still to take some passengers, for there were but few in it. I made a motion to descend, hoping that the Misses Schmaltz, who had, till that day, taken a great interest in our family, would allow us a place in their boat; but I was mistaken: those ladies, who had embarked in a mysterious incognito, had already forgotten us; and

M. Lachaumareys, who was still on the frigate, positively told me they would not embark along with us.

Nevertheless I ought to tell, what we learned afterwards, that that officer who commanded the pinnace had received orders to take us in, but, as he was already a great way from the frigate, we were certain he had abandoned us. My father however hailed him, but he persisted on his way to gain the open sea. A short while afterwards we perceived a small boat among the waves, which seemed desirous to approach the *Medusa*; it was the yawl.

When it was sufficiently near, my father implored the sailors who were in it to take us on board, and to carry us to the pinnace, where our family ought to be placed. They refused. He then seized a firelock, which lay by chance upon deck, and swore he would kill every one of them if they refused to take us into the yawl, adding that it was the property of the king, and that he would have advantage from it as well as another.

The sailors murmured, but durst not resist, and received all our family, which consisted of nine persons, viz. four children, our stepmother, my cousin, my sister Caroline, my father, and myself. A small box, filled with valuable papers, which we wished to save, some clothes, two bottles of ratafia, which we had endeavoured to preserve amidst our misfortunes, were seized and thrown overboard by the sailors of the yawl, who told us we would find in the pinnace everything which we could wish for our voyage.

We had then only the clothes which covered us, never thinking of dressing ourselves in two suits; but the loss which affected us most was that of several manuscripts at which my father had been labouring for a long while. Our trunks, our linen, and various chests of merchandise of great value, in a word, everything we possessed, was left in the *Medusa*. When we boarded the pinnace, the officer who commanded it began excusing himself for having set off without forewarning us, as he had been ordered, and said a thousand things in his justification. But without believing the half of his fine protestations, we felt very happy in having overtaken him; for it is most certain they had had no intention of encumbering themselves with our unfortunate family. I say encumber, for it is evident that four children, one of whom was yet at the breast, were very indifferent beings to people who were actuated by a selfishness without all parallel.

When we were seated in the longboat, my father dismissed the sailors with the yawl, telling them he would ever gratefully remember their services. They speedily departed, but little satisfied with the good action they had done. My father hearing their murmurs and the abuse they poured out against us, said, loud enough for all in the boat to hear: "We are not surprised sailors are destitute of shame, when their officers blush at being compelled to do a good action." The commandant of the boat feigned not to understand the reproaches conveyed in these words, and, to divert our minds from brooding over our wrongs, endeavoured to counterfeit the man of gallantry.

All the boats were already far from the *Medusa*, when they were brought to, to form a chain in order to tow the raft. The barge, in which was the governor of Senegal, took the first tow, then all the other boats in succession joined themselves to that. M. Lachaumareys embarked, although there yet remained upon the *Medusa* more than sixty persons. Then the brave and generous M. Espiau, commander of the shallop, quitted the line of boats, and returned to the frigate, with the intention of saving all the wretches who had been abandoned. They all sprung into the shallop; but as it was very much overloaded, seventeen unfortunates preferred remaining on board, rather than expose themselves as well as their companions to certain death.

But, alas! the greater part afterwards fell victims to their fears or their devotion. Fifty-two days after they were abandoned, no more than three of them were alive, and these looked more like skeletons than men. They told that their miserable companions had gone afloat upon planks and hen-coops, after having waited in vain forty-two days for the succour which had been promised them, and that all had perished.

The shallop, carrying with difficulty all those she had saved from the *Medusa*, slowly rejoined the line of boats which towed the raft. M. Espiau earnestly besought the officers of the other boats to take some of them along with them; but they refused, alleging to the generous officer that he ought to keep them in his own boat, as he had gone for them himself. M.

Espiau, finding it impossible to keep them all without exposing them to the utmost peril, steered right for a boat which I will not name.

Immediately a sailor sprung from the shallop into the sea, and endeavoured to reach it by swimming; and when he was about to enter it, an officer who possessed great influence, pushed him back, and, drawing his sabre, threatened to cut off his hands, if he again made the attempt. The poor wretch regained the shallop, which was very near the pinnace, where we were. Various friends of my father supplicated M. Lapérère, the officer of our boat, to receive him on board. My father had his arms already out to catch him, when M. Lapérère instantly let go the rope which attached us to the other boats, and tugged off with all his force. At the same instant every boat imitated our execrable example; and wishing to shun the approach of the shallop, which sought for assistance, stood off from the raft, abandoning in the midst of the ocean, and to the fury of the waves, the miserable mortals whom they had sworn to land on the shores of the Desert.

Scarcely had these cowards broken their oath, when we saw the French flag flying upon the raft. The confidence of these unfortunate persons was so great, that when they saw the first boat which had the tow removing from them, they all cried out, "The rope is broken! The rope is broken!" but when no attention was paid to their observation they instantly perceived the treachery of the wretches who had left them so basely. Then the cries of *Vive le Roi* arose from the raft, as if the poor fellows were calling to their father for assistance; or, as if they had been persuaded that, at that rallying word, the officers of the boats would return, and not abandon their countrymen. The officers repeated the cry of *Vive le Roi*, without a doubt, to insult them; but, more particularly, M. Lachaumareys, who, assuming a martial attitude, waved his hat in the air. Alas! what availed these false professions?

Frenchmen, menaced with the greatest peril, were demanding assistance with the cries of *Vive le Roi*; yet none were found sufficiently generous, nor sufficiently French, to go to aid them. After a silence of some minutes, horrible cries were heard; the air resounded with the groans, the lamentations, the imprecations of these wretched beings, and the echo of the sea frequently repeated, "Alas! how cruel you are to abandon us!!!" The raft already appeared to be buried under the waves, and its unfortunate passengers immersed.

The fatal machine was drifted by currents far behind the wreck of the frigate; without cable, anchor, mast, sail, oars; in a word, without the smallest means of enabling them to save themselves. Each wave that struck it, made them stumble in heaps on one another. Their feet getting entangled among the cordage, and between the planks, bereaved them of the faculty of moving. Maddened by these misfortunes, suspended, and adrift upon a merciless ocean, they were soon tortured between the pieces of wood which formed the scaffold on which they floated. The bones of their feet and their legs were bruised and broken, every time the fury of the waves agitated the raft; their flesh covered with contusions and hideous wounds, dissolved, as it were, in the briny waves, whilst the roaring flood around them was coloured with their blood.

As the raft, when it was abandoned, was nearly two leagues from the frigate, it was impossible these unfortunate persons could return to it: they were soon after far out at sea. These victims still appeared above their floating tomb; and, stretching out their supplicating hands towards the boats which fled from them, seemed yet to invoke, for the last time, the names of the wretches who had deceived them. O horrid day! a day of shame and reproach! Alas! that the hearts of those who were so well acquainted with misfortune, should have been so inaccessible to pity!

After witnessing that most inhuman scene, and seeing they were insensible to the cries and lamentations of so many unhappy beings, I felt my heart bursting with sorrow. It seemed to me that the waves would overwhelm all these wretches, and I could not suppress my tears. My father, exasperated to excess, and bursting with rage at seeing so much cowardice and inhumanity among the officers of the boats, began to regret he had not accepted the place which had been assigned for us upon the fatal raft. "At least," said he, "we would have died with the brave, or we would have returned to the wreck of the *Medusa*; and not have had the disgrace of saving ourselves with cowards." Although this produced no effect upon the officers, it proved very fatal to us afterwards; for, on our arrival at Senegal, it was reported to the governor, and very probably was the principal cause of all those evils and vexations which we endured in that colony.

Let us now turn our attention to the several situations of all those who were endeavouring to save themselves in the different boats, as well as to those left upon the wreck of the *Medusa*.

We have already seen, that the frigate was half sunk when it was deserted, presenting nothing but a hulk and wreck. Nevertheless, seventeen still remained upon it, and had food, which, although damaged, enabled them to support themselves for a considerable time; whilst the raft was abandoned to float at the mercy of the waves, upon the vast surface of the ocean. One hundred and fifty wretches were embarked upon it, sunk to the depth of at least three feet on its fore part, and on its poop immersed even to the middle. What victuals they had were soon consumed, or spoiled by the salt water; and perhaps some, as the waves hurried them along, became food for the monsters of the deep.

Two only of all the boats which left the *Medusa*, and these with very few people in them, were provisioned with every necessary; these struck off with security and despatch. But the condition of those who were in the shallop was but little better than those upon the raft; their great number, their scarcity of provisions, their great distance from the shore, gave them the most melancholy anticipations of the future. Their worthy commander, M. Espiau, had no other hope but of reaching the shore as soon as possible.

The other boats were less filled with people, but they were scarcely better provisioned; and, as by a species of fatality, the pinnace, in which were our family, was destitute of everything. Our provisions consisted of a barrel of biscuit, and a tierce of water; and, to add to our misfortunes, the biscuit being soaked in the sea, it was almost impossible to swallow one morsel of it. Each passenger in our boat was obliged to sustain his wretched existence with a glass of water, which he could get only once a day. To tell how this happened, how this boat was so poorly supplied, whilst there were abundance left upon the *Medusa*, is far beyond my power. But it is at least certain, that the greater part of the officers commanding the boats, the shallop, the pinnace, the Senegal boat, and the yawl, were persuaded, when they quitted the frigate, that they would not abandon the raft, but that all the expedition would sail together to the coast of Sahara; that when there, the boats would be again sent to the *Medusa* to take provisions, arms, and those who were left there; but it appears the chiefs had decided otherwise.

After abandoning the raft, although scattered, all the boats formed a little fleet, and followed the same route. All who were sincere hoped to arrive the same day at the coast of the Desert, and that everyone would get on shore; but MM. Schmaltz and Lachaumareys gave orders to take the route for Senegal. This sudden change in the resolutions of the chiefs was like a thunderbolt to the officers commanding the boats. Having nothing on board but what was barely necessary to enable us to allay the cravings of hunger for one day, we were all sensibly affected. The other boats, which, like ourselves, hoped to have got on shore at the nearest point, were a little better provisioned than we were; they had at least a little wine, which supplied the place of other necessaries.

We then demanded some from them, explaining our situation, but none would assist us, not even Captain Lachaumareys, who, drinking to a kept mistress, supported by two sailors, swore he had not one drop on board. We were next desirous of addressing the boat of the governor of Senegal, where we were persuaded were plenty of provisions of every kind, such as oranges, biscuits, cakes, comfits, plums, and even the finest liqueurs; but my father opposed it, so well was he assured we would not obtain anything.

We will now turn to the condition of those on the raft, when the boats left them to themselves.

If all the boats had continued dragging the raft forward, favoured as we were by the breeze from the sea, we would have been able to have conducted them to the shore in less than two days. But an inconceivable fatality caused the generous plan to be abandoned which had been formed.

When the raft had lost sight of the boats, a spirit of sedition began to manifest itself in furious cries. They then began to regard one another with ferocious looks, and to thirst for one another's flesh. Someone had already whispered of having recourse to that monstrous extremity, and of commencing with the fattest and youngest. A proposition so atrocious filled the brave Captain Dupont and his worthy lieutenant M. L'Heureux with horror; and that courage which had so often supported them in the field of glory, now forsook them.

Among the first who fell under the hatchets of the assassins, was a young woman who had been seen devouring the body of her husband. When her turn was come, she sought a little wine as a last favour, then rose, and without uttering one word, threw herself into the sea. Captain Dupont being proscribed for having refused to partake of the sacrilegious viands on which the monsters were feeding, was saved as by a miracle from the hands of the butchers. Scarcely had they seized him to lead him to the slaughter, when a large pole, which served in place of a mast, fell upon his body; and believing that his legs were broken, they contented themselves by throwing him into the sea. The unfortunate captain plunged, disappeared, and they thought him already in another world.

Providence, however, revived the strength of the unfortunate warrior. He emerged under the beams of the raft, and clinging with all his might, holding his head above water, he remained between two enormous pieces of wood, whilst the rest of his body was hid in the sea. After more than two hours of suffering, Captain Dupont spoke in a low voice to his lieutenant, who by chance was seated near the place of his concealment. The brave L'Heureux, with eyes glistening with tears, believed he heard the voice, and saw the shade of his captain; and trembling, was about to quit the place of horror; but, O wonderful! he saw a head which seemed to draw its last sigh, he recognised it, he embraced it, alas! it was his dear friend! Dupont was instantly drawn from the water, and M. L'Heureux obtained for his unfortunate comrade again a place upon the raft. Those who had been most inveterate against him, touched at what Providence had done for him in so miraculous a manner, decided with one accord to allow him entire liberty upon the raft.

The sixty unfortunates who had escaped from the first massacre, were soon reduced to fifty, then to forty, and at last to twenty-eight. The least murmur, or the smallest complaint, at the moment of distributing the provisions, was a crime punished with immediate death. In consequence of such a regulation, it may easily be presumed the raft was soon lightened. In the meanwhile the wine diminished sensibly, and the half rations very much displeased a certain chief of the conspiracy. On purpose to avoid being reduced to that extremity, the executive power decided it was much wiser to drown thirteen people, and to get full rations, than that twenty-eight should have half rations. Merciful Heaven! what shame! After the last catastrophe, the chiefs of the conspiracy, fearing doubtless of being assassinated in their turn, threw all the arms into the sea, and swore an inviolable friendship with the heroes which the hatchet had spared.

On the 17th of July, in the morning, Captain Parnajon, commandant of the *Argus* brig, still found fifteen men on the raft. They were immediately taken on board, and conducted to Senegal. Four of the fifteen are yet alive, viz. Captain Dupont, residing in the neighbourhood of Maintenon, Lieutenant L'Heureux, since Captain, at Senegal, Savigny, at Rochefort, and Corréard, I know not where.

On the 5th of July, at ten in the morning, one hour after abandoning the raft, and three after quitting the *Medusa*, M. Lapérère, the officer of our boat, made the first distribution of provisions. Each passenger had a small glass of water and nearly the fourth of a biscuit. Each drank his allowance of water at one draught, but it was found impossible to swallow one morsel of our biscuit, it being so impregnated with seawater. It happened, however, that some was found not quite so saturated. Of these we ate a small portion, and put back the remainder for a future day. Our voyage would have been sufficiently agreeable, if the beams of the sun had not been so fierce. On the evening we perceived the shores of the Desert; but as the two chiefs (MM. Schmaltz and Lachaumareys) wished to go right for Senegal, notwithstanding we were still one hundred leagues from it, we were not allowed to land.

Several officers remonstrated, both on account of our want of provisions and the crowded condition of the boats, for undertaking so dangerous a voyage. Others urged with equal force, that it would be dishonouring the French name, if we were to neglect the unfortunate people on the raft, and insisted we should be set on shore, and whilst we waited there, three boats should return to look after the raft, and three to the wrecks of the frigate, to take up the seventeen who were left there, as well as a sufficient quantity of provisions to enable us to go to Senegal by the way of Barbary. But MM. Schmaltz and Lachaumareys, whose boats were sufficiently well provisioned, scouted the advice of their subalterns, and ordered them to cast anchor till the following morning.

They were obliged to obey these orders, and to relinquish their designs. During the night, a certain passenger, who was doubtless no doctor, and who believed in ghosts and witches, was suddenly frightened by the appearance of flames, which he thought he saw in the waters of the sea, a little way from where our boat was anchored. My father, and some others, who were aware that the sea is sometimes phosphorated, confirmed the poor credulous man in his belief, and added several circumstances which fairly turned his brain. They persuaded him the Arabic sorcerers had fired the sea to prevent us from travelling along their deserts.

On the morning of the 6th of July, at five o'clock, all the boats were underway on the route to Senegal. The boats of MM. Schmaltz and Lachaumareys took the lead along the coast, and all the expedition followed. About eight, several sailors in our boat, with threats, demanded to be set on shore; but M. Lapérère not acceding to their request, the whole were about to revolt and seize the command; but the firmness of this officer quelled the mutineers.

In a spring which he made to seize a firelock which a sailor persisted in keeping in his possession, he almost tumbled into the sea. My father fortunately was near him, and held him by his clothes, but he had instantly to quit him, for fear of losing his hat, which the waves were floating away. A short while after this slight accident, the shallop, which we had lost sight of since the morning, appeared desirous of rejoining us. We plied all hands to avoid her, for we were afraid of one another, and thought that that boat, encumbered with so many people, wished to board us to oblige us to take some of its passengers, as M. Espiau would not suffer them to be abandoned like those upon the raft.

That officer hailed us at a distance, offering to take our family on board, adding, he was anxious to take about sixty people to the Desert. The officer of our boat, thinking that this was a pretense, replied, we preferred suffering where we were. It even appeared to us that M. Espiau had hid some of his people under the benches of the shallop. But, alas! in the end we deeply deplored being so suspicious, and of having so outraged the devotion of the most generous officer of the *Medusa*.

Our boat began to leak considerably, but we prevented it as well as we could, by stuffing the largest holes with oakum, which an old sailor had had the precaution to take before quitting the frigate. At noon the heat became so strong, so intolerable, that several of us believed we had reached our last moments. The hot winds of the Desert even reached us; and the fine sand with which they were loaded, had completely obscured the clearness of the atmosphere. The sun presented a reddish disk; the whole surface of the ocean became nebulous, and the air which we breathed, depositing a fine sand, an impalpable powder, penetrated to our lungs, already parched with a burning thirst.

In this state of torment we remained till four in the afternoon, when a breeze from the northwest brought us some relief. Notwithstanding the privations we felt, and especially the burning thirst which had become intolerable, the cool air which we now began to breathe, made us in part forget our sufferings. The heavens began again to resume the usual serenity of those latitudes, and we hoped to have passed a good night. A second distribution of provisions was made; each received a small glass of water, and about the eighth part of a biscuit. Notwithstanding our meagre fare, everyone seemed content, in the persuasion we would reach Senegal by the morrow. But how vain were all our hopes, and what sufferings had we yet to endure!

At half past seven, the sky was covered with stormy clouds. The serenity we had admired a little while before, entirely disappeared, and gave place to the most gloomy obscurity. The surface of the ocean presented all the signs of a coming tempest. The horizon on the side of the Desert had the appearance of a long hideous chain of mountains piled on one another, the summits of which seemed to vomit fire and smoke. Bluish clouds, streaked with a dark copper colour, detached themselves from that shapeless heap, and came and joined with those which floated over our heads. In less than half an hour the ocean seemed confounded with the terrible sky which canopied us. The stars were hid. Suddenly a frightful noise was heard from the west, and all the waves of the sea rushed to founder our frail bark.

A fearful silence succeeded to the general consternation. Every tongue was mute; and none durst communicate to his neighbour the horror with which his mind was impressed. At intervals the cries of the children rent our hearts. At that instant a weeping and agonized mother bared her breast to her dying child, but it yielded nothing to appease the thirst of the little innocent who pressed it in vain. O night of horrors! what pen is capable to paint thy terrible picture! How describe the agonizing fears of a father and mother, at the sight of their children tossed about and expiring of hunger in a small boat, which the winds and waves threatened to engulf at every instant! Having full before our eyes the prospect of inevitable death, we gave ourselves up to our unfortunate condition, and addressed our prayers to Heaven. The winds growled with the utmost fury; the tempestuous waves arose exasperated. In their terrific encounter a mountain of water was precipitated into our boat, carrying away one of the sails, and the greater part of the effects which the sailors had saved from the *Medusa*. Our bark was nearly sunk; the females and the children lay rolling in its bottom, drinking the waters of bitterness; and their cries, mixed with the roaring of the waves and the furious north wind, increased the horrors of the scene. My unfortunate father then experienced the most excruciating agony of mind. The idea of the loss which the shipwreck had occasioned to him, and the danger which still menaced all he held dearest in the world, plunged him into a deep swoon. The tenderness of his wife and children recovered him; but alas! his recovery was to still more bitterly deplore the wretched situation of his family. He clasped us to his bosom; he bathed us with his tears, and seemed as if he was regarding us with his last looks of love.

Every soul in the boat were seized with the same perturbation, but it manifested itself in different ways. One part of the sailors remained motionless, in a bewildered state; the other cheered and encouraged one another; the children, locked in the arms of their parents, wept incessantly. Some demanded drink, vomiting the salt water which choked them; others, in short, embraced as for the last time, intertwining their arms, and vowing to die together.

In the meanwhile the sea became rougher and rougher. The whole surface of the ocean seemed a vast plain furrowed with huge blackish waves fringed with white foam. The thunder growled around us, and the lightning discovered to our eyes all that our imagination could conceive most horrible. Our boat, beset on all sides by the winds, and at every instant tossed on the summit of mountains of water, was very nearly sunk in spite of our every effort in bailing it, when we discovered a large hole in its poop. It was instantly stuffed with everything we could find—old clothes, sleeves of shirts, shreds of coats, shawls, useless bonnets, everything was employed, and secured us as far as it was possible. During the space of six hours, we rowed suspended alternately between hope and fear, between life and death. At last towards the middle of the night, Heaven, which had seen our resignation, commanded the floods to be still. Instantly the sea became less rough, the veil which covered the sky became less obscure, the stars again shone out, and the tempest seemed to withdraw. A general exclamation of joy and thankfulness issued at one instant from every mouth. The winds calmed, and each of us sought a little sleep, whilst our good and generous pilot steered our boat on a still very stormy sea.

The day at last, the day so desired, entirely restored the calm; but it brought no other consolation. During the night, the currents, the waves, and the winds had taken us so far out to sea, that, on the dawning of the 7th of July, we saw nothing but sky and water, without knowing whither to direct our course; for our compass had been broken during the tempest. In this hopeless condition, we continued to steer sometimes to the right and sometimes to the left, until the sun arose, and at last showed us the east.

On the morning of the 7th of July, we again saw the shores of the Desert, notwithstanding we were yet a great distance from it. The sailors renewed their murmurings, wishing to get on shore, with the hope of being able to get some wholesome plants, and some more palatable water than that of the sea; but as we were afraid of the Moors, their request was opposed. However, M. Lapérère proposed to take them as near as he could to the first breakers on the coast; and when there, those who wished to go on shore should throw themselves into the sea, and swim to land. Eleven accepted the proposal; but when we had reached the first waves, none had the courage to brave the mountains of water which rolled between them and the beach.

Our sailors then betook themselves to their benches and oars, and promised to be more quiet for the future. A short while after, a third distribution was made since our departure from the *Medusa*; and nothing more remained than four pints of water, and one half dozen biscuits. What steps were we to take in this cruel situation? We were desirous of going on shore, but we had such dangers to encounter. However, we soon came to a decision, when we saw a caravan of Moors on the coast. We then stood a little out to sea. According to the calculation of our commanding officer, we would arrive at Senegal on the morrow.

Deceived by that false account, we preferred suffering one day more, rather than to be taken by the Moors of the Desert, or perish among the breakers. We had now no more than a small half glass of water, and the seventh of a biscuit. Exposed as we were to the heat of the sun, which darted its rays perpendicularly on our heads, that ration, though small, would have been a great relief to us; but the distribution was delayed to the morrow. We were then obliged to drink the bitter seawater, ill as it was calculated to quench our thirst.

Must I tell it! thirst had so withered the lungs of our sailors, that they drank water saltier than that of the sea! Our numbers diminished daily, and nothing but the hope of arriving at the colony on the following day sustained our frail existence. My young brothers and sisters wept incessantly for water. The little Laura, aged six years, lay dying at the feet of her mother. Her mournful cries so moved the soul of my unfortunate father, that he was on the eve of opening a vein to quench the thirst which consumed his child; but a wise person opposed his design, observing that all the blood in his body would not prolong the life of his infant one moment.

The freshness of the night-wind procured us some respite. We anchored pretty near to the shore, and, though dying of famine, each got a tranquil sleep. On the morning of the 8th of July at break of day, we took the route for Senegal. A short while after the wind fell, and we had a dead calm. We endeavoured to row, but our strength was exhausted. A fourth and last distribution was made, and, in the twinkling of an eye, our last resources were consumed. We were forty-two people who had to feed upon six biscuits and about four pints of water, with no hope of a further supply.

Then came the moment for deciding whether we were to perish among the breakers, which defended the approach to the shores of the Desert, or to die of famine in continuing our route. The majority preferred the last species of misery. We continued our progress along the shore, painfully pulling our oars. Upon the beach were distinguished several downs of white sand, and some small trees. We were thus creeping along the coast, observing a mournful silence, when a sailor suddenly exclaimed, "Behold the Moors!" We did, in fact, see various individuals upon the rising ground, walking at a quick pace, whom we took to be the Arabs of the Desert. As we were very near the shore, we stood farther out to sea, fearing that these pretended Moors, or Arabs, would throw themselves into the sea, swim out, and take us. Some hours after, we observed several people upon an eminence, who seemed to make signals to us. We examined them attentively, and soon recognised them to be our companions in misfortune. We replied to them by attaching a white handkerchief to the top of our mast.

Then we resolved to land, at the risk of perishing among the breakers, which were very strong towards the shore, although the sea was calm. On approaching the beach, we went towards the right, where the waves seemed less agitated, and endeavoured to reach it, with the hope of being able more easily to land. Scarcely had we directed our course to that point, when we perceived a great number of people standing near to a little wood surrounding the sand-hills. We recognised them to be the passengers of that boat, which, like ourselves, were deprived of provisions.

Meanwhile we approached the shore, and already the foaming surge filled us with terror. Each wave that came from the open sea, each billow that swept beneath our boat, made us bound into the air; so we were sometimes thrown from the poop to the prow, and from the prow to the poop. Then, if our pilot had missed the sea, we would have been sunk; the waves would have thrown us aground, and we would have been buried among the breakers. The helm of the boat was again given to the old pilot, who had already so happily steered us through the dangers of the storm.

He instantly threw into the sea the mast, the sails, and everything that could impede our proceedings. When we came to the first landing point, several of our shipwrecked companions, who had reached the shore, ran and hid themselves behind the hills, not to see us perish; others made signs not to approach at that place; some covered their eyes with their hands; others, at last despising the danger, precipitated themselves into the waves to receive us in their arms.

We then saw a spectacle that made us shudder. We had already doubled two ranges of breakers; but those which we had still to cross raised their foaming waves to a prodigious height, then sunk with a hollow and monstrous sound, sweeping along a long line of the coast. Our boat sometimes greatly elevated, and sometimes engulfed between the waves, seemed, at the moment, of utter ruin. Bruised, battered, tossed about on all hands, it turned of itself, and refused to obey the kind hand which directed it. At that instant a huge wave rushed from the open sea, and dashed against the poop; the boat plunged, disappeared, and we were all among the waves. Our sailors, whose strength had returned at the presence of danger, redoubled their efforts, uttering mournful sounds.

Our bark groaned, the oars were broken; it was thought aground, but it was stranded; it was upon its side. The last sea rushed upon us with the impetuosity of a torrent. We were up to the neck in water; the bitter sea-froth choked us. The grapnel was thrown out. The sailors threw themselves into the sea; they took the children in their arms, returned, and took us upon their shoulders; and I found myself seated upon the sand on the shore, by the side of my stepmother, my brothers and sisters, almost dead. Everyone was upon the beach except my father and some sailors; but that good man arrived at last, to mingle his tears with those of his family and friends.

Instantly our hearts joined in addressing our prayers and praises to God. I raised my hands to Heaven, and remained some time immoveable upon the beach. Everyone also hastened to testify his gratitude to our old pilot, who, next to God, justly merited the title of our preserver. M. Dumège, a naval surgeon, gave him an elegant gold watch, the only thing he had saved from the *Medusa*.

Let the reader now recollect all the perils to which we had been exposed in escaping from the wreck of the frigate to the shores of the Desert—all that we had suffered during our four days' voyage—and he will perhaps have a just notion of the various sensations we felt on getting on shore on that strange and savage land. Doubtless the joy we experienced at having escaped, as by a miracle, the fury of the floods, was very great; but how much was it lessened by the feelings of our horrible situation!

Without water, without provisions, and the majority of us nearly naked, was it to be wondered at that we should be seized with terror on thinking of the obstacles which we had to surmount, the fatigues, the privations, the pains and the sufferings we had to endure, with the dangers we had to encounter in the immense and frightful Desert we had to traverse before we could arrive at our destination? Almighty Providence! it was in Thee alone I put my trust.

TWELVE

Young Ironsides

James Fenimore Cooper

THE SECOND CRUISE OF OLD IRONSIDES COMMENCED IN AUGUST, 1799. Her orders were to go off Cayenne, in the first place, where she was to remain until near the close of September, when she was to proceed via Guadaloupe to Cape Francois, at which point, Talbot was to assume the command of all the vessels he found on the station. In the course of the season, this squadron grew to be six sail, three frigates and as many sloops, or brigs.

Two incidents occurred to Old Ironsides, while on the St. Domingo station, that are worthy of being noticed, the first being of an amicable, and the second of a particularly hostile character.

While cruising to windward the island, a strange sail was made, which, on closing, proved to be the English frigate, the _____. The commander of this ship and Com. Talbot were acquaintances, and the Englishman had the curiosity to take a full survey of the new Yankee craft. He praised her, as no unprejudiced seaman could fail to do, but insisted that his own ship could beat her on a wind. After some pleasantry on the subject, the English captain made the following proposition; he had touched at Madeira on his way out, and taken on board a few casks of wine for his own use. This wine stood him in so much a cask— now, he was going into port to refit, and clean his bottom, which was a little foul; but, if he could depend on finding the *Constitution* on that station, a few weeks later he would join her, when there should be a trial of speed between the two ships, the loser to pay a cask of the wine, or its price to the winner.

The bet was made, and the vessels parted. At the appointed time, the _____ reappeared; her rigging overhauled, new sails bent, her sides painted, her bottom cleaned. And, as Jack expressed it, looking like a new fiddle. The two frigates closed, and their commanders dined together, arranging the terms of the cartel for the next day's proceedings. That night, the vessels kept near each other, on the same line of sailing, and under short canvas. The following morning, as the day dawned, the *Constitution* and the _____ each turned up their hands, in readiness for what was to follow.

Just as the lower limb of the sun rose clear of the waves, each fired a gun, and made sail on a bowline. Throughout the whole of that day, did these two gallant ships continue turning to windward, on tacks of a few leagues in length, and endeavoring to avail themselves of every advantage which skill could confer on seamen. Hull sailed the *Constitution* on this interesting occasion and the admirable manner in which he did it, was long the subject of eulogy.

All hands were kept on deck all day, and there were tacks on which the people were made to place themselves to windward, in order to keep the vessel as near upright as possible, so as to hold a better wind. Just as the sun dipped, in the evening, the *Constitution* fired a gun, as did her competitor. At that moment the English frigate was precisely hull down dead to leeward; so much having Old Ironsides, or young Ironsides, as she was then, gained in the race, which lasted about eleven hours! The manner in which the *Constitution* beat her competitor out of the wind, was not the least striking feature of this trial, and it must in great degree be ascribed to Hull, whose dexterity in handling a craft under her canvas, was ever remarkable. In this particular, he was perhaps one of the most skilful seamen of his time, as he was also for coolness in moments of hazard. When the evening gun was fired and acknowledged, the *Constitution* put up her helm, and squared away to join her friend. The vessels joined a little after dark, the Englishman as the leeward ship, first rounding to. The *Constitution* passed under her lee, and threw her maintopsail to the mast. There was a boat out from the _____, which soon came alongside, and in it was the English Captain and his cask of wine; the former being just as prompt to "pay" as to "play."

The other occurrence was the cutting out of the *Sandwich*, a French letter of marque, which was lying in Port Platte, a small harbor on the Spanish side of St. Domingo. While cruising along the coast, the *Constitution* had seized an American sloop called the *Sally*, which had been selling supplies to the enemy.

Hearing that the *Sandwich*, formerly an English packet, but which had fallen into the hands of the French, was filling up with coffee, and was nearly full, Talbot determined to send Hull in, with the *Sally*, in order to cut her out. The sloop had not long before come out of that very haven, with an avowed intention to return, and offered every desirable facility to the success of the enterprise. The great and insuperable objection to its ultimate advantage, was the material circumstance that the Frenchman was lying in a neutral port, as respects ourselves, though watchful of the English who were swarming in those seas.

The *Constitution* manned the *Sally* at sea, near sunset, on the tenth of May, 1800, a considerable distance from Port Platte, and the vessels separated, Hull so timing his movements, as to reach his point of destination about midday of a Sunday, when it was rightly enough supposed many of the French, officers as well as men, would be ashore keeping holiday. Short sail was carried that night on board the *Sally*, and while she was quietly jogging along, thinking no harm, a gun was suddenly heard, and a shot came whistling over the sloop.

On looking around, a large ship was seen in chase, and so near, as to render escape impossible. The *Sally* rounded to, and presently, an English frigate ranged alongside. The boarding Officer was astonished when he found himself among ninety armed men, with officers in naval uniform at their head. On demanding an explanation, Hull told him his business, when the English lieutenant expressed his disappointment, candidly acknowledging that his own ship was waiting on the coast to let the *Sandwich* fill up, and get her sails bent, in order to send a party in, also, in order to cut her out! It was too late, however, as the *Sally* could not be, and would not be, detained, and Hull proceeded.

There have been many more brilliant exploits than this of the *Constitution* in sending in a party against the *Sandwich*, but very few that were more neatly executed, or ingeniously planned. The *Sally* arrived off the port, at the appointed hour, and stood directly in, showing the customary number of hands on deck, until coming near the letter of marque, she ran her aboard forward, and the *Constitution*'s clambered in over the *Sandwich*'s bows, led by Hull in person. In two minutes, the Americans had possession of their prize, a smart brig, armed with four sixes and two nines, with a pretty strong crew, without the loss of a man. A party of marines, led by Capt. Cormick, landed, drove the Spaniards from a battery that commanded the anchorage, and spiked the guns.

All this was against law and right, but it was very ingeniously arranged, and as gallantly executed. The most serious part of the affair remained to be achieved. The *Sandwich* was stripped to a girt line, and the wind blew directly into the harbor. As it was unsafe for the marines to remain in the battery any time, it was necessarily abandoned, leaving to the people of the place every opportunity of annoying their invaders by all the means they possessed.

The battery was reoccupied, and the guns cleared of the spikes as well and as fast as they could be, while the Americans set about swaying up topmasts and yards and bending sails. After some smart exertion, the brig got royal yards across, and, at sunset, after remaining several hours in front of the town, Hull scaled his guns, by way of letting it be known they could be used, weighed, and began to beat out of the harbor. The Spaniards fired a few shots after him, but with no effect.

Although this was one of the best executed enterprises of the sort on record, and did infinite credit to the coolness and spirit of all concerned, it was not quite an illustration of international law or of justice in general. This was the first victory of Old Ironsides in a certain sense, but all men must regret it was ever achieved, since it was a wrong act, committed with an exaggerated, if not an altogether mistaken notion of duty. America was not even at war with France, in the more formal meaning of the term, nor were all the legal consequences of war connected with the peculiar hostilities that certainly did exist; but with Spain she had no quarrel whatever, and the *Sandwich* was entitled to receive all the protection and immunities that of right belonged to her anchored in the neutral harbor of Port-au-Platte. In the end not only was the condemnation of the *Sandwich* resisted successfully, but all the other prize money made by Old Ironsides in the cruise went to pay damages. The reason why the exploit itself never received the public commendation to which, as a mere military achievement, it was so justly entitled, was connected with the illegality and recklessness of the enterprise in its inception.

It follows that this, which may be termed the *Constitution*'s earliest victory, was obtained in the face of law and right. Fortunately the old craft has lived long enough to atone for this error of her youth by many a noble deed achieved in defence of principles and rights that the most fastidious will not hesitate to defend. The *Constitution* returned to Boston in Aug. 1800, her cruise being up, not only on account of her orders, but on account of the short period for which men were then enlisted in the navy, which was one year.

On the 18th Nov., however, she was ordered to sail again for the old station, still wearing the broad pennant of Talbot. Nothing occurred of interest in the course of this cruise; and, early in the spring, orders were sent to recall all the cruisers from the West Indies, in consequence of an arrangement of the difficulties with France.

Thirteen

Rounding Cape Horn

Herman Melville

AND NOW, THROUGH DRIZZLING FOGS AND VAPORS, AND UNDER DAMP, double-reefed topsails, our wet-decked frigate drew nearer and nearer to the squally Cape. Who has not heard of it? Cape Horn, Cape Horn—a horn indeed, that has tossed many a good ship. Was the descent of Orpheus, Ulysses, or Dante into Hell, one whit more hardy and sublime than the first navigator's weathering of that terrible Cape?

Turned on her heel by a fierce West Wind, many an outwardbound ship has been driven across the Southern Ocean to the Cape of Good Hope—that way to seek a passage to the Pacific. And that stormy Cape, I doubt not, has sent many a fine craft to the bottom, and told no tales. At those ends of the earth are no chronicles. What signify the broken spars and shrouds that, day after day, are driven before the prows of more fortunate vessels? or the tall masts, imbedded in icebergs, that are found floating by? They but hint the old story—of ships that have sailed from their port, and never more have been heard of.

Impracticable Cape! You may approach it from this direction or that—in any way you please—from the East or from the West; with the wind astern, or abeam, or on the quarter; and still Cape Horn is Cape Horn. Cape Horn it is that takes the conceit out of freshwater sailors, and steeps in a still salter brine the saltest. Woe betide the tyro; the foolhardy, Heaven preserve!

Your Mediterranean captain, who with a cargo of oranges has hitherto made merry runs across the Atlantic, without so much as furling a t'-gallant-sail, oftentimes, off Cape Horn, receives a lesson which he carries to the grave; though the grave—as is too often the case—follows so hard on the lesson that no benefit comes from the experience.

Other strangers who draw nigh to this Patagonia termination of our Continent, with their souls full of its shipwrecks and disasters—top-sails cautiously reefed, and everything guardedly snug—these strangers at first unexpectedly encountering a tolerably smooth sea, rashly conclude that the Cape, after all, is but a bugbear; they have been imposed upon by fables, and founderings and sinkings hereabouts are all cock-and-bull stories.

"Out reefs, my hearties; fore and aft set t'gallantsails! stand by to give her the foretopmast stun'-sail!"

But, Captain Rash, those sails of yours were much safer in the sailmaker's loft. For now, while the heedless craft is bounding over the billows, a black cloud rises out of the sea; the sun drops down from the sky; a horrible mist far and wide spreads over the water.

"Hands by the halyard! Let go! Clew up!"

Too late.

For ere the ropes' ends can be cast off from the pins, the tornado is blowing down to the bottom of their throats. The masts are willows, the sails ribbons, the cordage wool; the whole ship is brewed into the yeast of the gale.

And now, if, when the first green sea breaks over him, Captain Rash is not swept overboard, he has his hands full to be sure. In all probability his three masts have gone by the board, and, raveled into list, his sails are floating in the air. Or, perhaps, the ship broaches to, or is brought by the lee. In either case, Heaven help the sailors, their wives and their little ones; and Heaven help the underwriters.

Familiarity with danger makes a brave man braver, but less daring. Thus with seamen: he who goes the oftenest round Cape Horn goes the most circumspectly. A veteran mariner is never deceived by the treacherous breezes which sometimes waft him pleasantly toward the latitude of the Cape. No sooner does he come within a certain distance of it—previously fixed in his own mind—than all hands are turned to setting the ship in storm trim; and never mind how light the breeze, down come his t'gallantyards. He "bends" his strongest storm-sails, and lashes everything on deck securely. The ship is then ready for the worst; and if, in reeling round the headland, she receives a broadside, it generally goes well with her. If ill, all hands go to the bottom with quiet consciences.

Among sea captains, there are some who seem to regard the genius of the Cape as a wilful, capricious jade, that must be courted and coaxed into complaisance. First, they come along under easy sails; do not steer boldly for the headland, but tack this way and that—sidling up to it. Now they woo the Jezebel with a t'-gallant studdingsail; anon, they deprecate her wrath with double-reefed topsails. When, at length, her unappeasable fury is fairly aroused, and all round the dismantled ship the storm howls and howls for days together, they still persevere in their efforts. First, they try unconditional submission; furling every rag and heaving to; laying like a log, for the tempest to toss wheresoever it pleases.

This failing, they set a spencer or trysail, and shift on the other tack. Equally vain! The gale sings as hoarsely as before. At last, the wind comes round fair; they drop the fore-sail; square the yards, and scud before it; their implacable foe chasing them with tornadoes, as if to show her insensibility to the last.

Other ships, without encountering these terrible gales, spend week after week endeavoring to turn this boisterous world-corner against a continual headwind. Tacking hither and thither, in the language of sailors they polish the Cape by beating about its edges so long.

Le Mair and Schouten, two Dutchmen, were the first navigators who weathered Cape Horn. Previous to this, passages had been made to the Pacific by the Straits of Magellan; nor, indeed, at that period, was it known to a certainty that there was any other route, or that the land now called Tierra del Fuego was an island. A few leagues southward from Tierra del Fuego is a cluster of small islands, the Diegoes; between which and the former island are the Straits of Le Mair, so called in honor of their discoverer, who first sailed through them into the Pacific. Le Mair and Schouten, in their small, clumsy vessels, encountered a series of tremendous gales, the prelude to the long train of similar hardships which most of their followers have experienced. It is a significant fact, that Schouten's vessel, the *Horne*, which gave its name to the Cape, was almost lost in weathering it.

The next navigator round the Cape was Sir Francis Drake, who, on Raleigh's Expedition, beholding for the first time, from the Isthmus of Darien, the "goodlie south Sea," like a trueborn Englishman, vowed, please God, to sail an English ship thereon; which the gallant sailor did, to the sore discomfiture of the Spaniards on the coasts of Chili and Peru.

But perhaps the greatest hardships on record, in making this celebrated passage, were those experienced by Lord Anson's squadron in 1736. Three remarkable and most interesting narratives record their disasters and sufferings. The first, jointly written by the carpenter and gunner of the *Wager*; the second by young Byron, a midshipman in the same ship; the third, by the chaplain of the *Centurion*. *White-Jacket* has them all; and they are fine reading of a boisterous March night, with the casement rattling in your ear, and the chimney-stacks blowing down upon the pavement, bubbling with raindrops.

But if you want the best idea of Cape Horn, get my friend Dana's unmatchable "Two Years Before the Mast." But you can read, and so you must have read it. His chapters describing Cape Horn must have been written with an icicle.

At the present day the horrors of the Cape have somewhat abated. This is owing to a growing familiarity with it; but, more than all, to the improved condition of ships in all respects, and the means now generally in use of preserving the health of the crews in times of severe and prolonged exposure.

Colder and colder; we are drawing nigh to the Cape. Now gregoes, pea jackets, monkey jackets, reefing jackets, storm jackets, oil jackets, paint jackets, round jackets, short jackets, long jackets, and all manner of jackets, are the order of the day, not excepting the immortal white jacket, which begins to be sturdily buttoned up to the throat, and pulled down vigorously at the skirts, to bring them well over the loins.

But, alas! those skirts were lamentably scanty; and though, with its quiltings, the jacket was stuffed out about the breasts like a Christmas turkey, and of a dry cold day kept the wearer warm enough in that vicinity, yet about the loins it was shorter than a ballet-dancer's skirts; so that while my chest was in the temperate zone close adjoining the torrid, my hapless thighs were in Nova Zembla, hardly an icicle's toss from the Pole.

Then, again, the repeated soakings and dryings it had undergone, had by this time made it shrink woefully all over, especially in the arms, so that the wristbands had gradually crawled up near to the elbows; and it required an energetic thrust to push the arm through, in drawing the jacket on.

I endeavored to amend these misfortunes by sewing a sort of canvas ruffle round the skirts, by way of a continuation or supplement to the original work, and by doing the same with the wristbands.

This is the time for oil-skin suits, dreadnaughts, tarred trousers and overalls, sea-boots, comforters, mittens, woolen socks, Guernsey frocks, Havre shirts, buffalo-robe shorts, and mooseskin drawers. Every man's jacket is his wigwam, and every man's hat his caboose.

Perfect license is now permitted to the men respecting their clothing. Whatever they can rake and scrape together they put on—swaddling themselves in old sails, and drawing old socks over their heads for night-caps. This is the time for smiting your chest with your hand, and talking loud to keep up the circulation.

Colder, and colder, and colder, till at last we spoke a fleet of icebergs bound North. After that, it was one incessant "cold snap," that almost snapped off our fingers and toes. Cold! It was cold as *Blue Flujin*, where sailors say fire freezes.

And now coming up with the latitude of the Cape, we stood south-ward to give it a wide berth, and while so doing were becalmed; ay, becalmed off Cape Horn, which is worse, far worse, than being becalmed on the Line.

Here we lay forty-eight hours, during which the cold was intense. I wondered at the liquid sea, which refused to freeze in such a temperature. The clear, cold sky overhead looked like a steel-blue cymbal, that might ring, could you smite it. Our breath came and went like puffs of smoke from pipebowls. At first there was a long gauky swell, that obliged us to furl most of the sails, and even send down t'gallantyards, for fear of pitching them overboard.

Out of sight of land, at this extremity of both the inhabitable and uninhabitable world, our peopled frigate, echoing with the voices of men, the bleating of lambs, the cackling of fowls, the grunting of pigs, seemed like Noah's old ark itself, becalmed at the climax of the Deluge.

There was nothing to be done but patiently to await the pleasure of the elements, and "whistle for a wind," the usual practice of seamen in a calm. No fire was allowed, except for the indispensable purpose of cooking, and heating bottles of water to toast Selvagee's feet. He who possessed the largest stock of vitality, stood the best chance to escape freezing. It was horrifying.

In such weather any man could have undergone amputation with great ease, and helped take up the arteries himself.

Indeed, this state of affairs had not lasted quite twenty-four hours, when the extreme frigidity of the air, united to our increased tendency to inactivity, would very soon have rendered some of us subjects for the surgeon and his mates, had not a humane proceeding of the Captain suddenly impelled us to rigorous exercise.

And here be it said, that the appearance of the Boatswain, with his silver whistle to his mouth, at the main hatchway of the gundeck, is always regarded by the crew with the utmost curiosity, for this betokens that some general order is about to be promulgated through the ship. What now? is the question that runs on from man to man. A short preliminary whistle is then given by "Old Yarn," as they call him, which whistle serves to collect round him, from their various stations, his four mates. Then Yarn, or Pipes, as leader of the orchestra, begins a peculiar call, in which his assistants join. This over, the order, whatever it may be, is loudly sung out and prolonged, till the remotest corner echoes again. The Boatswain and his mates are the towncriers of a man-of-war.

The calm had commenced in the afternoon: and the following morning the ship's company were electrified by a general order, thus set forth and declared: *"D'ye hear there, fore and aft! all hands skylark!"*

This mandate, nowadays never used except upon very rare occasions, produced the same effect upon the men that Exhilarating Gas would have done, or an extra allowance of "grog." For a time, the wonted discipline of the ship was broken through, and perfect license allowed. It was a Babel here, a Bedlam there, and a Pandemonium everywhere. The Theatricals were nothing compared with it. Then the faint-hearted and timorous crawled to their hiding places, and the lusty and bold shouted forth their glee. Gangs of men, in all sorts of outlandish habiliments, wild as those worn at some crazy carnival, rushed to and fro, seizing upon whomsoever they pleased—warrant-officers and dangerous pugilists excepted—pulling and hauling the luckless tars about, till fairly baited into a genial warmth. Some were made fast to and hoisted aloft with a will: others, mounted upon oars, were ridden fore and aft on a rail, to the boisterous mirth of the spectators, any one of whom might be the next victim. Swings were rigged from the tops, or the masts; and the most reluctant wights being purposely selected, in spite of all struggles, were swung from East to West, in vast arcs of circles, till almost breathless. Hornpipes, fandangoes, Donnybrook-jibs, reels, and quadrilles, were danced under the very nose of the most mighty captain, and upon the very quarterdeck and poop. Sparring and wrestling, too, were all the vogue; Kentucky bites were given, and the Indian hug exchanged. The din frightened the seafowl, that flew by with accelerated wing.

It is worth mentioning that several casualties occurred, of which, however, I will relate but one. While the "skylarking" was at its height, one of the foretopmen—an ugly-tempered devil of a Portuguese, looking on—swore that he would be the death of any man who laid violent hands upon his inviolable person. This threat being overhead, a band of desperadoes, coming up from behind, tripped him up in an instant, and in the twinkling of an eye the Portuguese was straddling an oar, borne aloft by an uproarious multitude, who rushed him along the deck at a railroad gallop. The living mass of arms all round and beneath him was so dense, that every time he inclined one side he was instantly pushed upright, but only to fall over again, to receive another push from the contrary direction. Presently, disengaging his hands from those who held them, the enraged seaman drew from his bosom an iron belaying pin, and recklessly laid about him to right and left. Most of his persecutors fled; but some eight or ten still stood their ground, and, while bearing him aloft, endeavored to wrest the weapon from his hands. In this attempt, one man was stuck on the head, and dropped insensible. He was taken up for dead, and carried below to Cuticle, the surgeon, while the Portuguese was put under guard. But the wound did not prove very serious; and in a few days the man was walking about the deck, with his head well bandaged.

This occurrence put an end to the "skylarking," further headbreaking being strictly prohibited. In due time the Portuguese paid the penalty of his rashness at the gangway; while once again the officers shipped their quarterdeck faces.

Ere the calm had yet left us, a sail had been discerned from the foretop-masthead, at a great distance, probably three leagues or more. At first it was a mere speck, altogether out of sight from the deck. By the force of attraction or something else equally inscrutable, two ships in a calm, and equally affected by the currents, will always approximate, more or less. Though there was not a breath of wind, it was not a great while before the strange sail was described from our bulwarks; gradually, it drew still nearer.

What was she, and whence? There is no object which so excites interest and conjecture, and, at the same time, baffles both, as a sail, seen as a mere speck on these remote seas off Cape Horn.

A breeze! a breeze! for lo! the stranger is now perceptibly nearing the frigate; the officer's spyglass pronounced her a full-rigged ship, with all sail set, and coming right down to us, though in our own vicinity the calm still reigns.

She is bringing the wind with her. Hurrah! Ay, there it is! Behold how mincingly it creeps over the sea, just ruffling and crisping it.

Our top-men were at once sent aloft to loose the sails, and presently they faintly began to distend. As yet we hardly had steerage-way. Toward sunset the stranger bore down before the wind, a complete pyramid of canvas. Never before, I venture to say, was Cape Horn so audaciously insulted. Stun'-sails alow and aloft; royals, moonsails, and everything else. She glided under our stern, within hailing distance, and the signal quartermaster ran up our ensign to the gaff.

"Ship ahoy!" cried the Lieutenant of the Watch, through his trumpet.

"Halloa!" bawled an old fellow in a green jacket, clapping one hand to his mouth, while he held on with the other to the mizzenshrouds.

"What ship's that?"

"The *Sultan*, Indiaman, from New York, and bound to Callao and Canton, sixty days out, all well. What frigate's that?"

"The United States ship *Neversink*, homeward bound."

"Hurrah! hurrah! hurrah!" yelled our enthusiastic countryman, transported with patriotism.

By this time the *Sultan* had swept past, but the Lieutenant of the Watch could not withhold a parting admonition.

"D'ye hear? You'd better take in some of your flying-kites there. Look out for Cape Horn!"

But the friendly advice was lost in the now increasing wind. With a suddenness by no means unusual in these latitudes, the light breeze soon became a succession of sharp squalls, and our sailproud braggadocio of an Indiaman was observed to let everything go by the run, his t'-gallant stun'sails and flying jib taking quick leave of the spars; the flying-jib was swept into the air, rolled together for a few minutes, and tossed about in the squalls like a football. But the wind played no such pranks with the more prudently managed canvas of the *Neversink*, though before many hours it was stirring times with us.

About midnight, when the starboard watch, to which I belonged, was below, the boatswain's whistle was heard, followed by the shrill cry of "All hands take in sail! Jump, men, and save ship!"

Springing from our hammocks, we found the frigate leaning over to it so steeply, that it was with difficulty we could climb the ladders leading to the upper deck.

Here the scene was awful. The vessel seemed to be sailing on her side. The maindeck guns had several days previous been run in and housed, and the portholes closed, but the lee carronades on the quarter-deck and forecastle were plunging through the sea, which undulated over them in milk-like billows of foam. With every lurch to leeward the yard-arm-ends seemed to dip in the sea, while forward the spray dashed over the bows in cataracts, and drenched the men who were on the foreyard. By this time the deck was alive with the whole strength of the ship's company, five hundred men, officers and all, mostly clinging to the weather bulwarks. The occasional phosphorescence of the yeasting sea cast a glare upon their uplifted faces, as a night fire in a populous city lights up the panic-stricken crowd.

In a sudden gale, or when a large quantity of sail is suddenly to be furled, it is the custom for the First Lieutenant to take the trumpet from whoever happens then to be officer of the deck. But Mad Jack had the trumpet that watch; nor did the First Lieutenant now seek to wrest it from his hands. Every eye was upon him, as if we had chosen him from among us all, to decide this battle with the elements, by single combat with the spirit of the Cape; for Mad Jack was the saving genius of the ship, and so proved himself that night. I owe this right hand, that is this moment flying over my sheet, and all my present being to Mad Jack. The ship's bows were now butting, battering, ramming, and thundering over and upon the head seas, and with a horrible wallowing sound our whole hull was rolling in the trough of the foam. The gale came athwart the deck, and every sail seemed bursting with its wild breath.

All the quartermasters, and several of the forecastlemen, were swarming round the double-wheel on the quarter-deck. Some jumping up and down, with their hands upon the spokes; for the whole helm and galvanized keel were fiercely feverish, with the life imparted to them by the wild tempest.

"Hard up the helm!" shouted Captain Claret, bursting from his cabin like a ghost in his nightdress.

"Damn you!" raged Mad Jack to the quarter-masters; "hard down—hard down, I say, and be damned to you!"

Contrary orders! but Mad Jack's were obeyed. His object was to throw the ship into the wind, so as the better to admit of close-reefing the topsails, but though the halyards were let go, it was impossible to clew down the yards, owing to the enormous horizontal strain on the canvas. It now blew a hurricane. The spray flew over the ship in floods. The gigantic masts seemed about to snap under the world-wide strain of the three entire topsails.

"Clew down! Clew down!" shouted Mad Jack, husky with excitement, and in a frenzy, beating his trumpet against one of the shrouds.

But, owing to the slant of the ship, the thing could not be done. It was obvious that before many minutes something must go—either sails, rigging, or sticks; perhaps the hull itself, and all hands.

Presently a voice from the top exclaimed that there was a rent in the main topsail. And instantly we heard a report like two or three muskets discharged together; the vast sail was rent up and down like the Veil of the Temple. This saved the main-mast; for the yard was now clewed down with comparative ease, and the topmen laid out to stow the shattered canvas. Soon, the two remaining top-sails were also clewed down and close-reefed.

Above all the roar of the tempest and the shouts of the crew, was heard the dismal tolling of the ship's bell—almost as large as that of a village church—which the violent rolling of the ship was occasioning. Imagination cannot conceive the horror of such a sound in a night tempest at sea.

"Stop that ghost!" roared Mad Jack; "away, one of you, and wrench off the clapper!"

But no sooner was this ghost gagged, than a still more appalling sound was heard, the rolling to and fro of the heavy shot, which, on the gundeck, had broken loose from the gunracks, and converted that part of the ship into an immense bowling alley. Some hands were sent down to secure them; but it was as much as their lives were worth. Several were maimed; and the midshipmen who were ordered to see the duty performed reported it impossible, until the storm abated.

The most terrific job of all was to furl the main-sail, which, at the commencement of the squalls, had been clewed up, coaxed and quieted as much as possible with the bunt-lines and slab lines. Mad Jack waited some time for a lull, ere he gave an order so perilous to be executed. For to furl this enormous sail, in such a gale, required at least fifty men on the yard; whose weight, superadded to that of the ponderous stick itself, still further jeopardized their lives. But there was no prospect of a cessation of the gale, and the order was at last given.

At this time a hurricane of slanting sleet and hail was descending upon us; the rigging was coated with a thin glare of ice, formed within the hour.

"Aloft, main-yard-men! and all you maintop-men! and furl the main-sail!" cried Mad Jack.

I dashed down my hat, slipped out of my quilted jacket in an instant, kicked the shoes from my feet, and, with a crowd of others, sprang for the rigging. Above the bulwarks (which in a frigate are so high as to afford much protection to those on deck) the gale was horrible. The sheer force of the wind flattened us to the rigging as we ascended, and every hand seemed congealing to the icy shrouds by which we held.

"Up—up, my brave hearties!" shouted Mad Jack; and up we got, some way or other, all of us, and groped our way out on the yard-arms.

"Hold on, every mother's son!" cried an old quartergunner at my side. He was bawling at the top of his compass, but in the gale, he seemed to be whispering; and I only heard him from his being right to windward of me.

But his hint was unnecessary; I dug my nails into the Jack-stays, and swore that nothing but death should part me and them until I was able to turn round and look to windward. As yet, this was impossible; I could scarcely hear the man to leeward at my elbow; the wind seemed to snatch the words from his mouth and fly away with them to the South Pole.

All this while the sail itself was flying about, sometimes catching over our heads, and threatening to tear us from the yard in spite of all our hugging. For about three quarters of an hour we thus hung suspended right over the rampant billows, which curled their very crests under the feet of some four or five of us clinging to the lee yard-arm, as if to float us from our place.

Presently, the word passed along the yard from windward that we were ordered to come down and leave the sail to blow; since it could not be furled. A midshipman, it seemed, had been sent up by the officer of the deck to give the order, as no trumpet could be heard where we were.

Those on the weather yard-arm managed to crawl upon the spar and scramble down the rigging; but with us, upon the extreme leeward side, this feat was out of the question; it was, literally, like climbing a precipice to get to windward in order to reach the shrouds; besides, the entire yard was now encased in ice, and our hands and feet were so numb that we dared not trust our lives to them. Nevertheless, by assisting each other, we contrived to throw ourselves prostrate along the yard, and embrace it with our arms and legs. In this position, the stun'sail booms greatly assisted in securing our hold. Strange as it may appear, I do not suppose that, at this moment, the slightest sensation of fear was felt by one man on that yard. We clung to it with might and main; but this was instinct. The truth is, that, in circumstances like these, the sense of fear is annihilated in the unutterable sights that fill all the eye, and the sounds that fill all the ear. You become identified with the tempest; your insignificance is lost in the riot of the stormy universe around.

Below us, our noble frigate seemed thrice its real length—a vast black wedge, opposing its widest end to the combined fury of the sea and wind.

At length the first fury of the gale began to abate, and we at once fell to pounding our hands, as a preliminary operation to going to work; for a gang of men had now ascended to help secure what was left of the sail; we somehow packed it away, at last, and came down.

About noon the next day, the gale so moderated that we shook two reefs out of the top-sails, set new courses, and stood due east, with the wind astern.

Thus, all the fine weather we encountered after first weighing anchor on the pleasant Spanish coast, was but the prelude to this one terrific night; more especially, that treacherous calm immediately preceding it. But how could we reach our long promised homes without encountering Cape Horn? by what possibility avoid it? And though some ships have weathered it without these perils, yet by far the greater part must encounter them. Lucky it is that it comes about midway in the homeward-bound passage, so that the sailors have time to prepare for it, and time to recover from it after it is astern.

But, sailor or landsman, there is some sort of a Cape Horn for all. Boys! beware of it; prepare for it in time. Greybeards! thank God it is passed. And ye lucky livers, to whom, by some rare fatality, your Cape Horns are placid as Lake Lemans, flatter not yourselves that good luck is judgment and discretion; for all the yolk in your eggs, you might have foundered and gone down, had the Spirit of the Cape said the word.

Fourteen

The Last Cruise of the *Saginaw*

George H. Read

As Doctor Frank, our surgeon, and myself were walking down the beach to the last boat off to the ship, there occurred an incident which I will relate here for psychological students.

He remarked, as we loitered around the landing, that he felt greatly depressed without being able to define any cause for it and that he could not rid himself of the impression that some misfortune was impending. I tried to cheer him up; told him that the "blues" were on him, when he ought to be rejoicing instead; that we had a fair wind and a smooth sea to start us on a speedy return to the old friends in San Francisco. It was in vain, however; he expressed a firm belief that we should meet with some disaster on our voyage and I dropped the subject with a "pooh pooh."

As soon as we reached the open sea, the captain ordered the ship headed to the westward and the pressure of steam to be reduced, as with topsails set we sailed along to a light easterly breeze. It was his intention, he stated, to come within sight of Ocean Island about daylight and to verify its location by steaming around it before heading away for San Francisco.

The evening following the departure passed quietly in our wardroom quarters and in fact all over the ship. Officers and men were more than usually fatigued after the preparations for sea both on shore and on board. There was none of the general hilarity accompanying a homeward cruise. There was also a prevailing dread of a long and tedious journey of over three thousand miles, mostly to be made under sail, and we all knew the tendency of the old *Saginaw* in a headwind to make "eight points to leeward," or, as a landlubber would say, to go sideways. We occupied ourselves in stowing and securing our movables, and after the bugle sounded "Out lights" at 9 p.m. the steady tramp of the lookouts and their half-hour hail of "All's well" were all that disturbed the quiet of the night.

The night was dark, but a few stars were occasionally visible between the passing clouds. The sea continued smooth and the ship on an even keel. When I turned in at ten o'clock I had the comforting thought that by the same time tomorrow night we should be heading for San Francisco. We were making about three knots an hour, which would bring Ocean Island in sight about early dawn, so that there would be plenty of time to circumnavigate the reef and get a good offing on our course before dark.

How sadly, alas! our intentions were frustrated and how fully our surgeon's premonitions were fulfilled! My pen falters at the attempt to describe the events of the next few hours. I was suddenly awakened about three o'clock in the morning by an unusual commotion on deck; the hurried tramping of feet and confusion of sounds. In the midst of it I distinguished the captain's voice sounding in sharp contrast to his usual moderate tone, ordering the taking in of the topsails and immediately after the cutting away of the topsail halliards. Until the latter order was given I imagined the approach of a rain squall, a frequent occurrence formerly, but I knew now that some greater emergency existed, and so I hastily and partly dressed myself sufficiently to go on deck.

Just before I reached the top of the wardroom ladder I felt the ship strike something and supposed we were in collision with another vessel. The shock was an easy one at first, but was followed immediately by others of increasing force, and, as my feet touched the deck, by two severe shocks that caused the ship to tremble in every timber. The long easy swell that had been lifting us gently along in the open sea was now transformed into heavy breakers as it reached and swept over the coral reef, each wave lifting and dropping with a frightful thud the quaking ship. It seemed at each fall as though her masts and smokestack would jump from their holdings and go by the board.

To a landsman or even a professional seaman who has never experienced the sensation it would be impossible to convey a realizing sense of the feelings aroused by our sudden misfortune. There is a something even in the air akin to the terror of an earthquake shock—a condition unnatural and uncanny. The good ship that for years has safely sailed the seas or anchored in ports with a free keel, fulfilling in all respects the destiny marked out for her at her birth, suddenly and without warning enters upon her death-struggle with the rocks and appeals for help. There is no wonder that brave men—men having withstood the shock of battle and endured the hardships of the fiercest storms—should feel their nerves shaken from their first glance at the situation.

The captain had immediately followed his orders, to take in the sails that were forging us on towards the reef, by an order to back engines. Alas! the steam was too low to give more than a few turns to the wheels, and they could not overcome the momentum of the ship. In less than an hour of the fierce pounding the jagged rock broke through the hull and tore up the engine and fireroom floor; the water rushed in and reached the fires; the doom of our good ship was now apparent and sealed.

I hastily returned to my stateroom, secured more clothing, together with some of the ship's papers, then ascended to the hurricane deck to await developments or to stand by to do rescue work as ordered. I had participated in the past in drills that are called in Navy Regulations "abandoned ship." In these drills everyone on board is supposed to leave the vessel and take station as assigned in one of the ship's boats. I had only taken part in these drills during calm weather at sea, and thought it a pretty sight to see all the boats completely equipped and lying off in view of the deserted vessel. Here, however, no programme could help us. Our captain's judgment and quickness of decision must control events as they developed.

The night was clear and starlit, but we could see nothing of any land. Perhaps we had struck on some uncharted reef, and while strenuously employed in getting the boats over the side opposite the sea we waited anxiously for daylight. The scene was one for a lifelong remembrance and is beyond my power adequately and calmly to describe.

There was at first some confusion, but the stern and composed attitude of the captain and his sharp, clear orders soon brought everyone to his senses, and order was restored.

A few of the more frightened ones had at first, either through a misunderstanding or otherwise, rushed to our largest boat—the launch—hanging at the starboard quarter and partly lowered it before the act was noticed. A large combing sea came along and tore it from their hold, smashing it against the side of the ship and then carrying its remnants away with its tackles and all its fittings. This was a great loss, we felt, if we should have to take to the boats, for we did not know at that time where we were.

The same wave also carried off one of the crew, a member of the Marine Guard, who had been on the bulwarks; and whisking him seaward, returned him miraculously around the stern of the ship to the reef, where his struggles and cries attracted the notice of others. He was hauled over the lee side, somewhat bruised and water-soaked, but, judging from his remarks, apparently not realizing his wonderful escape from death.

As the night wore on, the wind increased and also the size of the breakers. The ship, which had first struck the reef "bows on," was gradually swung around until she was at first broadside to the reef, and then further until the after part, to which we were clinging, was lifted over the jagged edge of the perpendicular wall of rock. She was finally twisted around until the bow hung directly to seaward, with the middle of the hull at the edge. Thus the ship "seesawed" from stem to stern with each coming wave for an hour or more and until the forward part broke away with a loud crash and disappeared in the deep water outside. Our anchors, which had been "let go," apparently never touched bottom until the bow went with them.

All that was left of our good ship now heeled over towards the inner side of the reef; the smokestack soon went by the board and the mainmast was made to follow it by simply cutting away the starboard or seaward shrouds. Over this mast we could pass to the reef, however, and there was comparative quiet in the waters under our lee. This helped us in passing across whatever we could save from the wreck, and in this manner went three of our boats, the captain's gig, one of the cutters, and the dinghy, without much damage to them. We also secured in this way an iron lifeboat belonging to the contractor.

As the first gray streaks of dawn showed us a small strip of terra firma in the smooth water of the lagoon and not far from the reef, many a sigh of relief was heard, and our efforts were redoubled to provide some means of prolonging existence there. At any rate, we knew now where we were and could at least imagine a possible relief and plan measures to secure it.

Although the sea had robbed us of the larger part of our provisions, in the forward hold there were still some of the most important stowed within the fragment we were clinging to, which contained the bread and clothing storerooms. With daylight our task was made easier.

A line was formed across the reef and everything rescued was passed over the side and from hand to hand to the boats in the lagoon, for transfer to the island. Thus we stood waist-deep in the water, feet and ankles lacerated and bleeding, stumbling about the sharp and uneven coral rock, until five in the afternoon, and yet our spirits, which had been low in the dark, were so encouraged by a sight of a small portion of dry land and at least a temporary escape from a watery grave that now and then a jest or a laugh would pass along the line with some article that suggested a future meal.

At five o'clock in the afternoon the order was given to abandon the wreck (which was done while hoping that it would hold together until tomorrow), and as the sun went down on the "lone barren isle," all hands were "piped" by the boatswain's whistle to supper.

A half-teacup of water, half a cake of hardtack, and a small piece of boiled pork constituted our evening meal, to which was added a piece of boiled mutton that had been intended for the wardroom table.

After this frugal meal all hands were mustered upon the beach to listen to a prayer of thankfulness for our deliverance and then to a few sensible and well-timed remarks from the captain enjoining discipline, good nature, and economy of food under our trying circumstances. He told us that by the Navy Regulations he was instructed, as our commanding officer, to keep up, in such sad conditions as we were thrown into, the organization and discipline of the Service so far as applicable; that he would in the event of our rescue be held responsible for the proper administration of law and order; that officers and crew should fare alike on our scanty store of food, and that with care we should probably make out, with the help of seal meat and birds, a reduced ration for some little time. He would detail our several duties tomorrow. Then we were dismissed to seek "tired nature's sweet restorer" as best we could.

With fourteen hours of severe labor, tired, wet, and hungry, we were yet glad enough to sink to rest amid the bushes with but the sky for a canopy and a hummock of sand for a pillow. In my own case sleep was hard to win. For a long time I lay watching the stars and speculating upon the prospects of release from our island prison. Life seemed to reach dimly uncertain into the future, with shadow pictures intervening of famished men and bereaved families.

I could hear the waves within a few rods of our resting-places—there was no music in them now—lapping the beach in their restlessness, and now and then an angry roar from the outside reef, as though the sea was in rage over its failure to reach us. I realized that for more than a thousand miles the sea stretched away in every direction before meeting inhabited shores and for treble that distance to our native land; that our island was but a small dot in the vast Pacific—a dot so small that few maps give it recognition. Truly it was a dismal outlook that "tired nature" finally dispelled and that sleep transformed into oblivion; for I went to sleep finally while recalling old stories of family gatherings where was always placed a vacant chair for the loved absent one should he ever return.

ON THE ISLAND

Sunday, October 30. No pretensions to the official observance of the Sabbath were made today. We always had religious services on board the ship when the weather permitted on Sunday, but today every effort has been made to further the safety of our condition.

The captain, executive officer, and many of the crew went off early to the wreck in order to make further search for supplies and equipment. The wreck appears from the island to be about as we left it, for the wind has been light and the sea calm during the night.

I remained on shore with a few men to assist in sorting out and making a list of the articles rescued yesterday and to assemble them in the best place suitable for their preservation. We spread out in the sun the bread, bags of flour, and other dry foodstuffs, even to the smallest fragments, and it was early apparent that unless much more food was secured we shall be compelled to live upon a greatly reduced ration and that our main source of food would be the seal and brown albatross (or "goonies," as they are commonly called). Both of these seemed plentiful and are easily captured.

The seal succumb quickly to a blow upon the head, a fact we discovered early in our first visit to the Midway Islands. One of the boat's crew, when pushing off from the beach, carelessly and without intent to kill, struck a nearby seal on the head with an oar, and the next morning it was found dead, apparently not having moved from the spot. Its mate had found it and was nosing it about, while moaning in a most humanlike voice.

These seal are quite different from the Alaska fur seal, of such great value for their fur. These have a short lustreless hair, and their principal value is in the oil that is extracted by the few seal hunters who seek them. They frequently exceed two hundred pounds in weight, and are savage fighters if one can judge by the many scars found upon them. We never thought, when, a few months ago, we amused ourselves on the verandas of the Cliff House at San Francisco in watching their disporting about Seal Rock, that we should make such a close acquaintance with them.

The "goonies" also are easy to capture, although they are large and strong and a blow from the wing would break a man's limb. I measured one of them from tip to tip of wing, and it was over seven feet. They are, however, very awkward on their feet, and, having a double-jointed wing (that is, a joint in it like an elbow) can only rise from the ground when the wind is in their faces. Owing to this fact one only needs to get to the windward of them with a club and look out for the wings. We should like to add some of their eggs to our bill of fare, but dare not for fear of driving the birds away. I imagine it would take but a few of the eggs, if eatable, to go around, for I saw one at the Midways that was as large as those of the ostrich.

Fresh water will, however, apparently be our greatest cause for anxiety; for we have secured but a small supply, considering our number—ninety-three. A few breakers or kegs only, which were stowed in the boats, were secured. Rain, of course, we count upon; but to conserve our scanty supply until it comes is most necessary. Today several wells have been dug in various parts of the island, but the water found in them is near the surface and is too brackish for any use.

The old timbers of a former wreck, probably of the *Gledstanes*—the "bones" as sailors call them—lie near on the beach and look as though they would yield us fuel for a long time. Our fire, which was started last evening by a match that Mr. Bailey, the chief diver, had fortunately kept dry, has been constantly going for lack of more lighting material.

Evening. The reef party returned at sundown, reporting a strenuous day on the wreck. We all had a supper of "scouse" (a dish of pork, potato, and hardtack), and before sleeping the campsite was laid out, the sails and awnings which had come on shore temporarily set up, to our greater comfort. Besides the sails and awnings, more food supplies were captured from the after storeroom and a particularly fortunate prize secured in a small portable boiler that had been lashed to the after deck. This had been used by the contractor's party in hoisting to the scow the blasted coral from the reef, at Midway Islands.

There were also in one of the wheelhouses of the wreck some distilling coils, which the engineer's force with our chief engineer successfully rescued after hard labor, for the sea was washing through the wheelhouse with terrible force. The boiler, suspended between two boats, was successfully landed on the beach, and we are greatly encouraged at the promise of fresh water tomorrow. We secured a barrel, also, partly filled with sperm oil, and a lantern in good condition. These two articles ensure us a supply of lighting material for the cooking-fire, which can now be put out at night and much fuel saved. Considerable clothing was secured from the officers' staterooms, and I was fortunate enough to find some of mine rolled up in one of the large wet bundles; and a few soaked mattresses and blankets were also brought in. The carpenter's chest, too, came ashore intact, and altogether we feel our situation greatly improved.

Mr. Talbot tells me that they are literally "stripping" the wreck, and nothing movable will be left on it if the weather will but hold good long enough. No one stops to question the utility of an article found adrift;

it is seized hastily and thrown out on the reef to be transported later to the island. Pieces of rigging, boxes of tinned coffee, canned goods, tools, crockery, sails, awnings, etc., all come to the beach in a promiscuous mass to be sorted out later.

Tuesday, November 1. The crew was formed into several messes today, and also into watches. Each mess was provided with a tent, that for our mess (the wardroom) being made from the *Saginaw's* quarter-deck awning. Such of our dry goods and bedding as had been rescued were removed to them, and our little camp begins to take on the appearance of comfort.

The duties of every member of the ship's company have been so arranged that it is hoped and expected that no one will have much time to brood over our situation or the future.

Wednesday, November 2. The bad weather we have feared has arrived. It came on suddenly this morning from the southeast with a high wind and a heavy rainfall, and before we had been able firmly to secure the tents. After strenuous exertion, however, we saved them from being blown over, but were wet to the skin when they were finally safe in place.

Fortunately the wreck on the reef has been thoroughly explored and there is very little material there now that could be of use to us, unless it may be the timbers themselves, to help us in building a seaworthy boat should it be necessary to do so in a final effort to get away. The idea of sending a boat to the Sandwich Islands for relief has been already revolving in our minds, and today was revealed by an order from the captain to the senior officers. After a consultation singly with us, he has directed each one to file with him an opinion on the feasibility and necessity of doing so—each written opinion to be without knowledge of the others.

It is probable that the hulk will be considerably broken up before the wind and sea go down, for one can see it rise and fall with the breakers, and occasionally a piece is detached and floated across the reef into the lagoon. As soon as it is safe to launch the boats, the work of securing these pieces will be started.

The boats are now resting at the highest part of the island in the centre of the camp, for even with the protecting reef the sea in the lagoon has been so rough that combers have reached within a few feet of our tents. As I write my journal we are a wet and sad party of unfortunates.

Our captain and his boat's crew must be having an experience worse than ours, however. They left this morning in the cutter for the sandspit near and to the west of us, to collect driftwood, and are "marooned" there in the storm. They can be seen, with the glasses, huddled together beneath the upturned boat. They do not, however, seem to be in imminent danger, and have made no signals of distress; so we expect them to return as soon as the sea abates.

Thursday, November 3. It has been still too rough today to launch the boats for work in the lagoon. We have, however, busied ourselves in erecting a storehouse for the better preservation of our food supplies, and tonight have them safely under cover. Last night the rats robbed us of a box of macaroni, and, therefore, we have put our storehouse on posts and two feet above the ground with inverted pans upon the posts.

We made the acquaintance of the rats last night in our tent when a noisy fight over a piece of candle disturbed our sleep. We had seen a few of them before, but did not suppose them to be so very numerous—as on first thought there seemed to be so very little for them to eat. We now found them to have good lungs and appetites, however, and a good deal of thrashing around with boots, etc., was necessary to expel them. We discussed them before we went to sleep again in the light of a future food supply—an addition to our one-quarter ration—and the opinion was general that should the seal and goonies desert us the rats would become more valuable. At any rate, they would thrive on the refuse of the food we had now.

The captain returned this morning from his expedition and gives a sad story of their luck. They had to literally bury themselves to the neck in the sand and lie under the boat to prevent being drenched by the rain. During the height of the storm they had one streak of good luck. They found some companions that the rough sea had induced to seek the shelter of the lagoon and beach. They were large sea turtles, and he and his crew turned them on their backs to prevent their escape. Today we have them added to our food supply and they are very welcome, notwithstanding the sad plight of their captors when they returned.

We have also added to our fresh water a supply of about fifty gallons caught in the rainstorm of yesterday, and doled out an extra cupful to each person.

Sunday, November 6. We were mustered for divine service today, and it being the first Sunday of the month the roll was called and each man answered "Here" as his name was called. After that prayers were read by the captain and an extra cup of water served out from the quantity caught during the recent gale. Work was suspended so far as possible, but the lagoon being so quiet it was thought necessary to launch two of the boats and tow in some of the floating timbers. We were overjoyed thus to receive and haul up free of the water a large fragment of the old hurricane deck. We can imagine some value in almost any piece of timber, but in this particular we are confident of securing much material for the building of our future boat, it being of three-inch-thick narrow planking. We believe we can make one-and-a-half-inch stuff from it by rigging up a staging and converting our one bucksaw into a jigsaw with a man above and one below. The blacksmith believes that he can extract a good supply of nails, and in many ways it is evident that we are not going to wait supinely for the relief we hope for from our brave comrades' voyage.

Today we killed our first goonies and had some for supper. They were very tough and "fishy," and Solomon Graves, once the *Saginaw*'s cabin cook, but now "King of the Galley" on Ocean Island, says that he cooked them all day. Only a portion of the bird could be masticated. However, it was voted superior to seal, the latter being so tough that Graves has to parboil it overnight and fry it in the morning. The hardtack is exhausted, but so much of the flour has been found good that we are to have a table-spoonful every other day and the same quantity of beans on the alternate days as substitutes for the hardtack. A cup of coffee or tea every day for the morning meal. Supper we have at five.

We had a luxury after supper. There are nine of us in the wardroom mess who smoke, and each of us was generously supplied with a cigar by Passed Assistant Engineer Blye, whose chest was rescued the second day; it contained a box of five hundred Manila cigars.

Monday, November 7. The mainmast is ready to raise tomorrow. An excavation has been made at the highest point of the island, near the captain's tent, and the mast rolled up to it with the rope guys ready to hold it upright. The carpenter's gang have been busy all day in sorting out material for the gig's deck and for raising her sides eight inches.

While the weather is fine, there seems to be a considerable swell at sea from the late storm, and the wreck is gradually, as it were, melting away. Today a piece of the hull floated towards us and a boat was sent after it. When it reached the beach I recognized the remains of my stateroom, with twisted bolts protruding from the edge where it had been wrenched away from the rest of the hull. I viewed mournfully the remnant of my longtime home and reflected how it had once been my protection and that now fate had turned me out of its shelter.

Many of the hopes that were bred within its wooden walls have been shattered by its destruction, and I thought it would be appropriate to bury it on the beach with an epitaph above it showing the simple words "Lights out" which I had so often heard at its door when the ship's corporal made his nightly rounds at the "turning-in" hour. However, it was valuable even in its ruin for building and burning material. Besides, we are not ready yet to think of anything like a funeral.

Sunday, November 13. Ship breaking up rapidly and boats out to pick up driftwood. Had prayers (read by Captain S.) at 3 p.m., and he addressed us with remarks as to necessity in our situation of working on Sabbath. Thousands of rats about. Put extra night watch on storehouse, for fear of further depredations.

Thursday, November 17. Blowing hard from north. Tea at 7 a.m. The gig anchored off shore. Mr. Bailey and I fixed up the well where fresh water was found when mast fell; goodbye to the old condenser. "The little cherub that sits up aloft" doing good work for us all.

THE SAILING OF THE GIG

Friday, November 18. The weather has been fine since the breaking up of the storm of the second.

As to work, everyone has had his duties portioned out to him, and there is no doubt of the captain's wisdom in providing thus an antidote to homesickness or brooding. Faces are—some of them—getting "peaked," and quite a number of the party have been ill from lack of power to digest the seal meat; but there are no complaints, we all fare alike. Medicines are not to hand, but a day or two of abstinence and quiet generally brings one around again. In the evenings, when we gather around the smoking lamp after supper, there are frequent discussions over our situation and prospects.

They are, however, mostly sanguine in tone, and it is not uncommon to hear the expression "when we get home." No one seems to have given up his hope of eventual relief. It has been very noticeable, too, at such times that no matter where the conversation begins it invariably swings around, before the word is passed to "douse the glim," to those things of which we are so completely deprived—to narratives of pleasant gatherings—stories of banquets and festival occasions where toothsome delicacies were provided. It would seem as though these reminiscences were given us as a foil to melancholy, and they travel along with us into our dreams.

Upon one point we are all agreed, that we are very fortunate in being wrecked in so agreeable a climate, where heavy clothing is unnecessary. The temperature has been, aside from the storm we had soon after the landing, between seventy and seventy-five degrees during the day and around fifty degrees at night. We are very sensible of the discomforts that would be ours if tumbled upon some of the islands of the northern ocean in winter.

The moonlit nights have been grand, and calculated to foster romance in a sailor's thoughts were the surroundings appropriate. As it is, the little cheer we extract from them is in the fact that we see the same shining face that is illuminating the home of our loved ones.

Often in my corner of the tent, Mr. Foss and I pass what would be a weary hour otherwise, over a game of chess, the pieces for which he has fashioned from gooney bones and blocks of wood.

Mr. Main has made a wonderful nautical instrument—a sextant—from the face of the *Saginaw*'s steam gauge, together with some broken bits of a stateroom mirror and scraps of zinc. Its minute and finely drawn scale was made upon the zinc with a cambric needle, and the completed instrument is the result of great skill and patience. Mr. Talbot has tested it and pronounces it sufficiently accurate for navigating purposes.

Another officer has made a duplicate of the official chart of this part of the Pacific, and still another has copied all the Nautical Almanac tables necessary for navigation.

I have been directed by the captain to make a selection from the best-preserved supplies in the storehouse most suitable for boat service, and calculate that Talbot will have the equivalent of thirty-five days' provender at one-half rations, although many of the articles are not in the regular ration tables.

This morning the boat was surrounded by many men and carried bodily into water that was deep enough to float her. There she was anchored and the stores carried out to her. Mr. Butterworth, standing waist-deep in the water, put on the last finishing touches while she was afloat by screwing to the gunwales the rowlocks for use in calm weather.

There was expended from store-book the following articles: ten breakers (a small keg) of water, five days' rations of hardtack sealed in tin, ten days of the same in canvas bags, two dozen small tins of preserved meat, five tins (five pounds each) of desiccated potato, two tins of cooked beans, three tins of boiled wheaten grits, one ham, six tins of preserved oysters, ten pounds of dried beef, twelve tins of lima beans, about five pounds of butter, one gallon of molasses, twelve pounds of white sugar, four pounds of tea and five pounds of coffee. A small tin cooking apparatus for burning oil was also improvised and furnished.

I had intended putting on board twenty-five pounds of boiled rice in sealed tins, but discovered one of the tins to be swollen just before the provisions were started off. Hastily the tins were opened and the rice found unfit for use. The desiccated potatoes were at once served out in place of the rice, the cans scalded and again sealed.

With the navigating instruments and the clothing of the voyagers on board, the boat was pronounced ready and we went to dinner. There was little conversation during the meal. The impending departure of our shipmates hung like a pall of gloom over us at the last and was too thought-absorbing for speech. Talbot seemed to be the most unconcerned of all, but as I watched him I felt that the brave fellow was assuming it to encourage the rest of us. I had a long friendly talk with him, last evening, during which he seemed thoroughly to estimate the risk he was to take, and entrusted to me his will to be forwarded to his parents in Kentucky in case he should not survive the journey.

The hour set for the boat's departure (four o'clock) arrived and we were all mustered upon the beach. Prayers were read by the captain, after which final farewells were said and the brave men who were to peril their lives for us waded off to the gig and climbed on board. They quickly stepped the little masts, spread the miniature sails, raised their anchor, and slowly gaining headway stood off for the western channel through the reef. With full hearts and with many in tears, we gave them three rousing cheers and a tiger, which were responded to with spirit, and we watched them until the boat faded from sight on the horizon to the northward.

As I write this by the dim light of a candle the mental excitement due from the parting with our shipmates seems still to pervade the tent and no one is thinking of turning in.

Mr. Bailey, the foreman of the contractor's party, came into the tent soon after we had gathered for the evening. He had in his hand a small book and on his face a smile as he passed it around, showing each one an open page of the book; when he reached me I saw it was a pocket Bible opened at the fifty-first chapter of Isaiah, where Mr. B.'s finger rested under the words, "The isles shall wait upon me and on my arm shall they trust." He did not speak until I had read, and then said he had opened the Bible by chance, as was his habit every evening. Poor Bailey! We all feel very sorry for him. He is a fine character, well advanced in years; and having by economy accumulated considerable money, had bought himself a home, before coming out, to which he was intending to retire when this contract was completed.

By invitation from the captain I accompanied him in walking around the entire island, avoiding, however, the extreme point to the westward, where albatross were nesting. He talked but little, and I saw that his eyes often turned to the spot where the gig had disappeared from view. As we separated in front of his little tent he said with a voice full of pathos to me, "Goodnight, Paymaster; God grant that we see them again."

I find that I have so far omitted to give the personnel of Talbot's crew. As stated before there were many volunteers, but the surgeon was ordered to select from a list given him four of the most vigorous and sturdy of the applicants and report their names to the captain. There was considerable rivalry among them. In fact I was accidentally a witness to a hard-fought wrestling match between two of the crew who sought the honor of going and risking their lives. The defeated one, I was told, was to waive his claim in favor of the victor.

WAITING

Thursday, November 24. Thanksgiving Day—at home; the noble bird, roast turkey, has not graced our tarpaulin-covered table. He has been replaced by a tough section of albatross. Nor was there any expression of thanks at the mess table until one of the officers, having finished the extra cup of coffee served in honor of the day, said, "Say, fellows, let's be thankful that we are alive, well and still with hope."

Last evening about nine o'clock we were given another flurry of excitement over expected relief. The storehouse sentry reported a light to the eastward and in a "jiffy" our tent was empty. Sure enough, there was a bright light close to the horizon which, as we watched, appeared to grow larger and nearer. The captain was called, and I joined him with Mr. Cogswell (our new executive officer since Talbot left) in front of his tent. After watching the light for a few minutes, the captain turned to us and said, "Gentlemen, it is only a star rising and the atmosphere is very clear. Better turn in again"; and he entered the tent.

Sunday, November 27. Last Sunday and today we have had divine service led by the captain reading the prayers of the Episcopal ritual.

Work has been steadily pushed on the schooner. The keel has been hewed out of the *Saginaw's* late topmast and is blocked up on the beach. We are ripping the old deck planks in two with our old bucksaw and one handsaw, and while it is slow work we can see our boat planking ahead of us when the frame is ready. The schooner is to be forty feet long, of centre-board, flat-bottomed type, and the captain has settled upon her shape and dimensions after experimenting with a small model in company with the contractor's carpenter, who has had experience in boat-building.

This morning about sunrise the camp was roused to excitement by the loud cry of "Sail ho!" I found on joining the crowd at the landing that the captain had ordered a boat launched and her crew were already pulling away in a northerly direction.

I could see nothing from the crow's nest at the masthead, but the statement of one of the crew that he had seen a sail was positive; and the camp was full of a nervous expectancy until nine o'clock, when the boat returned with the disappointing news that the alleged sail was only a large white rock on the north end of the reef that had reflected the sun's rays. As the sun rose to a greater angle the reflection disappeared. An order was at once given out that no one should again alarm the camp before permission from the captain was obtained.

Sunday, December 25. Christmas Day!! Merry Christmas at home, but dreary enough here! Still the salutation was passed around in a halfhearted manner. It is the first day since the wreck that depression of spirit has been so contagious and camp-wide. The religious services, as we stood in the sand bareheaded (some barefooted also), hardly seemed to fit our situation, and the voice of the captain was subdued and occasionally tremulous. I had donned my best uniform coat, which had come ashore when the wreck was stripped, and tried also to put on a cheerful face. No use; I could not keep up the deceit, and I slipped out of line before the service was ended, to change back to the blue sailor shirt and working clothes. I felt that I had been "putting on airs." It has been my first really blue day, for the pictures in my mind of the Christmas festivities at home but emphasized the desolation of the life here.

Strangely enough, Doctor Frank has seemed to a certain extent to be more cheerful than usual. It seems queer that he, pessimist as he appeared to me when he predicted disaster before we sailed from the Midway Islands, should now be the optimist and attempt to dispel our gloom. Some expert in psychical research may be able to discern, as I cannot, why the doctor's belief in Talbot's success should now have influence enough to change my melancholy into a firmer hope than ever.

We borrowed the chart from the captain and followed in pure imagination the course of the gig; and when we folded it, the doctor said that he believed Talbot had arrived at the end of his journey and we should be relieved. Talbot has now been away thirty-seven days, and our several estimates of the time he would consume have been between thirty and forty.

Every afternoon, when work is suspended for the day and we have repaired to the tent, the expression of Talbot's whereabouts is the first note of discussion; as though it had not been in our minds all the long weary day of work.

As the possible failure of Talbot's brave effort begins to enter our calculations, the greater is the exertion to provide in the near future another avenue of escape. So, with gradually weakened strength, owing to lack of sustaining food, the labor we find arduous and exhausting; I, being included in the carpenter's gang on the schooner, realize that fact thoroughly. Yesterday the captain and myself made another circuit of the island, and both were glad to rest on the return to the camp.

The captain has ordered the cutter to be also fitted for a voyage to the Midway Islands. There he intends to have a sign erected stating, briefly, our situation; to serve in case the Navy Department should send (as we expect it will) a searching vessel for us. Twice every day I have climbed the rope ladder on the mast and searched with anxious eyes through my rescued opera glasses the shipless horizon; sometimes with such a strain of nerves and hope that phantom vessels plague my vision. The loneliness and solitude of the vast expanse of water surrounding us is beyond expression. Truly, it is the desert of the Pacific Ocean, and more dangerous than that upon the land, for there are no trails or guideposts for the weary traveler when the sky is obscured. One might easily fancy that beyond the line of the horizon there exists only infinite space. As the Prince of the Happy Valley observes in "Rasselas," after an ocean voyage, "There is no variety but the difference between rest and motion."

I do not remember the cry of "Sail ho!" during all of our cruising between the Hawaiian and Midway Islands save in the vicinity of the former.

The rats are more in evidence of late. At first small and timid, they are now growing larger and bolder; running about and over us in the tents during the night. We are getting quite accustomed to their visits, however, and, rolling ourselves in blankets or whatever covering we have, pay small attention to them. If we stay here, though, our attention will become more acute; for they begin to loom up in importance as a food supply.

The seal, on the contrary, are growing less in numbers, although great care has been taken not to frighten them away. Also, we have not lately attempted fishing on the reef, for fear of reducing their food. We have been prevented from trying the eggs of the albatross, that their nesting may continue without interruption. They will probably leave, too, when the hatching season is over and the young have been taught to fly.

So far as our present ration is concerned, with the exception of beans, flour, and coffee from which our small daily issue is made, we are situated as though no provisions had been rescued from the wreck; for the captain has wisely ordered that all the rest must be held intact to provision the schooner. So, with all the nerve we can muster, the work on the schooner is being pushed. Today the frame stands ready for the planking, and the captain thinks that in another week her mast can be ready for stepping.

Sunday, January 1, 1871. New Year's Day—"Happy New Year"! I think no one but the marine sentry at the storehouse saw the birth of the new year or cared to see the new year come in. For myself I hope there will be no more holidays to chronicle here except it may be the one that liberates us from these surroundings. They have—the three we have had here—aroused too many sombre reflections in contrasting those of the past with the present.

Talbot has now been away forty-three days and it seems almost beyond probability that he should have reached the Sandwich Islands before the food was exhausted. There is a lingering hope, however, that some delay in starting relief for us may have occurred or that he may have reached some island other than Oahu, where Honolulu is situated, and that communication with Oahu may be limited. We are "threshing out" the whole situation tonight in earnest discussion between the sanguine and non-sanguine members of the mess.

RESCUED

Tuesday, January 3. At midnight. It is near an impossibility sanely and calmly to write up my journal tonight—my nerves are shaken and my pencil falters. I have climbed into the storehouse to get away from the commotion in the tent and all over the camp. No one can possibly sleep, for I can see through a rent in the canvas men dancing around a huge fire on the highest point of the island, and hear them cheering and singing while feeding the fire with timbers that we have been regarding as worth their weight in coin. To a looker-on the entire camp would seem to have gone crazy. I will tell what I can now and the rest some other time.

At half-past three this afternoon I was working on the schooner near Mr. Mitchell, one of the carpenters of the contractor's party. I was handing him a nail when I noticed his eyes steadily fixed on some point seaward. He paid no attention to me, and his continued gaze induced me to turn my eyes in the same direction to find what was so attractive as to cause his ignoring me. I saw then, too, something that held my gaze. Far off to the northeast and close to the horizon there was something like a shadow that had not been there when I had last visited the lookout. It appeared as a faintly outlined cloud, and as we both watched with idle tools in our hands it seemed to grow in size and density. Very soon he spoke in a low voice, as though not wishing to give a false alarm: "Paymaster, I believe that is the smoke of a steamer," and after another look, "I am sure of it"; and then arose a shout that all could hear: "Sail ho!"

The order concerning alarms was forgotten in his excitement, but as the captain stood near and his face beamed with his own joy, no notice was taken of the violation. He directed me at once to visit the lookout, and I did so, rapidly securing my glasses. By the time I reached the top of the mast I could see that the shadow we had watched was developing into a long and well-marked line of smoke and that a steamer was headed to the westward in front of it. I notified the eager, inquiring crowd at the foot of the mast and still kept my glasses trained on the steamer until her smokestack came into view. She was not heading directly for us, and I cannot describe the anxiety with which I watched to see if she was going to pass by—my heart was thumping so that one could hear it. I could not believe she would fail to see our signal of distress that waved above me, and pass on to leave us stricken with despair.

When she arrived at a point nearly to the north of us, I saw her change her course until her masts were in line, and then I shouted the fact to those below, for it was evident she was bound for Ocean Island.

The long dreary suspense was over; our relief was near, and I slid down the Jacob's ladder, pale and speechless. The few moments of tense watchfulness had seemed to me like hours of suspense, and it is slight wonder that it took some time to recover my speech. When I did so I acquainted the captain with all I had seen. By the time I had completed my statement the steamer was in view from the ground, and then I witnessed such a scene as will never be forgotten.

Rough-looking men—many of them having faced the shocks of storm and battle—all of them having passed through our recent misfortunes without a murmur of complaint—were embracing each other with tears of joy running down their cheeks, while laughing, singing, and dancing.

By the time we had finished supper she was very near and was recognized as the *Kilauea*, a vessel belonging to the King of the Sandwich Islands. She came within half a mile of the reef where the *Saginaw* was wrecked and dipped her flag and then slowly steamed away in a southerly direction. This manoeuvre we understood, for, as it was getting late in the day, our rescuers were evidently intending to return tomorrow.

Thursday, January 5. The *Kilauea* appeared at daybreak and anchored near the west entrance of the lagoon, and very soon after her captain came to our landing-place in a whaleboat. I recognized in him an old Honolulu friend—Captain Thomas Long, a retired whaling captain—and as he stepped from his boat, we gave him three rousing cheers while we stood at attention near the fringe of bushes around the camp. Captain Sicard went down the beach alone to receive him, and after a cordial greeting, they conferred together for a few minutes. Together they came towards us apparently in sober thought, and Captain Sicard held up his hand as a signal for silence. He uncovered his head and said, in a tremulous voice, "Men, I have the great sorrow to announce to you that we have been saved at a great sacrifice. Lieutenant Talbot and three of the gig's crew are dead. The particulars you will learn later; at present, Captain Long is anxious for us to remove to the *Kilauea* as quickly as possible."

He bowed his head and a low murmur of grief passed along our line. From a cheering, happy crowd we were as in an instant changed to one of mourning. All the dreary waiting days we had passed seemed to fade into insignificance in the face of this great sorrow.

* * *

Our captain has made the following report to the Secretary of the Navy, which adds to and confirms the story of the lone survivor of the gig:

> *Honolulu, Hawaiian Islands,*
> *January 18, 1871.*
>
> *Sir: I forward herewith the brief report called for by regulation of the death of Lieutenant J.G. Talbot (and also three of the crew of the United States Steamer Saginaw) at the island of Kauai (Hawaiian Group).*
>
> *I feel that something more is due to these devoted and gallant friends, who so nobly risked their lives to save those of their shipmates, and I beg leave to report the following facts regarding their voyage from Ocean Island and its melancholy conclusion.*
>
> *The boat (which had been the Saginaw's gig and was a whaleboat of very fine model) was prepared for the voyage with the greatest care. She was raised on the gunwale eight inches, decked over, and had new sails, etc.*
>
> *The boat left Ocean Island November 18, 1870. The route indicated by me to Lieutenant Talbot was to steer to the northward "by the wind" until he got to the latitude of about 32 degrees north, and then to make his way to the eastward until he could "lay" the Hawaiian Islands with the northeast trade winds. He seems to have followed about that route. The boat lost her sea anchor and oars in a gale of wind and a good deal of her provision was spoiled by salt water. The navigation instruments, too, were of but little use, on account of the lively motions of the boat. When she was supposed to be in the longitude of Kauai she was really about one and one half degrees to the westward; thus, instead of the island of Kauai she finally sighted*

the rock Kauhulaua (the southwestern point of land in the group) and beat up from thence to the island of Kauai. She was hove off the entrance of Hanalei Bay during part of the night of Monday, December 19, and in attempting to run into the bay about 2:30 a.m. she got suddenly into the breakers (which here made a considerable distance from the shore) and capsized.

I enclose herewith a copy of the deposition of William Halford, coxswain, the only survivor of this gallant crew; his narrative being the one from which all accounts are taken. I have not seen him, personally, as he left here before my arrival.

Peter Francis, quartermaster, and John Andrews, coxswain, were washed overboard at once and disappeared. Lieutenant Talbot was washed off the boat, and when she capsized he clung to the bottom and tried to climb up on it, going to the stern for that purpose; the boat gave a plunge and Halford thinks that the boat's gunwale or stern must have struck Mr. Talbot in the forehead as he let go his hold and went down.

James Muir was below when the boat struck the breakers, and does not appear to have come out of her until she had rolled over once. He must have suffered some injury in the boat, as he appears to have been out of his mind and his face turned black immediately after his death. As will be seen by Halford's statement, Muir reached shore, but died of exhaustion on the way to the native huts.

The body of John Andrews did not come on shore until about December 20. All clothes had been stripped from it. The body of Peter Francis has never been recovered.

The bodies are buried side by side at Hanalei (Kauai). The service was read over them in a proper manner. Suitable gravestones will be erected over them by subscription of the officers and crew of the Saginaw.

As soon as we had gotten on Ocean Island after the Saginaw's wreck, Lieutenant Talbot volunteered to take this boat to Honolulu, and the rest volunteered as soon as it was known that men might perhaps be wanted for such service.

Mr. Talbot was a very zealous and spirited officer. I had observed his excellent qualities from the time of his joining the Saginaw (September 23, 1870) in Honolulu. During the wreck and afterwards he rendered me the greatest assistance and service by his fine bearing, his cheerfulness, and devotion to duty. His boat was evidently commanded with the greatest intelligence, fortitude, and gallantry and with the most admirable devotion. May the Service always be able to find such men in the time of need.

The men were fine specimens of seamen—cool and brave, with great endurance and excellent physical strength. They were, undoubtedly, those best qualified in the whole party on Ocean Island to perform such a service. Both Lieutenant Talbot and his men had very firm confidence in their boat and looked forward with cheerfulness to the voyage. Such men should be the pride of the Navy, and the news of their death cast a deep gloom over the otherwise cheerful feelings with which the Kilauea was welcomed at Ocean Island.

I do not know that I sufficiently express my deep sense of their devotion and gallantry; words seem to fail me in that respect.

Previous to the sailing of the boat from Ocean Island I had enlisted John Andrews and James Muir as seamen for one month. Since I have ascertained their fate I have ordered them to be rated as petty officers (in ratings allowed to most of the "fourth rates"), as I have thought that all the crew of that boat should have stood on equal footing as regards the amount they might be entitled to in case of disaster, as they all incurred the same risk.

Andrews and Muir belonged to the party of Mr. G.W. Townsend (the contractor at Midway Islands), and it was made a condition, by them, of their enlistment that it should not interfere with their contract with Mr. Townsend. It was intended as the security of their families against the risk incurred while performing the great service for the shipwrecked party. I have forwarded their enlistment papers to the Bureau of Equipment and Recruiting.

I am very respectfully,
Your obedient Servant,

Montgomery Sicard,
Lieut. Comdr. U.S.N. Comd'g.
Hon. George M. Robeson,
Secretary of the Navy.

Fifteen

Loss of a Man—Superstition

Richard Henry Dana

Monday, Nov. 19th. This was a black day in our calendar. At seven o'clock in the morning, it being our watch below, we were aroused from a sound sleep by the cry of "All hands ahoy! a man overboard!" This unwonted cry sent a thrill through the heart of every one, and hurrying on deck we found the vessel hove flat aback, with all her studdingsails set; for the boy who was at the helm left it to throw something overboard, and the carpenter, who was an old sailor, knowing that the wind was light, put the helm down and hove her aback.

The watch on deck were lowering away the quarter-boat, and I got on deck just in time to heave myself into her as she was leaving the side; but it was not until out upon the wide Pacific, in our little boat, that I knew whom we had lost. It was George Ballmer, a young English sailor, who was prized by the officers as an active lad and willing seaman, and by the crew as a lively, hearty fellow, and a good shipmate.

He was going aloft to fit a strap round the main topmasthead, for ringtail halyards, and had the strap and block, a coil of halyards and a marlinespike about his neck. He fell from the starboard futtock shrouds, and not knowing how to swim, and being heavily dressed, with all those things round his neck, he probably sank immediately. We pulled astern, in the direction in which he fell, and though we knew that there was no hope of saving him, yet no one wished to speak of returning, and we rowed about for nearly an hour, without the hope of doing anything, but unwilling to acknowledge to ourselves that we must give him up. At length we turned the boat's head and made towards the vessel.

Death is at all times solemn, but never so much so as at sea. A man dies on shore; his body remains with his friends, and "the mourners go about the streets"; but when a man falls overboard at sea and is lost, there is a suddenness in the event, and a difficulty in realizing it, which give to it an air of awful mystery. A man dies on shore—you follow his body to the grave, and a stone marks the spot. You are often prepared for the event. There is always something which helps you to realize it when it happens, and to recall it when it has passed.

A man is shot down by your side in battle, and the mangled body remains an object, and a real evidence; but at sea, the man is near you—at your side—you hear his voice, and in an instant he is gone, and nothing but a vacancy shows his loss. Then, too, at sea—to use a homely but expressive phrase—you miss a man so much. A dozen men are shut up together in a little bark, upon the wide, wide sea, and for months and months see no forms and hear no voices but their own, and one is taken suddenly from among them, and they miss him at every turn. It is like losing a limb. There are no new faces or new scenes to fill up the gap. There is always an empty berth in the forecastle, and one man wanting when the small night watch is mustered. There is one less to take the wheel, and one less to lay out with you upon the yard. You miss his form, and the sound of his voice, for habit had made them almost necessary to you, and each of your senses feels the loss.

All these things make such a death peculiarly solemn, and the effect of it remains upon the crew for some time. There is more kindness shown by the officers to the crew, and by the crew to one another. There is more quietness and seriousness. The oath and the loud laugh are gone. The officers are more watchful, and the crew go more carefully aloft. The lost man is seldom mentioned, or is dismissed with a sailor's rude eulogy—"Well, poor George is gone! His cruise is up soon! He knew his work, and did his duty, and was a good shipmate." Then usually follows some allusion to another world, for sailors are almost all believers; but their notions and opinions are unfixed and at loose ends. They say,—"God won't be hard upon the poor fellow," and seldom get beyond the common phrase which seems to imply that their sufferings and hard treatment here will excuse them hereafter,—"To work hard, live hard, die hard, and go to hell after all, would be hard indeed!" Our cook, a simple-hearted old African, who had been through a good deal in his day, and was rather seriously inclined, always going to church twice a day when on shore, and reading his Bible on a Sunday in the galley, talked to the crew about spending their Sabbaths badly, and told them that they might go as suddenly as George had, and be as little prepared.

Yet a sailor's life is, at best, but a mixture of a little good with much evil, and a little pleasure with much pain. The beautiful is linked with the revolting, the sublime with the commonplace, and the solemn with the ludicrous. We had hardly returned on board with our sad report, before an auction was held of the poor man's clothes. The captain had first, however, called all hands aft and asked them if they were satisfied that everything had been done to save the man, and if they thought there was any use in remaining there longer. The crew all said that it was in vain, for the man did not know how to swim, and was very heavily dressed. So we then filled away and kept her off to her course.

The laws regulating navigation make the captain answerable for the effects of a sailor who dies during the voyage, and it is either a law or a universal custom, established for convenience, that the captain should immediately hold an auction of his things, in which they are bid off by the sailors, and the sums which they give are deducted from their wages at the end of the voyage. In this way the trouble and risk of keeping his things through the voyage are avoided, and the clothes are usually sold for more than they would be worth on shore.

Accordingly, we had no sooner got the ship before the wind, than his chest was brought up upon the forecastle, and the sale began. The jackets and trowsers in which we had seen him dressed but a few days before, were exposed and bid off while the life was hardly out of his body, and his chest was taken aft and used as a store-chest, so that there was nothing left which could be called his. Sailors have an unwillingness to wear a dead man's clothes during the same voyage, and they seldom do so unless they are in absolute want.

As is usual after a death, many stories were told about George. Some had heard him say that he repented never having learned to swim, and that he knew that he should meet his death by drowning. Another said that he never knew any good to come of a voyage made against the will, and the deceased man shipped and spent his advance and was afterwards very unwilling to go, but not being able to refund, was obliged to sail with us. A boy, too, who had become quite attached to him, said that George talked to him during most of the watch on the night before, about his mother and family at home, and this was the first time that he had mentioned the subject during the voyage.

The night after this event, when I went to the galley to get a light, I found the cook inclined to be talkative, so I sat down on the spars, and gave him an opportunity to hold a yarn. I was the more inclined to do so, as I found that he was full of the superstitions once more common among seamen, and which the recent death had waked up in his mind. He talked about George's having spoken of his friends, and said he believed few men died without having a warning of it, which he supported by a great many stories of dreams, and the unusual behavior of men before death. From this he went on to other superstitions, the Flying Dutchman, etc., and talked rather mysteriously, having something evidently on his mind.

At length he put his head out of the galley and looked carefully about to see if any one was within hearing, and being satisfied on that point, asked me in a low tone—

"I say! you know what countryman 'e carpenter be?"

"Yes," said I; "he's a German."

"What kind of a German?" said the cook.

"He belongs to Bremen," said I.

"Are you sure o' dat?" said he.

I satisfied him on that point by saying that he could speak no language but the German and English.

"I'm plaguy glad o' dat," said the cook. "I was mighty 'fraid he was a Fin. I tell you what, I been plaguy civil to that man all the voyage."

I asked him the reason of this, and found that he was fully possessed with the notion that Fins are wizards, and especially have power over winds and storms. I tried to reason with him about it, but he had the best of all arguments, that from experience, at hand, and was not to be moved. He had been in a vessel at the Sandwich Islands, in which the sail-maker was a Fin, and could do anything he was of a mind to. This sailmaker kept a junk bottle in his berth, which was always just half full of rum, though he got drunk upon it nearly every day. He had seen him sit for hours together, talking to this bottle, which he stood up before him on the table. The same man cut his throat in his berth, and everybody said he was possessed.

He had heard of ships, too, beating up the gulf of Finland against a head wind, and having a ship heave in sight astern, overhaul and pass them, with as fair a wind as could blow, and all studdingsails out, and find she was from Finland.

"Oh ho!" said he; "I've seen too much of them men to want to see 'em 'board a ship. If they can't have their own way, they'll play the d—l with you."

As I still doubted, he said he would leave it to John, who was the oldest seaman aboard, and would know, if anybody did. John, to be sure, was the oldest, and at the same time the most ignorant, man in the ship; but I consented to have him called. The cook stated the matter to him, and John, as I anticipated, sided with the cook, and said that he himself had been in a ship where they had a head wind for a fortnight, and the captain found out at last that one of the men, whom he had had some hard words with a short time before, was a Fin, and immediately told him if he didn't stop the head wind he would shut him down in the fore peak, and would not give him anything to eat. The Fin held out for a day and a half, when he could not stand it any longer, and did something or other which brought the wind round again, and they let him up.

"There," said the cook, "what do you think o' dat?"

I told him I had no doubt it was true, and that it would have been odd if the wind had not changed in fifteen days, Fin or no Fin.

"Oh," said he, "go 'way! You think, 'cause you been to college, you know better than anybody. You know better than them as 'as seen it with their own eyes. You wait till you've been to sea as long as I have, and you'll know."

Sixteen

Treacherous Passage

Douglas A. Campbell

THE SEA WAS IN A FURIOUS MOOD. PILED ON ITS SURFACE WERE GREAT, gray waves, living monsters who could humble even the greatest warships. Yet, the USS *Flier* was but a submarine, at about 300 feet, one of the smaller vessels in the navy. Even when submerged, it pitched and rolled like a slender twig. But inside *Flier* were no ordinary sailors. They were submariners: men—most of them quite young—selected from the ranks for their virtues of fearlessness and its companion trait, optimism. Their mood was bright. Despite the beastly roar and hiss of the sea above them, none believed that on this day his death was at hand.

The Reaper might come later, when their boat reached the actual battle lines in this, the third year of World War II. And probably not then, either, they thought. The momentum of the conflict had turned in their favor. There was a sense, pervasive on board, that destiny was with the Allies. Everyone expected to be around for the final victory. These were young men—many of them green—led by a handful of sailors creased by the experience of having survived at sea. Death was for someone else, the enemy, even on January 16, 1944, even on the Pacific Ocean, the greatest naval battleground in history, a place where tens of thousands of Americans had already died.

But the men aboard the *Flier* could not ignore the thrashing as she bucked and twisted. For the one young cowboy in the crew, it had to make him think rodeo bull. He and his mates joked uneasily about the sobriety of the welders who had built the submarine back in Groton, Connecticut.

In these angry seas they approached the atoll known as Midway, one of the navy's refueling depots. Once beyond Midway, their first wartime patrol aboard *Flier* would begin, and their record—distinguished or dreadful—would be tallied in tons of enemy shipping sunk. With young hearts and a sense of invincibility, they knew that the slamming of their submarine by the sea was only a tune-up for the coming combat. And they had no fear.

* * *

On August 12, the now-battle-tested *Flier* approached Sibutu Passage like a slugger stepping into the batter's box. On the far side of this strait was the Sulu Sea, nearly 90,000 square miles of unbroken blue water shaped roughly like a baseball diamond. Sibutu Passage was home plate. The opposing team—the Japanese soldiers and sailors—had taken all the land around that diamond two years earlier. They were scattered along the first-base line, a string of islands called the Sulu Archipelago that ended in Mindanao, more than 200 miles to the northeast. More Japanese troops were strung along the islands from first base to second—Mindoro, at the top of the diamond, 500 miles due north. The enemy also held third base—the small island of Balabac to the northwest. And the huge island nation of Borneo, due west of Sibutu Passage, was thick with supporting troops, like the bench-dwellers in the dugout. Throughout the more than 7,000 Philippine Islands and their Indonesian and Malaysian neighbors, the Japanese navy and army were arrayed in what until now had been an almost impenetrable defense. Americans entered the Sulu Sea only by submarine, and when they did, they knew it was kill or be killed in this deadly World Series.

Flier's general orders, drafted back in Fremantle, became specific on the evening of August 13 as the submarine approached Balabac Strait. The word came around dinnertime, the normal hour for submarine headquarters in Australia to broadcast the war news along with any special instructions for the submarines on patrol. On this, the third consecutive day, there was a message for the submarine *Robalo*, which was scheduled to return from its most recent patrol. The message asked for the boat's location and estimated time of arrival in Fremantle. There was no urgency in the transmission. A returning submarine could easily be a few days late.

There was a message for *Flier*, as well. When the radioman's message was typed into the machine, the officer informed the captain, Commander John Crowley, of the new orders. The submarine *Puffer*, which had been patrolling in the northern Philippines, had encountered a Japanese convoy heading south. *Puffer* had sent torpedoes into several of the ships in the convoy and was now trailing "cripples," the message said. The rest of the convoy, thwarted by *Puffer* from entering Mindoro Strait on the northern end of Palawan, was now traveling southwest, along the western shore of Palawan in the South China Sea. Until now, *Flier's* assignment had been to patrol the South China Sea, looking particularly for four Japanese submarines making supply runs from Vietnam. The new orders directed *Flier* to go after *Puffer's* convoy. There was no need for Crowley to change course. *Flier* was already headed for Balabac Strait, and that would take the submarine right into the path of the approaching convoy.

Crowley was energized. The patrol had just begun and already there were targets. The word was passed along by intercom, and the crew knew they were back in the war.

On this evening, Baumgart had lookout duty after dinner, so he donned a pair of red glasses after seven o'clock and wore them for a half hour before he went to the control room. At eight o'clock, he climbed the ladders up through the conning tower. The glasses, filtering the harsh incandescent submarine lighting, prepared his eyes for scanning the darkened ocean. He was wearing his navy denims and boots. The warmth of a night in the tropics required nothing else. And despite his continuing anger over the way he had been assigned this duty, he was beginning to enjoy the hours he spent standing on the A-frame above the deck, cooled by the breeze as *Flier* made eighteen knots across the surface.

The conning tower was crowded with its usual complement of officers and crew. Jim Liddell, the executive and navigation officer, stood at the foot of the ladder leading to the bridge so that he could talk with

Crowley, whose stool was on deck beside the hatch. Jim Russo stood beside Liddell, helping him with the charts. Arthur Gibson Howell was at the rear of the compartment, operating the radar. Beside him, Charles Pope, the hero who nearly drowned on the trip between Midway and Hawaii, ran the sonar.

Howell's radar presented him with an image of the nearest shoreline, many miles away. They had traveled on the surface throughout the day and had seen neither Japanese ships nor aircraft, and the radar screen still showed no enemy threats. The night was going as easily as had the day.

Admiral Ralph W. Christie's orders directing *Flier* through Balabac Strait remained unchanged by the message the radioman had transcribed earlier. Crowley was to take the deepest water route through the strait that he could. In deep water, it was assumed, mines could not be anchored. Specifically, the orders directed Crowley to use the Nasubata Channel, one of eight channels between the Sulu and South China seas allowing east and westbound ships to pass through the reef-strewn Balabac Strait. Nasubata Channel was the deepest—more than 500 feet deep in spots—and the broadest, with about five miles' leeway between Roughton Island's reefs to the north and Comiran Island to the south.

As he approached the channel, Crowley had several concerns, as would any skipper. While the ability to navigate safely around natural obstructions such as reefs was always a consideration, in wartime a captain had two more problems to solve. He had to give himself enough room to maneuver if an enemy ship attacked, and he also had to be wary of shallow water where mines could be anchored. Roughton Island's extensive reefs to the north took away maneuvering room and presented a navigation problem. Crowley, talking the matter over with Liddell through the conning-tower hatch, decided he would try a more southerly route through the channel. If he stayed in fifty fathoms—300 feet or more of water—Crowley believed *Flier* would pass through the channel untroubled.

Mines were the only military threat Crowley felt he faced that night. He trusted his radar and its operator, Chief Howell, and felt the device could find a target the size of a surfaced submarine—with the possible exception of a midget submarine—at a range of more than three miles. Unless the Japanese had developed a superior night periscope, he believed that on a night as dark as this one, a submarine could not make a submerged attack. And as Howell reported from below what he was seeing on the radar, Crowley was convinced the only things out there in the dark were islands and mountains, a few of which he knew harbored enemy soldiers. *Flier* could make it through.

Chief Howell relayed a constant stream of radar readings to Liddell, who passed them along to the skipper. And Chief Pope, watching the sonar, gave depth readings. With the radar showing the nearest land about 5,000 yards away, *Flier* was traveling in sixty-five to ninety fathoms of ocean when Pope reported a reading of forty-one fathoms. *Flier* wasn't about to scrape bottom, but the depth was shallow enough to raise Crowley's concern about mines. He asked Liddell, a veteran of a Philippine tour before the war, what he thought.

Crowley was standing in the forward end of the bridge, leaning over the open hatch in the bridge floor to talk with Liddell about taking a new course, when the explosion came. The blast caused the entire submarine to whip to one side and then snap back like an angry stallion trying to throw its rider. Crowley felt the violent motion, but the concussion was without sound, like the thunder from the electrical storms that played their lightning fingers across those distant mountains.

Jim Russo's job had been simply to help handle the conning-tower charts. He was at Liddell's side when the explosion rocked the boat as if it had rammed a wall. Instinctively, he looked down at the hatch to the control room. Something slammed into his cheek below his eye, ripping his flesh like a bullet. A shaft of air was venting straight up from below, blasting out through the hatch to the bridge. Blood was draining down his cheek when Russo felt himself lifted by the column of air, along with Liddell, the 200-pound ex-football-player, straight up to the bridge. Once above deck, Russo—by instinct and without hesitation—followed Liddell, whose shirt had been ripped off by the blast, to the rear of the bridge where, at the railing, they dove into the ocean. When Russo turned around in the water, *Flier* was gone.

Wesley Miller, standing on the A-frame above the bridge, was nearly thrown from his watch but managed to hook his legs over a railing to avoid falling. He was confused. Somehow, he had lost his binoculars, and he was concerned about the discipline that would result. Then there was screaming coming from below and air was blasting out of the hatch in the bridge floor under him. He stood frozen on the A-frame for an instant, although it seemed longer, until he heard someone yell, "Abandon ship!" and saw the bow of the submarine go under. Then the ocean was swelling around him, dragging him down into its darkness. The radio antenna had snagged him. Miller struggled to free himself and then swam and swam, reaching for the surface. Then he was alone in the water, and the submarine was gone.

Al Jacobson, lost in his reverie, watching the lightning and the mountains silhouetted in the darkness, felt the blast of air and, curiously, found Lieutenant Reynolds standing on the deck beside him, complaining that his side hurt. Jacobson told Reynolds to lie down, and then he crouched over the lieutenant, hoping to help him. He assumed that an air bank, used to store compressed air for use in diving and surfacing, had blown, and he told Reynolds to lie still. But as he talked with Reynolds, he saw Ensign Mayer and Ed Casey diving over *Flier*'s side. Just then, water rose around Jacobson and Reynolds, and the submarine sank below them, sucking them down with it. The image in Jacobson's mind was of the two huge propellers at the rear of the boat, still spinning as they passed him, slicing him to bits. He struggled to swim up and away from his death. It took a few seconds before he surfaced in a slick of diesel fuel that floated on warm, calm seas. Baseball-size chunks of cork from inside *Flier* floated around him. He could feel them. But there was no light to see what, or who, else was there.

Crowley, who had been standing to port at the front of the bridge, saw a geyser shoot toward the sky from the forward starboard side of the submarine. The next thing he knew, he was standing against the aft railing of the bridge, near Jacobson. He ran forward to trigger the collision alarm that was mounted on the bulwark just above the conning-tower hatch. When he got there, he smelled diesel fuel. He looked down into the conning tower but it was dark. There was no time!

"Abandon ship!"

The skipper's yell carried across the deck and perhaps a short way down into the submarine, where many in the crew already were being thrown about by the air blast and the flooding that made Crowley's command superfluous. On the bridge, the skipper felt the shaft of air rising from within the submarine, carrying with it the sounds of rushing water and the screams of seventy-one men trapped inside. Some men were climbing the ladder from the conning tower, just in time because the deck was heading under, *Flier's* engines still driving it like a train entering a tunnel. Crowley found himself in a raging stream of water as the sea poured around the bulwark and into the bridge, and then he was washed out the rear of the bridge, into the sea. In a matter of just twenty seconds after the blast, *Flier* was gone, and after its passing, the ocean was calm. The dead sailors of the Japanese minelayer *Tsugaru* had once again struck from their graves.

A mine had touched the side of *Flier*—just a glancing blow, but enough to trigger its explosives. The geyser had appeared near the rear of the forward torpedo room. The explosion, quiet as it was on the surface, would have been enough to punch a hole through the submarine's superstructure and one or more of its watertight welded-steel compartments.

On this night, Crowley had ordered battle stations for the conning tower, but the rest of the crew was not on alert. If the watertight doors were not dogged in place with their big handles, the blast from the mine—having opened a huge hole in *Flier's* side—flooded the forward torpedo room and the officers' quarters immediately. At the same time, a rupturing of the tanks full of compressed air sent a shockwave through the submarine's ductwork ahead of the flooding water. The seawater raced to the rear, in seconds reaching the control room where Ensign Behr would have been among the first in its path, followed by the bow and stern planesmen and the other sailors handling the various controls.

The flooding would stop the engines as *Flier* sank deeper, and the darkness that Crowley had seen when he looked down the hatch from the bridge would spread throughout the submarine. For the men in the rear, there was but one hope to temper their panic—an escape hatch in the aft torpedo room. If they could get the hatch open, then for the first time since their submarine training in Connecticut, they would strap on their air tanks and take their chances floating to the surface.

But what then?

The darkness was nearly absolute. Al Jacobson could see nothing, but he could taste diesel fuel on his lips. His body felt the warm, wet embrace of the sea as his uniform clung to his arms and legs. All was quiet except for the lapping of small waves. And then there were shouts, the sounds of a human voice, the first indication to the ensign that he was not alone. He began swimming toward the voice, floating easily in the salt water but slowed by the weight of his shirt, trousers, and boots. The strap of his binoculars was still around his neck, and the glasses floated harmlessly by his chest as he did the sidestroke. *Flier*'s sinking had disoriented him, snatching him from the tranquility of a warm ocean night, plunging him into a struggle for survival.

Several of the men had responded to the same yell that had drawn Jacobson. Once they were all together, they shared what they knew and tried to decide what had happened to *Flier*. It could have been an explosion in the batteries, but Crowley discarded that notion. The diesel engines had been running and the batteries were idle, not a situation in which they were likely to explode. The other topic concerned who else from the crew had escaped the boat. With almost no light, they could not expect to see other survivors, so their only choice was to call for them. Soon they had gathered more men into their group. A headcount was taken.

A total of fifteen men responded to the roll call. All fifteen men were already getting a lift. Not only had they escaped the terrifying death of their seventy-one trapped shipmates, but they had also surfaced on a sea that was unusually docile for this time of year. Summer is monsoon season, and storms can whip the Sulu Sea into a froth. Swimming in those waves would have been exhausting, and the chances of all these men finding and communicating with each other would have been slim. With lightning flashing on distant islands, there remained the possibility that a storm could still come, funneling winds between the mountains. But for now, here on the open ocean, the wind was light and the waves were gentle. As long as each man could stay afloat, he could remain with the group. For the next two hours, as they assessed their situation and developed a plan, that is what the men did.

There was a sense shared by most of the men that they were still part of a military unit. Perhaps this was because of their training, to always follow the lead of Crowley and Liddell. It took time for them to realize that there was no longer a formal chain of command, and that neither Crowley nor Liddell was in charge. They needed a plan.

The first thing to consider was their location. The skipper and his executive officer knew where they were. Liddell began explaining the options, most of which everyone already understood. There was land on three sides, Liddell said. To the west was Balabac, the largest chunk of land in the vicinity, roughly ten miles away. Each man was aware that the Japanese occupied that mountainous island. They could swim in that direction and with some certainty, due to the island's long shoreline, land on Balabac's beaches. To the south was Comiran Island, less than two miles away. Every survivor could probably swim that far, despite injuries. But there was a problem with this option: Comiran was tiny—only a few hundred feet across. If, swimming in this opaque darkness, they missed Comiran, they would have another forty miles of ocean before they reached land on one of northern Borneo's islands.

The lightning occasionally lit a mountainside to the northwest, but judging from what they could see, that land was about thirty miles away.

And if they headed toward any eastern quadrant, the Sulu Sea threatened them, with hundreds of miles of ocean, uninterrupted by land of any kind.

These were the options Liddell presented, each one unpromising. And none was worth even attempting right now. There were no stars to guide the men, and no moon, only clouds overhead. And the occasional lightning flash on the horizon, while it gave them something of a beacon, left their eyes blinded for several minutes.

But even the strongest among the survivors could not expect to stay afloat forever, and so they adopted two rules. First, they would turn so that the waves were lapping their left cheeks, and then they would swim in the direction they were facing. The course was randomly chosen, as far as young Jacobson knew. Perhaps Crowley and Liddell had a reason. But the skipper and his second in command did not share their thinking.

The second rule was Crowley's idea—a death sentence for several of the men, and everyone knew it: It would be each man for himself. The cruel reality was that wherever they were going, it was a long way off. Some of the injured men could not swim the distance without help, and if the whole group waited for the injured, the chances were overwhelming that no one would survive.

Crowley was uninjured, but he was the oldest, at thirty-five, and his physique after several years of sedentary submarine service was not particularly athletic. Crowley could be among the first victims of his edict. But like their skipper, all of the men agreed to the pact, and all fifteen began swimming across the waves, which lifted them a few inches and then gently lowered them in a mesmerizing rhythm.

* * *

Chief Pope called out in the night for Jim Liddell, asking the distance that lay ahead of them before they reached shore. After two hours in the water, the men still felt little wind. Liddell, pondering Pope's question, knew the entire swim could be fifteen miles or more, but he wanted to be encouraging.

"About nine miles, Chief," Liddell replied.

"Oh, fuck it!" Pope said in disgust. With that, the chief stopped swimming and said no more, his faint image dissolving forever behind the swimmers in the night.

It was not much later that Jacobson, keeping pace with Ed Casey, saw him veer. Instead of calling him back to the course as he had done before, the ensign swam over to his mentor.

"Ed, rest a minute, and then just float on your back and put your feet on my shoulders and I'll push you back," Jacobson offered.

"Remember, we agreed every man for himself?" Casey said, refusing his young friend's gesture. But the two of them swam back toward the group, talking as they went. They were joking about a blowout party they had planned to throw in Perth when the patrol was over, and as they talked, they reached the wake of the others.

Ten minutes later, Casey disappeared in the darkness. When Jacobson and the others called to him, there was no response. The lieutenant had chosen not to burden his shipmates any longer.

Paul Knapp had been struggling like Casey, but was keeping in line with the others. Jacobson saw him swim off to the side without a word. The ensign thought little of it until Knapp did not return. Then he realized the courage it had taken for Knapp to separate himself.

As the night wore on, one after another of the men, when they felt they could swim no more, silently turned to the side and disappeared, each man choosing for himself when his time had come.

If anyone among the survivors were thinking about the beasts that swam below them, none gave voice to the image. But the reefs of the Sulu Sea were habitat for a vast assortment of large animals. Sharks of every description shared the water with barracudas and rays. Some were harmless, like the white and blacktip sharks, and the guitar sharks. But others were legendary, like the hammerheads and bull sharks, predators that would eat another shark as quickly as they would consume a human being.

If the swimmers were ignoring the carnivores beneath them, it may have been because their minds were filled with the death they had just dodged, and, not that far below the sea, their shipmates already dead inside *Flier*. The thing that now would keep these men alive was their determination to keep swimming.

The overcast sky that had kept the stars hidden was overcome at about three o'clock that morning by the moon, rising grudgingly in the east to give the swimmers a navigational beacon. By now, only nine of the original fifteen survivors remained in the group. Wesley Miller straggled far behind the main pack but could hear their voices in the dark, and shouted to them to maintain contact.

At about five o'clock, when the first hint of daybreak was tingeing the sky from black to gray, helmsman Gerald Madeo began to panic. He fell below the surface, and after seven hours in the water, no one had the strength to help him. They simply continued swimming, led away from Madeo by the moon toward an unknown destination.

The trio of Howell, Baumgart, and Jacobson kept pace with each other throughout the morning, cooled under the blazing tropical sun by the same glass-clear sea that had warmed them during the night. Slowly, they drew toward their island, probably helped by a change in the tide or the currents.

It was one o'clock in the afternoon when Jacobson checked his wristwatch. The approaching drone of an airplane came from a distance, and when the men stopped swimming to look, they saw a low-flying Japanese craft, coming directly toward them. A half-dozen heads on the surface of the Sulu Sea were too tiny for the pilot to notice, however, and the plane kept going. The swimmers resumed their strokes, their luck apparently intact.

CHOOSING FREEDOM

The jungle island floated in the distance like a thin, green wafer. Little about its shore could be determined, but it was closer than anything else, and distance was important for the men, who by now had been in the water for nearly seventeen hours. Jacobson, Howell, and Baumgart had managed to stay close to each other since they had first spotted land. If there was anyone else afloat, they were no longer in sight. There were only the three and the island.

And then ahead, almost on a line toward the island, the men noticed something else in the water. It was long, and above it rose some perpendicular objects. Perhaps it was a native fishing boat, they thought, and the objects were the fishermen. They waved, but there was no response, so they decided to avoid the unfriendly thing. Swimming the straight route toward the island, they nevertheless drew closer to it and discovered it was a bamboo tree, its buoyant trunk riding lightly on the surface, its limbs rising toward the afternoon sun. Eager for some rest, they swam to the tree and Jacobson climbed up to have a look at the surrounding area. Howell and Baumgart struggled up beside where he balanced as the ensign scanned the sea. A short way off he saw more swimmers. He began shouting, joined by his mates, their voices carrying across the now-choppy water, their arms waving in excited arcs.

* * *

At daylight, when the island was first spotted, Crowley had given the order to anyone within shouting distance: Swim toward land at your own pace. He soon fell behind the rest, alternately swimming and resting when exhaustion overcame him. That he continued to swim is indisputable. What kept him moving is less certain.

Early in the afternoon, Crowley had seen Liddell ahead of him, clinging to another floating tree. The skipper and his executive officer stayed together then until they heard the shouts from Jacobson's group. It took a few minutes for Crowley and Liddell to reach the larger tree and cling to it. For the skipper, the plant had become a lifesaver. Exhausted, he had felt—even with the island in sight—that he could no longer swim. But the shouts from Jacobson and the others gave him new energy.

Breaking branches from the tree for paddles, the five men now straddled its trunk, urging the tree toward the shore. Off to one side, they saw Don Tremaine, swimming alone. He waved back when they shouted and gestured, but he avoided them. Tremaine had seen them but he could not hear them, and had assumed they were natives. If they were too unfriendly to pick him up, he reasoned, he would not chance swimming toward them.

The water changed from dark blue to a pale aqua a few hundred yards from the shore where the coral reef began, and then it was shallow enough for the men to walk. Their feet were wrinkled and white from seventeen hours in the water, and the entire seabed on which they stepped was coral. It was like walking on crushed glass rather than gravel, and the sharp coral edges sliced into the soft soles of their feet. Abandoning their bamboo tree, they stumbled, trying to keep their balance as the hot afternoon sun dried the salt water from their backs. The pain in their feet was numbed by their eagerness to feel dry land. And up on the beach stood Jim Russo, urging them on.

Staggering ashore, the men could, for the first time since *Flier* sank, see each other from head to toe. The sight was shocking. In the ten hours since daylight, the unrelenting rays of the sun had bombarded their water-softened white flesh. Now, where their skin was exposed, they were scalded red. Baumgart alone had long trousers on, saving his legs from the scorching. But like the others, he waded from the ocean with his face and arms as red as if the seawater had been boiling, and his blood drained into the sand from the coral slashes in his feet.

Byan Island, roughly triangular in shape, is just east of Mantangule Island. In 1944, the island was uninhabited by humans. A few hours after the men reached Byan Island, the sun settled beyond sprawling Mantangule and the air grew cool. Crowley and Liddell believed they were on Mantangule Island and that the big land to the west was Balabac, which they knew was occupied by the Japanese. To build a fire, if they could manage it, might attract attention, so they faced a night of cold. Even before sunset had cooled the air, they were swept alternately by fever chills and sweats. In the dusk, they huddled together for warmth, lying directly on the sand and, having successfully outlasted death for a day, sought the peace of sleep.

Neither sleep nor peace was to be theirs, however. Roused by their fevers, they would seek a more comfortable position, only to have the grains of the beach rasp across their sunburned flesh like sandpaper. At times, they were awakened by rats nibbling on their feet. Young Jacobson lay awake, his body shaking, the watch on his wrist slowly ticking off the seconds and minutes. He wanted nothing more than for the hours to pass and the day to come, bringing with it warmth.

They had learned, in their first stumbling hours ashore before nightfall, that the area near the beach offered neither food nor water. So when the sun rose on August 15, 1944, the *Flier* survivors knew they had to begin a search. Crowley directed Tremaine, Russo, and Howell, who had injured his knee when he had jumped off the submarine, to stay on the beach and improve the lean-to shelter. Jacobson and Baumgart were to head east and scout out the island, while Crowley and Liddell would head the other way.

Howell, Russo, and Tremaine started gathering scraps of wood and palm leaves in the hope of creating some real shelter for the coming night. At the same time, Ensign Jacobson and Baumgart hobbled along the shore, Baumgart in trousers and an undershirt, the ensign in his underwear with binoculars dangling from his neck. There were coconuts everywhere along the water's edge. They picked up the ones that looked whole and opened them with their bare hands by smashing them on coral. But each one that broke open left them disappointed, its meat rotten, its milk spoiled.

They trudged for hours without success. And then they rounded a point and ahead they saw a string of islands. Still ankle-deep in the sea and standing on coral beds, they splashed forward until they found a sandy beach where driftwood had gathered. Then they decided it was time to head back. Realizing that if they crossed the island, it should take less time than circling the beach, they tried to climb ashore up the coral cliffs. But the thorns and vines repelled them, and they waded once more into the shallows across the coral, retracing their painful steps toward their shipmates.

They had another reason for leaving together: They wanted to talk about the prospects for the group's ultimate survival. Crowley was familiar with the territory only from having studied nautical charts. Liddell had served on a submarine in the Philippines before the war and had a deeper understanding of the locale. What they had found so far was that they could not stay on this island. It was little more than a coral reef with no food or shelter. The jungle that began at the shore was a tangle of thorns and vines.

Rounding the western tip of the island, they saw their two options: To the northeast, beyond two more small islands, lay a large island. Liddell identified it as Bugsuk Island. And behind them, to the southwest, beyond long, flat Mantangule Island, was the mountainous mass they knew had to be Balabac. Intelligence reports that they had reviewed back in Australia said the Japanese were on Balabac. But from what they remembered, there was less chance of finding the enemy on Bugsuk, which was about five or six miles away. The trip could be made manageable by hopping only to the next island, Gabung, and resting before going on. There were tremendous currents that funneled between islands like these, Crowley and Liddell knew, and in their weakened state, the survivors could easily be swept out to sea if they tried to swim across. They needed another plan, and more information. So Liddell decided to leave Crowley behind and explore a bit further on the northern side of the island. The lieutenant had walked some distance when, coming around a curve in the shore, he saw a man ahead on the beach—a white man, clothed only in underwear.

* * *

At daybreak the day before, Wesley Miller had lost contact with the other swimmers. But he saw several islands on the horizon, and, since it was in the direction he had been swimming, he kept going for the closest one. As the afternoon wore on, however, he found that the current was sweeping him to his left, past the nearest island. He would never be able to reach it, he knew, so he began to swim for the next island. But when he was perhaps two miles from his target, the current increased, carrying him fast along the beach. Still he swam toward the shore, cutting the distance in half when, to his left, he saw the end of the island approaching. After that, there was nothing, and Miller believed that his long trek from the Oregon ranch to the middle of the Pacific Ocean was at an end.

It was startling when, letting his feet fall below him, Miller felt his toes touch the bottom. He began walking now, and soon the water was only waist-deep, and to his left the coral actually rose above the surface. Then the sun set over Balabac and Mantangule and the water grew deeper. He no longer could wade, but although he must swim again to survive, his arms and legs were unwilling to move. So he willed himself toward the beach, and when he could touch bottom again, he was too tired to stand. Sand and coral rose beneath him, and he leaned forward in the water so that his knees, not his feet, propelled him ashore while his body and arms floated listlessly. Crawling as an infant might, he worked his way out of the sea and partway up the sand, where his thoughts and his will ceased and he fell asleep. Awakened in the middle of the night by rising water, he dragged himself to higher ground, up against the coral cliff, and slept once more.

In the morning, Miller began to walk along the shore, looking for a way to scale the cliffs. As he stumbled on, he searched for clams in the sand. In a mile of hiking, he had found only solid rock cliffs along the bank, with jungle growth snarling out of their cracks.

Then Liddell found Miller and led him back to Crowley. The skipper, perhaps noticing the sailor for the first time, realized that this crewman was little more than a boy, a child who was pathetically grateful to find that he was not a sole survivor.

Later that afternoon, everyone assembled at the beach and reported on their work. Jacobson and Baumgart had found neither food nor water, but they told of locating a pile of driftwood on the northern beach. Howell, Russo, and Tremaine, when they were not working on the lean-to, had set out seashells to collect water should it rain. And they had found water seeping out of the coral cliffs. They had set some shells below the cliffs and collected some water, one drip at a time—three shells full, in all. Everyone shared it, each person drinking a couple of teaspoons. It was merely seawater that had splashed onto the coral at high tide, but their thirst convinced the men they were getting fresh water.

If they continued to wet their lips with this water for long, they were going to be doomed. The human body, in order to rid itself of excess salt, passes the salt through the kidneys where it is washed away in urine. That means that the body is losing water as well as salt. The more salt in the system, the more water must be expelled. In a short time, the consumption of salt water will actually dehydrate the body, increasing the level of salt in the bloodstream and damaging bodily tissues. Soon, the drinker will die. But first, normal body functions will be damaged. Saliva will dry up, leaving the mouth and tongue without lubrication, exposing them to infection. Drying of the tongue may cause it to swell and split. Death might be preferable.

With the other reports submitted, Crowley told the men of their options. They could head west, eventually reaching Balabac where there was food and shelter—and Japanese soldiers. They would probably be captured and become prisoners of war. (Earlier in the afternoon, another Japanese patrol plane had flown low over the island, the red rising sun insignia on its wings easily seen by the survivors.) Or, they could use the driftwood that Jacobson and Baumgart had found, build a raft, try to reach Bugsuk, and, accepting the uncertainty of finding food and water there, remain free men.

To a man, they chose freedom. They would begin work in the morning.

Sunrise brought all the men back onto their feet. The agony of standing on those festering cuts was not enough to keep them on the beach, and soon the eight were hobbling in the shallow water, where vegetation coated some of the coral, making it less sharp. Splashing up Byan Island's eastern shore, they could see Gabung Island in the distance. When they reached the place where the two islands were closest together—just under a mile separated them—they began building their raft. Liddell and Russo, both strong men, reached into the jungle from the edge of the beach and tore out vines. As some of the men used the vines to lash the bamboo driftwood logs together, Chief Howell sat on the beach, improvising two paddles by splitting slender bamboo poles partway, inserting small pieces in the split crossway, and then tying them in place with thin vines. Occasionally, he would lick moisture that he found on leaves.

Crowley saw how his men slowed in their work as the day wore on, their movements becoming uncoordinated, their attention wandering. Thirst was on everyone's mind. But even though they scoured the coast looking for edible coconuts, they found none all day.

It was about two-thirty that afternoon, just before slack tide, when the eight men surrounded their little raft and pushed it out toward Gabung. Ahead of them was a crossing of slightly less than one mile. The water was the pale blue of reef water out for several hundred yards off the beach, and the reef resumed on the far side of the channel, where dark blue water indicated a depth that no one would be able to wade. They had brought two long poles with them, and for the first quarter of the voyage, the younger men took turns poling the raft, on which Crowley was the only permanent passenger. The rest of the men leaned on the raft for support as they walked in the shallows across another long bed of razor-sharp coral. Crowley paddled.

Before they had made it halfway across the channel, they saw the daily patrol plane coming in low. Crowley and the man poling slipped into the water, and everyone tried to hide under the raft. The plane kept going, and the men, clinging to the sides, kicked in the deeper water, slowly moving the raft across the channel.

Now on the open water, they found themselves directly in the path of an oncoming squall. Abruptly, they were pelted with large, pure droplets, delicious on their lips, and everyone tipped his head back and opened his mouth. But while the raindrops splattered off their foreheads and cheeks, none of it seemed to find their tongues before the squall passed on into the ocean, leaving the scorching sun in its wake.

They had not yet reached the reef on the far side of the channel when the tide seemed to shift and a new current swept between the islands. With only a quarter of a mile to go, they suddenly seemed unable to make any progress, and the raft appeared to be drifting away from Gabung Island. The men on the sides kicked with all their feeble power and Crowley, feeling like a very elderly thirty-five, paddled, and the raft circled the end of the island and settled in its lee, the current having deposited the men close enough to shore that they could swim the final leg.

It was seven o'clock, more than four hours since they had left Byan Island a mile to the south, and the sun had already set. They found a sandy beach and were content to collapse where they could find room. The little slivers of coconut they had eaten earlier had done little to curb their appetite, and their thirst was only growing. But no one had the energy to forage. More than food and water, right now they wanted sleep.

Sunrise the following morning—August 17—brought with it relief from the tremors of the night and hope that this would be the day the men would eat. Before launching their raft, they gathered to discuss their next steps. It would be another nine or ten hours until the tides allowed them to leave this island for the next one in the chain. Crowley and Liddell took suggestions, and the group decided that their time could be best used by traveling around the island the long way. There would be more chance of finding food if they were covering a longer shoreline. It meant more walking, but now empty stomachs and parched mouths were overpowering the screaming pains from their feet and the swollen and blistered burns on their backs and arms. They pushed the raft into the shallow water and began circling the island to the west. Once more, the coral in the shallow water was softened by plants that grew on it, so wading was less painful than it might have been. But there was a trade-off, because when they were not swimming, their burns were always exposed to the sun as it rose high above the island.

On the eastern side of Byan the day before, the men had walked along the beach with open ocean to their right. Now, walking along Gabung's western coast, they felt surrounded by islands. Mantangule's long, low bulk stretched out to the southwest, and Bugsuk's broad sweep consumed the view to the northeast, only three or so miles across the reef-strewn water. To the north, another large island—Pandanan—was indistinguishable from Bugsuk. And to the northwest, more, smaller islands rose above the reefs to hide the horizon. With their goal of Bugsuk in sight, the men could think of food and water and let those images draw them ahead. But there were distractions. Swarms of stinging insects flew around them, and their thoughts drifted uncontrollably, clouded by the lack of food and water.

Apo Island was on the far side of a strait nearly two miles across, with the dark blue of deep water again in the middle, between the two shores. The men had about two hours to wait for slack tide and the passing of the next enemy patrol plane, and they gathered more coconuts from the beach, but as so often before, none was edible. Surrounded by a sea full of fish and water, they were dying of thirst and starvation.

The airplane arrived on schedule and continued south over the island. Certain the danger had passed, the men pushed their raft back to sea. The water was shallow enough for them to wade and to keep the weight off their feet as they leaned on the raft. Pushing and splashing, they moved their craft into the dark blue of the deeper water.

They were midway between the two islands, with no retreat possible, when someone noticed the fins. Two sharks cruised just beneath the surface, looking for food. The men kept paddling, splashing and kicking, and the sharks, perhaps sensing the hunger that drove these eight beings, stayed clear.

Aided by the shallow water, the raft crossed between Gabung and Apo islands in only three hours, and the men found a sandy beach just before dark. By now, they knew what to expect. They posted their rat guards and waited in troubled dreams and fitful sleep for the morning.

Sunrise was again their alarm clock, but they lingered until about eight o'clock before gathering around their sole possession, the raft, and heading to the west. Apo is a small, round island, but in all other ways it seemed no different from Byan and Gabung. Again, the men had chosen to take the long way, and each grudging step along the curved shoreline revealed some new aspect of the land ahead. Before noon, they had found the first indication of humans—a dugout canoe abandoned on the beach. The boat was riddled with holes and useless, so they left it and went on. Then they saw a trail leading up over the coral cliff, and Jacobson and Baumgart decided to explore. A trail like this meant human activity. But after a few hundred yards of walking on the coral pavement of the path, the men turned back, leaving the place to the monkeys that chattered and scampered in the trees around them. Joining the other men, they continued north along the shore.

Ancient trees, their trunks varicose and black, their roots writhing like serpents, the weave of their arched branches creating darkened tunnels, grew out from the coral cliffs along the northwestern shore of Apo Island. The men walked under the trees, hidden from observation, until, in the distance, they saw the green shoreline of their destination, Bugsuk Island.

They stood transfixed, for there, under coconut trees that swayed like tall, slender dancers lining the edge of a broad, sandy beach, were houses. There were no Japanese launches on the shore and no sign of activity around the buildings. That did not mean there were no risks. So they would wait and watch.

But not for long.

SPIRITS OF THE LAND

There were eyes behind the towering coconut trees that swayed in the sea breeze along Bugsuk Island's sparkling beach in a gentle hula. The eyes were watching the *Flier* survivors.

All that Crowley and his men saw when they looked toward the island were the apparently tranquil settlement of houses and, in their imaginations, food and water. But they were cautious. With their raft in tow, they worked their way around the northern edge of Apo Island to a point on the beach where they could no longer see the houses. Their plan was to arrive on Bugsuk just before sunset and to use the half-light of dusk to sneak toward the settlement. By now, their starvation and thirst had robbed the men of whatever athletic ability they had once possessed, so when they swam across the narrow channel between the islands, they would lack the strength to swim against the flow. But if they judged the current correctly, they would land about a mile and a half from the houses. Then they would have enough cover to sneak closer, undetected. There was no more than a half-mile between Apo and the far shore, and all of it was the pale blue of reef water. They expected no problems.

Late that afternoon, they pushed the raft off the sandy beach. Most of the men waded at its side, and when they reached the far shore, they climbed out of the water, not on coral but with another long stretch of white-sand beach under their tender feet. Stowing the raft, they walked west toward the setting sun. They were on a narrow peninsula, on the far side of which was the tidal mouth of a saltwater stream. Crossing the peninsula with a wary eye toward the far shore, they waded into the stream. When they climbed the far bank, they were on the same beach that, to the west, passed in front of the Bugsuk houses. Here a grove of baring trees, a species that, like mangroves, sinks its roots in salt water, blocked their view of the settlement. The men worked their way through the shallow water under the trees, with the low rays of the sun slanting between the tree trunks, and then moved ashore, peering through the grove at what appeared to be a once-thriving but now-abandoned village. The houses that they had seen from Apo Island were surrounded by a coconut grove, and between the survivors and those houses were the remnants of bamboo and palm-leaf native huts.

Jacobson and Baumgart were the last to arrive, their arms filled with coconuts. For the first time in five days, the *Flier* survivors would have unspoiled food to eat, and apparently a place to sleep. The main building in this settlement—well built of bamboo and lumber, with a thatched roof—looked like the home of a person of wealth. But the home had been ransacked, the furniture carted out of its now-barren rooms, and any remnant of the former owners' presence stripped from the now-naked walls.

The house had a good wooden floor for sleeping, probably free from rats and certainly protected from sand crabs. But weary as they were, the men were also excited by their discoveries and were not yet ready for rest. They wanted to explore. Standing in front of the main house and looking south, they could imagine that they were in an exotic resort. A lawn fifty yards deep or more and shaded by the high canopy of coconut trees led to the beach of pure, white, soft sand, framed in this view by drooping coconut palm fronds. Beyond the beach was an island paradise. Stretching out to the left was the chain of islands the men had spent the last four days hopping, and between the last—Byan—and Mantangule, on the right, rose the distant blue mountains of Balabac. A good-size wooden boat—Al Jacobson guessed it was thirty-eight feet long—was beached in front of the house and looked like it had been intentionally destroyed. Nearby was another launch of about the same size that appeared to have been under construction. On either side of the house and inland from it were several clearings, which suggested that the owners had raised vegetables. And farther inland, some of the men reported, there was a stream. In its clear water swam schools of fish, meals for days to come.

Exploring by himself, Earl Baumgart found a curious concrete structure just behind the main house. It stood about five feet high and was another six feet long, and when he climbed atop it he was elated. Someone had built a cistern to collect rainwater, probably from the roof of the house. There was all the water the men would ever need, and more! He called out his discovery to the others, who came running.

Once more, the skipper lived up to his reputation for cautiousness outside of the realm of battle. He told his gathered crew that they should drink sparingly from the cistern. They wanted to guzzle to their thirst's content, he knew. But having gone without water for five days, and with almost no food in the same period, their bodies could not handle much. When he had explained this, each man took a small sip from Baumgart's pool and then went away. Only Chief Howell ignored Crowley's caution. He drank until his belly was full, and then he drank some more.

Now Jacobson and some others set about opening the good coconuts. They found a sharp rock in the ground and smashed each nut against it until they had removed the soft green outer shell. Then they punched out the eye of the inner, hard brown shell, drained the milk, and crushed the nut into pieces that could be chewed.

With these small pleasures, the men began to settle in for the night. Jacobson found a bamboo door that he laid on the floor as his mattress, and he stretched out on it, content. Images filled his head as the palm leaves rustled above him. There were fish and coconuts to eat, a roof over his head, water to drink, and, it appeared, no enemies within miles. There was no more need to walk, so his feet could heal. There was shade from the sun, so his blisters would dry and disappear. This was a place where a man could wait out the war, if he had to.

Not long after Chief Howell drank his fill from the cistern, he began to feel ill. His condition worsened during the night, but there was no help for him. If some of the others showed little sympathy, it may have been because they knew his sickness, self-inflicted as it was, was not lethal. In time, his body would acclimate. The little bit of coconut in their stomachs had satisfied their appetites, and they knew there would be more meals to come. With a home around them to keep away the chilling breezes, they succumbed to their exhaustion, dreaming untroubled dreams.

Once more, they arose with the sun and began planning their day. There was work to be done, and Crowley and Liddell started organizing teams. One group would catch some fish, while another would build a fire for cooking. They had no matches, but Jacobson still had his binoculars, and their lenses would make perfect magnifying glasses for focusing the sun's rays in an incendiary beam on dry tinder. Someone needed to scout the area, and the group would need more coconuts.

Jacobson was the first one up, and he was standing looking out a window toward the rear of the house and the jungle beyond when he saw two small boys—they might have been thirteen or fourteen—emerge from the trees. Jacobson told his shipmates what he saw, and they all were quickly on their feet. It was obvious to them that the boys knew the sailors were there, so they filed out of the house and approached the visitors. The boys were wearing ragged shorts and tattered shirts, and their feet were bare, like the sailors'. Crowley stepped forward.

"Americans or Japanese?" he asked.

"Americanos!" one of the boys, Oros Bogata, said, smiling. "Japanese!" he said, drawing a finger across his throat as if slitting it.

The men felt a collective wave of relief. Then the boy pointed to the cistern by the house.

"Don't drink water," he said.

Perhaps they misunderstood his puzzling words, they thought. But with their *Que sera, sera* attitude, they disregarded that comment and asked whether the boys had any food. Oros patted his small stomach.

"Rice," he said, and he motioned for the men to follow him and his silent friend back into the jungle. Stepping in line behind the boys, they found themselves on a narrow path. The boys, seeing that they were being followed, scampered ahead to a spot where they had left poles with small packs tied at the ends. Each balanced his pole on his shoulder, and then Oros led the file of hobbling, nearly naked men while his friend followed, sweeping the trail behind them to camouflage evidence of the group's passing.

In a short distance, they reached an abandoned sugarcane field. Oros motioned for the men to sit down, and he and his companion cut sections of cane a yard long and offered each man his own piece. The heart of the cane was a sweet and juicy bundle of fibers, and for the next half-hour, Crowley and his men chewed in bliss, until they simply had no more strength left in their jaws.

Back on the path, the boys led the men a short way to a clearing about the size of a football field. In one corner of the field was a raised wooden platform with a thatched roof supported on bamboo poles, but with no walls. Again the boys motioned for the men to sit and rest. Then they dropped their poles and opened their packs. One took a stick, sharpened at one end, and placed the tip in a notched piece of wood that he drew from his pack. He spun the stick between his palms, and in less time than it would have taken to remove a match from a box and strike it, he had some tinder smoldering. Jacobson, the Eagle Scout who had been taught to start fires with a bow and a stick, was impressed.

Then the boys produced a small pot, and one left and got water from a nearby stream. They poured rice from their pack into the pot, and while the fire brought the water to a boil, they cut leaves from a banana tree and made plates for their guests. Now the same boy who had cautioned them against drinking from the cistern gave them a cup of muddy water and, by sign language, told them they should drink it. The men hesitated, so the boy drank some himself. *Que sera, sera!* The men drank, as well.

When the rice was cooked, the boys spread it on the banana leaves. Then they produced three dried fish from the bounty of their packs and divided them among the men. There was enough for everyone.

Four days earlier, Crowley and his men had chosen survival with freedom over survival with food when they had elected to head away from Balabac. Theirs was a decision that prolonged the pain of hunger and thirst, which might easily have been cut short had they allowed themselves instead to come under Japanese control. Now, without hesitation, they had turned themselves over to the authority of two small boys whose friendship they accepted as a stray dog does that of a man with a scrap of food. Led by their stomachs, the sailors had followed the boys into the jungle with only the promise of rice, and now, with the smell of steaming hot rice and fish rising to their nostrils, they attacked their meal.

Their focus changed abruptly when, looking up from his food, one of them saw nine men, bristling with weapons—rifles, blowguns, and bolos—stepping into the clearing from every point of the compass.

They were surrounded!

The *Flier* survivors had traded their safety for scraps of sustenance. The price of their meal now stared across the clearing at them.

Wesley Miller was ready to bolt like a startled fawn, but these fierce-looking warriors were everywhere. No one budged. The shredded soles of their feet precluded it.

"Hello!" one of the armed men called. His voice was cheerful and a smile lit his face. He dashed across the grass to the platform where the sailors still sat. Crowley struggled to his feet, as did his men at the approach of this stranger. When the man reached them, he grasped the hands of the *Flier* crew, shaking them vigorously.

"Welcome to Bugsuk Island," the man said. "I am Pedro Sarmiento."

Sarmiento said he was the leader of the local bolo battalion, indicating the men who were with him. Sarmiento had instructions from the guerrilla headquarters that if he found any Allied survivors, he was to ship them to a guerrilla outpost on Palawan's southern tip.

Crowley and Liddell were becoming comfortable with Sarmiento, and they were prepared to follow his instructions. At this point someone recalled the earlier direction from the two boys, to not drink the water in the cistern, and asked Sarmiento to explain.

Oh, he replied, earlier in the war, when the Japanese had driven the owner of the home from his property, Sarmiento had poisoned the water with arsenic in hopes that Japanese soldiers would drink it!

Everyone looked at Chief Howell. The man had a cast-iron gut!

A Japanese patrol would reach Bugsuk later in the morning, Sarmiento told the sailors, so they could not remain at the schoolhouse. The Japanese soldiers would inspect the area and then would spend the night in the house where Crowley and his men had slept so peacefully the night before. So the sailors would have to hike at least a mile inland to be safe. The Japanese were afraid to penetrate the center of the island, Sarmiento said.

Sarmiento reported that his instructions from the guerrilla leader, Captain Mayor, were to take any survivors all the way north across the center of Bugsuk and then to bring them by boat to the guerrilla outpost at Cape Buliluyan, the southernmost tip of Palawan. He told Crowley and Liddell that it was important to begin the hike soon. When they said they were ready, he told them to finish their breakfast. Then he sent the two boys back to the beach to make sure the Americans had not left any evidence behind.

A few minutes later, the boys returned with the lens that the sailors had removed from Jacobson's binoculars, with which they had planned to start their fires. Someone offered the lens to Sarmiento, who produced a pipe and tobacco that he lit with the lens. He smiled with gratitude. Then, seeing that they had finished their rice and fish, he invited the men from *Flier* to begin their cross-island trek.

Seventeen

The Merchant's Cup

David W. Bone

"Fatty" Reid burst into the half-deck with a whoop of exulta-tion. "Come out, boys," he yelled. "Come out and see what luck! The *James Flint* comin' down the river, loaded and ready for sea! Who-oop! What price the *Hilda* now for the Merchants' Cup?"

"Oh, come off," said big Jones. "Come off with your Merchants' Cup. Th' *James Flint*'s a sure thing, and she wasn't more than half-loaded when we were up at Crockett on Sunday!"

"Well, there she comes anyway! *James Flint*, sure enough! Grade's house-flag up, and the Stars and Stripes!"

We hustled on deck and looked over by the Sacramento's mouth. "Fatty" was right. A big barque was towing down beyond San Pedro. The *James Flint*! Nothing else in 'Frisco harbour had spars like hers; no ship was as trim and clean as the big Yankee clipper that Bully Nathan commanded. The sails were all aloft, the boats aboard. She was ready to put to sea.

Our cries brought the captain and mate on deck, and the sight of the outward-bounder made old man Burke's face beam like a nor'west moon.

"A chance for ye now, byes," he shouted. "An open race, bedad! Ye've nothin' t' be afraid of if th' *James Flint* goes t' sea by Saturday!"

Great was our joy at the prospect of the Yankee's sailing. The 'Frisco Merchants' Cup was to be rowed for on Saturday. It was a mile-and-half race for ships' boats, and three wins held the Cup for good. Twice, on previous years, the *Hilda's* trim gig had shot over the line—a handsome winner. If we won again, the Cup was ours for keeps! But there were strong opponents to be met this time. The *James Flint* was the most formidable. It was open word that Bully Nathan was keen on winning the trophy. Every one knew that he had deliberately sought out boatmen when the whalers came in from the north. Those who had seen the Yankee's crew at work in their snaky carvel-built boat said that no one else was in it. What chance had we boys in our clinker-built against the thews and sinews of trained whalemen? It was no wonder that we slapped our thighs at the prospect of a more open race.

Still, even with the Yankee gone, there were others in the running. There was the *Rhondda* that held the Cup for the year, having won when we were somewhere off the Horn; then the *Hedwig Rickmers*—a Bremen four-master—which had not before competed, but whose green-painted gig was out for practice morning and night. We felt easy about the *Rhondda* (for had we not, time and again, shown them our stern on the long pull from Green St. to the outer anchorage?), but the Germans were different. Try as we might, we could never pull off a spurt with them. No one knew for certain what they could do, only old Schenke, their skipper, and he held his tongue wisely.

The *James Flint* came around the bend, and our eager eyes followed her as she steered after the tug. She was making for the outer anchorage, where the laden ships lie in readiness for a good start off.

"Th' wind's 'bout west outside," said Jones. "A 'dead muzzler'! She'll not put t' sea tonight, even if she has all her 'crowd' aboard."

"No, worse luck! mebbe she'll lie over till Saturday after all. They say Bully's dead set on getting th' Cup. He might hang back. . . . Some excuse—short-handed or something!" Gregson was the one for "croaking."

"No hands?" said Fatty. "Huh! How could he be short-handed when everybody knows that Daly's boardin'-house is chock-full of fightin' Dutchmen? No, no! It'll be the sack for Mister Bully B. Nathan if he lets a capful o' fair wind go by and his anchor down. Gracie's agents 'll watch that!"

"Well! He's here for th' night, anyway. . . . There goes her mudhook!"

We watched her great anchor go hurtling from the bows and heard the roar of chain cable as she paid out and swung round to the tide.

"Come roun', yo' boys dere! Yo' doan' want no tea, eh?" The cook, beating tattoo on a saucepan lid, called us back to affairs of the moment, and we sat down to our scanty meal in high spirits, talking—all at one time—of our chances of the Cup.

The *Hilda* had been three months at San Francisco, waiting for the wheat crop and a profitable charter. We had come up from Australia, and most of our crew, having little wages due to them, had deserted soon after our arrival. Only we apprentices and the sail-maker remained, and we had work enough to set our muscles up in the heavy harbour jobs. Trimming coal and shovelling ballast may not be scientific training, but it is grand work for the back and shoulders.

We were in good trim for rowing. The old man had given us every opportunity, and nothing he could do was wanting to make us fit. Day and daily we had set our stroke up by the long pull from the anchorage to the wharves, old Burke coaching and encouraging, checking and speeding us, till we worked well together. Only last Sunday he had taken us out of our way, up the creek, to where we could see the flag at the *Rhondda*'s masthead. The old man said nothing, but well we knew he was thinking of how the square of blue silk, with Californian emblem worked in white, would look at his trim little *Hilda*'s fore-truck! This flag accompanied the Cup, and now (if only the Yankee and his hired whale-men were safely at sea) we had hopes of seeing it at our masthead again.

Tea over—still excited talk went on. Some one recalled the last time we had overhauled and passed the *Rhondda*'s gig.

"It's all very well your bucking about beating the *Rhondda*," said Gregson; "but don't think we're going to have it all our own way! Mebbe they were 'playing 'possum' when we came by that time!"

"Maybe," said Jones. "There's Peters and H. Dobson in her crew. Good men! Both rowed in the Worcester boat that left the Conways' at the start, three years ago. . . . And what about the *Rickmers*? . . . No, no! It won't do to be too cocksure! . . . Eh, Takia?"

Takia was our cox-n, a small wiry Jap. Nothing great in inches, but a demon for good steering and timing a stroke. He was serving his apprenticeship with us and had been a year in the *Hilda*. Brute strength was not one of his points, but none was keener or more active in the rigging than our little Jap.

He smiled,—he always smiled,—he found it the easiest way of speaking English. "Oh, yes," he said. "Little cocksu'—good! Too much cocksu'—no good!"

We laughed at the wisdom of the East.

"Talk about being cocky," said Gregson; "you should hear Captain Schenke bragging about the way he brought the *Hedwig Rickmers* out. I heard 'em and the old man at it in the ship-chandler's yesterday. Hot . . . Look here, you chaps! I don't think the old man cares so much to win the Cup as to beat Schenke! The big 'squarehead' is always ramming it down Burke's throat how he brought his barque out from Liverpool in a hundred and five days, while the *Hilda* took ten days more on her last run out!"

"That's so, I guess," said Jones. (Jones had the Yankee "touch.") "Old Burke would dearly love to put a spoke in his wheel, but it'll take some doing. They say that Schenke has got a friend down from Sacramento—gym.-instructor or something to a college up there. He'll be training the 'Dutchy' crew like blazes. They'll give us a hot time, I'll bet!"

Gregson rose to go on deck. "Oh, well," he said, "it won't be so bad if the *James Flint* only lifts his hook by Saturday. Here's one bloomin' *hombre* that funks racin' a fancy whaler! . . . An' doesn't care who knows it, either!"

Thursday passed—and now Friday—still there was no sign of the wind changing, and the big Yankee barque lay quietly at anchor over by the Presidio.

When the butcher came off from the shore with the day's stores, we eagerly questioned him about the prospects of the *James Flint*'s sailing. "*Huh!* I guess yew're nat the only 'citizens' that air concarned 'bout that!" he said. "They're talkin' 'bout nuthin' else on every 'lime-juicer' in the Bay! . . . An' th' *Rickmers*! Gee! Schenkie's had his eye glued ter th' long telescope ever since daybreak, watchin' fer th' *Flint* heavin' up anchor!"

The butcher had varied information to give us. Now it was that Bully Nathan had telegraphed to his New York owners for permission to remain in port over Sunday. Then again, Bully was on the point of being dismissed his ship for not taking full advantage of a puff of nor'-west wind that came and went on Thursday night.

. . . The *Flint* was short of men! . . . The Flint had a full crew aboard! Rumours and rumours! "All sorts o' talk," said the butcher; "but I know this fer certain—she's got all her stores aboard. Gosh! I guess—she—has! I don't like to wish nobody no harm, byes, but I hope Bully Nathan's first chop 'll choke him, fer th' way he done me over the beef! . . . Scorch 'im!"

In the forenoon we dropped the gig and put out for practice. Old Burke and the mate came after us in the dinghy, the old man shouting instruction and encouragement through his megaphone as we rowed a course or spurted hard for a furious three minutes. Others were out on the same ploy, and the backwaters of the Bay had each a lash of oars to stir their tideless depths. Near us the green boat of the *Rickmers* thrashed up and down in style. Time and again we drew across—"just for a friendly spurt"—but the "Dutchies" were not giving anything away, and sheered off as we approached. We spent an hour or more at practice and were rowing leisurely back to the ship when the green boat overhauled us, then slowed to her skipper's orders.

"How you vass, Cabtin Burke?" said Schenke, an enormous fair-headed Teuton, powerful-looking, but run sadly to fat in his elder years. "You t'ink you get a chanst now, *hein?*. . . Now de Yankee is goin' avay!" He pointed over to the Presidio, where the *Flint* lay at anchor. We followed the line of his fat forefinger. At anchor, yes, but the anchor nearly a-weigh. Her flags were hoisted, the blue peter fluttering at the fore, and the *Active* tug was passing a hawser aboard, getting ready to tow her out. The smoke from the tugboat's funnel was whirling and blowing over the low forts that guard the Golden Gates. Good luck! A fine nor'-west breeze had come that would lift our dreaded rival far to the south'ard on her way round Cape Horn!

Schenke saw the pleased look with which old Burke regarded the Yankee's preparations for departure.

"Goot bizness, eh?" he said. "You t'ink you fly de flack on de *Hilda* nex' *Sonndag*, Cabtin? Veil! Ah wish you goot look, but you dond't got it all de same!"

"Oh, well, Captain Schenke, we can but thry," said the old man. "We can but thry, sorr! . . . Shure, she's a foine boat—that o' yours. . . . An' likely-looking lads, too!" No one could but admire the well-set figures of the German crew as they stroked easily beside us.

"*Schweinehunden*," said Schenke brutally. We noticed more than one stolid face darkling as they glanced aside. Schenke had the name of a "hard case." "*Schweinehunden*," he said again. "Dey dond't like de hard vork, Cabtin. . . . Dey dond't like it—but ve takes der Coop, all de same! Dey pulls goot und strong, oder"—he rasped a short sentence in rapid Low German—"Shermans dond't be beat by no durn lime-juicer, *nein*!"

Old Burke grinned. "Cocky as ever, Captain Schenke! Bedad now, since ye had the luck of ye're last passage there's no limit to ye!"

"Luck! Luck! Alvays de luck mit you, Cabtin!"

"An' whatt ilse?. . . Sure, if I hadn't struck a bilt of calms an' had more than me share of head winds off the Horn, I'd have given ye a day or two mesilf!"

"Ho! Ho! Ho! *Das ist gut*!" The green boat rocked with Schenke's merriment. He laughed from his feet up—every inch of him shook with emotion. "Ho! Ho! Hoo! *Das ist ganz gut*. You t'ink you beat de *Hedwig Rickmers* too, Cabtin? You beat 'm mit dot putty leetle barque? You beat 'm mit de *Hilda*, *nichtwahr*?"

"Well, no," said our old man. "I don't exactly say I beat the *Rickmers*, but if I had the luck o' winds that ye had, bedad, I'd crack th' *Hilda* out in a hundred an' five days too!"

"Now, dot is not drue, Cabtin! *Aber ganz und gar nicht*! You know you haf bedder look von de vind as Ah got. Ah sail mein sheep! Ah dond't vait for de fair winds nor not'ings!"

"No," said Burke, "but ye get 'em, all the same. Everybody knows ye've th' divil's own luck, Schenke!"

"Und so you vas! Look now, Cabtin Burke. You t'nk you got so fast a sheep as mein, eh? Veil! Ah gif you a chanst to make money. Ah bet you feefty dollars to tventig, Ah take mein sheep home quicker as you vass!"

"Done wit' ye," said stout old 'Paddy' Burke, though well he know the big German barque could sail round the little *Hilda*. "Fifty dollars to twenty, Captain Schenke, an' moind y've said it!"

The green boat sheered off and forged ahead, Schenke laughing and waving his hand derisively. When they had pulled out of earshot, the old man turned ruefully to the mate: "Five pounds clean t'rown away, mister! Foine I know the *Rickmers* can baate us, but I wasn't goin' t' let that ould 'squarehead' have it all his own way! Divil th' fear!"

We swung under the *Hilda*'s stern and hooked on to the gangway. The old man stepped out, climbed a pace or two, then came back.

"Look ye here, byes," he said, "I'll give ye foive dollars a man—an' a day's 'liberty' t' spind it—if ye only baate th' 'Dutchmen.' . . . Let th' Cup go where it will!"

The Bay of San Francisco is certainly one of the finest natural harbours in the world, let Sydney and Rio and Falmouth all contest the claim. Land-locked to every wind that blows, with only a narrow channel open to the sea, the navies of the world could lie peacefully together in its sheltered waters. The coast that environs the harbour abounds in natural beauties, but of all the wooded creeks—fair stretches of undulating downs—or stately curves of winding river, none surpasses the little bay formed by the turn of Benita, the northern postern of the Golden Gates. Here is the little township of Sancilito, with its pretty white houses nestling among the dark green of the deeply wooded slopes. In the bay there is good anchorage for a limited number of vessels, and fortunate were they who manned the tall ships that lay there, swinging ebb and flood, waiting for a burthen of golden grain.

On Saturday the little bay was crowded by a muster of varied craft. The ships at anchor were "dressed" to the mastheads with gaily-coloured flags. Huge ferryboats passed slowly up and down, their tiers of decks crowded with sightseers. Tug-boats and launches darted about, clearing the course, or convoying racing boats to the starting lines. Ships' boats of all kinds were massed together close inshore: gigs and pinnaces, lean whaleboats, squat dinghys, even high-sided ocean lifeboats with their sombre broad belts of ribbed cork. A gay scene of colour and animation. A fine turn-out to see the fortune of the Merchants' Cup.

At two the Regatta began. A race for longshore craft showed that the boarding-house "crimps" were as skillful at boatman's work as at inducing sailormen to desert their ships. Then two outriggers flashed by, contesting a heat for a College race. We in the *Hilda*'s gig lay handily at the starting line and soon were called out. There were nine entries for the Cup, and the judges had decided to run three heats. We were drawn in the first, and, together with the *Ardlea*'s and *Compton*'s gigs, went out to be inspected. The boats had to race in sea-service conditions, no lightening was allowed. At the challenge of the judges we showed our gear. "Spare oar—right! Rowlocks—right! Sea-anchor—right! Bottom boards and stern grating—right. Painter, ten fathoms; hemp. . . . A bit short there, *Compton*! Eh? . . . Oh—all right," said the official, and we manoeuvred into position, our sterns held in by the guard-boats. Some of the ships' captains had engaged a steam-launch to follow the heats, and old Burke was there with his trumpet, shouting encouragement already.

"Air yew ready?"

A pause: then, pistol shot! We struck water and laid out! Our task was not difficult. The *Ardlea*'s gig was broad-bowed and heavy; they had no chance; but the *Compton*'s gave us a stiff pull to more than midway. Had they been like us, three months at boat-work, we had not pulled so easily up to the mark, but their ship was just in from Liverpool, and they were in poor condition for a mile and a half at pressure. We won easily, and scarce had cheered the losers before the launch came fussing up.

"Come aboard, Takia," shouted old Burke. "Ye come down wit' me an' see what shape the German makes. He's drawn wit' th' *Rhondda* in this heat!"

Takia bundled aboard the launch and we hauled inshore to watch the race. There was a delay at the start. Schenke, *nichts verstehen*, as he said, was for sending his boat away without a painter or spare gear. He was pulled up by the judges, and had to borrow.

Now they were ready. The *Rickmers* outside, *Rhondda* in the middle berth, and the neat little *Slieve Donard* inshore. At the start the Rhonddas came fair away from the German boat, but even at the distance we could see that the "Dutchmen" were well in hand. At midway the *Rhondda* was leading by a length, still going strong, but they had shot their bolt, and the green boat was surely pulling up. The *Slieve Donard*, after an unsteady course, had given up. Soon we could hear old Schenke roaring oaths and orders, as his launch came flying on in the wake of the speeding boats.

The Germans spurted.

We yelled encouragement to the Rhonddas. "Give 'em beans, old sons! . . ."

"*Rhondda*! *Rhondda*! . . . Shake 'er up" Gallantly the white boat strove to keep her place, but the greens were too strong. With a rush, they took the lead and held it to the finish, though two lengths from the line their stroke faltered, the swing was gone, and they were dabbling feebly when the shot rang out.

"A grand race," said every one around. "A grand race"—but old Burke had something to say when he steamed up to put our cox'n among us. "Byes, byes," he said, "if there had been twinty yards more the *Rhondda* would have won. Now d'ye moind, Takia, ye divil . . . d'ye moind! Keep th' byes in hand till I give ye th' wurrd! . . . An' whin ye get th' wurrd, byes! . . . Oh, Saints! Shake her up when ye get th' wurrd!"

The third heat was closely contested. All three boats, two Liverpool barques and a Nova Scotiaman, came on steadily together. A clean race, rowed from start to finish, and the *Tuebrook* winning by a short length.

The afternoon was well spent when we stripped for the final, and took up our positions on the line. How big and muscular the Germans looked! How well the green boat sat the water! With what inward quakings we noted the clean fine lines of stem and stern! . . . Of the *Tuebrook* we had no fear. We knew they could never stand the pace the Germans would set. Could we?

Old Burke, though in a fever of excitement when we came to the line, had little to say. "Keep the byes in hand, Takia—till ye get th' wurrd," was all he muttered. We swung our oar-blades forward.

"Ready?" The starter challenged us.

Suddenly Takia yelped! We struck and lay back as the shot rang out! A stroke gained! Takia had taken the flash; the others the report!

The Jap's clever start gave us confidence and a lead. Big Jones at stroke worked us up to better the advantage. The green boat sheered a little, then steadied and came on, keeping to us, though nearly a length astern. The *Tuebrook* had made a bad start, but was thrashing away pluckily in the rear.

So we hammered at it for a third of the course, when Takia took charge. Since his famous start he had left us to take stroke as Jones pressed us, but now he saw signs of the waver that comes after the first furious burst—shifting grip or change of foothold.

"'*Trok!*—'*trok!*—'*trok!*'" he muttered, and steadied the pace. "'*Troke!*—'*troke!*—'*troke!*'" in monotone, good for soothing tension.

Past midway the green boat came away. The ring of the German's rowlocks rose to treble pitch. Slowly they drew up, working at top speed. Now they were level—level! and Takia still droning "'*troke!*—'*troke!*—'*troke!*'"—as if the lead was ours!

Wild outcry came from the crowd as the green boat forged ahead! Deep roars from Schenke somewhere in the rear! Now, labouring still to Takia's '*troke!*—'*troke!* we had the foam of the German's stern wash at our blades! "Come away, *Hilda's!*" . . . "*Shake her up, there!*" . . . "*Hilda-h! Hilda-h!*"—Takia took no outward heed of the cries. He was staring stolidly ahead, bending to the pulse of the boat. No outward heed—but '*troke!*—'*troke!* came faster from his lips. We strained, almost holding the Germans' ensign at level with our bow pennant.

Loud over the wild yells of the crowd we heard the voice we knew— old Burke's bull-roar: "Let 'er rip, Taki'! Let 'er rip, bye!"

Takia's eyes gleamed as he sped us up—up—up! '*Troke* became a yelp like a wounded dog's. He crouched, standing, in the sternsheets, and lashed us up to a furious thrash of oars! Still quicker! . . . The eyes of him glared at each of us, as if daring us to fail! The yelp became a scream as we drew level—the Germans still at top speed. "Up! Up! Up!" yells Takia, little yellow devil with a white froth at his lips! "*Up! Up! Up!*" swaying unsteadily to meet the furious urging.

The ring of the German rowlocks deepens—deepens—we see the green bow at our blades again. Her number two falters—jars—recovers again—and pulls stubbornly on. Their "shot" is fired! They can do no more! Done!

And so are we! Takia drops the yoke ropes and leans forward on the gunwale! Oars jar together! Big Jones bends forward with his mouth wide—wide! Done!

But not before a hush—a solitary pistol shot—then roar of voices and shrilling of steamer syrens tell us that the Cup is ours!

A month later there was a stir in the western seaports. No longer the ships lay swinging idly at their moorings. The harvest of grain was ready for the carriers, and every day sail was spread to the free wind outside the Golden Gates, and laden ships went speeding on their homeward voyages. The days of boat-races and pleasant time-passing harbour jobs were gone; it was now work—work—to get the ship ready for her burden, and, swaying the great sails aloft, to rig harness for the power that was to bear us home. From early morning till late evening we were kept hard at it; for Captain Burke and the mate were as keen on getting the *Hilda* to sea after her long stay in port as they were on jockeying us up to win the Cup. Often, when we turned to in the morning, we would find a new shipmate ready to bear a hand with us. The old man believed in picking up a likely man when he offered. Long experience of Pacific ports had taught him how difficult it is to get a crew at the last moment.

So when at length the cargo was stowed, we were quite ready to go to sea, while many others—the *Hedwig Rickmers* among them—were waiting for men.

On the day before sailing a number of the ship captains were gathered together in the chandler's store, talking of freights and passages, and speculating on the runs they hoped to make. Burke and Schencke were the loudest talkers, for we were both bound to Falmouth "for orders," and the *Rickmers* would probably sail three days after we had gone.

"Vat 'bout dot bett you make mit me, Cabtin?" said Schenke. "Dot is all recht, no?"

"Oh, yess," answered the old man, but without enthusiasm. "That stands."

"Hoo! Hoo! Hoo! Tventig dollars to feefty—dot you goes home quicker as me, no?" Schencke turned to the other men. "Vat you tinks, yenthelmen? Ah tinks Ah sbend der tventig dollars now—so sure Ah vass."

The others laughed. "Man, man," said Findlayson of the *Rhondda*. "You don't tell me Burke's been fool enough to take that bet. Hoo! You haven't the ghost of a chance, Burke."

"Och, ye never know," said the now doleful sportsman. "Ye never know ye're luck."

"Look here, Cabtin," said Schencke (good-humoured by the unspoken tribute to his vessel's sailing powers)—"Ah gif you a chanst. Ah make de bett dis vay—look. Ve goes to Falmouth—you *und* me, *hein?* Now, de first who comes on de shore vins de money. Dot vill gif you t'ree days' start, no?"

"That's more like it," said the other captains. "I wish you luck, Burke," said Findlayson. "Good luck—you'll need it too—if you are to be home before the big German."

So the bet was made.

At daybreak next morning we put out to sea. The good luck that the *Rhondda* wished us came our way from the very first. When the tug left us we set sail to a fine fair wind, and soon were bowling along in style. We found the nor'-east Trades with little seeking; strong Trades, too, that lifted us to the Line almost before the harbour dust was blown from our masts and spars. There calms fell on us for a few days, but we drifted south in the right current, and in less than forty days had run into the "westerlies" and were bearing away for the Horn.

Old Burke was "cracking on" for all the *Hilda* could carry canvas. Every morning when he came on deck the first question to the mate would be: "Any ships in sight, mister?" . . . "Any ships astern," he meant, for his first glance was always to where the big green four-master might be expected to heave in sight. Then, when nothing was reported, he would begin his day-long strut up and down the poop, whistling "Garryowen" and rubbing his hands.

Nor was the joy at our good progress his alone. We in the half-deck knew of the bet, and were keen that the ship which carried the Merchants' Cup should not be overhauled by the runner-up! We had made a fetish of the trophy so hardly won. The Cup itself was safely stowed in the ship's strong chest, but the old man had let us have custody of the flag. Big Jones had particular charge of it; and it had been a custom while in 'Frisco to exhibit it on the Saturday nights to admiring and envious friends from other ships. This custom we continued when at sea. True, there were no visitors to set us up and swear what lusty chaps we were, but we could frank one another and say, "If you hadn't done this or that, we would never have won the race."

On a breezy Saturday evening we were busy at these rites. The *Hilda* was doing well before a steady nor'-west wind, but the weather—though nothing misty—was dark as a pall. Thick clouds overcast the sky, and there seemed no dividing line between the darkling sea and the windy banks that shrouded the horizon. A dirty night was in prospect; the weather would thicken later; but that made the modest comforts of the half-deck seem more inviting by comparison; and we came together for our weekly "sing-song"—all but Gregson, whose turn it was to stand the lookout on the fo'c'sle-head.

The flag was brought out and hung up—Jones standing by to see that no pipe-lights were brought near—and we ranted at "Ye Mariners of England" till the mate sent word that further din would mean a "work-up" job for all of us.

Little we thought that we mariners would soon be facing dangers as great as any we so glibly sang about. Even as we sang, the *Hilda* was speeding on a fatal course! Across her track the almost submerged hull of a derelict lay drifting. Black night veiled the danger from the keenest eyes.

A frenzied order from the poop put a stunning period to our merriment. "Helm up, f'r God's sake! . . . *Up!—oh God!—Up! Up!*" A furious impact dashed us to the deck. Staggering, bruised, and bleeding, we struggled to our feet. Outside the yells of fear-stricken men mingled with hoarse orders, the crash of spars hurtling from aloft vied with the thunder of canvas, as the doomed barque swung round broadside to the wind and sea.

Even in that dread moment Jones had heed of his precious flag. As we flew to the door, he tore the flag down, stuffing it in his jumper as he joined us at the boats.

There was no time to hoist out the lifeboats—it was pinnace and gig or nothing. Already the bows were low in the water. "She goes. She goes!" yelled some one. "Oh, Christ! She's going!"

We bore frantically on the tackles that linked the gig, swung her out, and lowered by the run; the mate had the pinnace in the water, men were swarming into her. As the gig struck water, the barque heeled to the rail awash. We crowded in, old Burke the last to leave her, and pushed off. Our once stately *Hilda* reeled in a swirl of broken water, and the deep sea took her!

Sailor work! No more than ten minutes between "Ye Mariners" and the foundering of our barque!

We lay awhile with hearts too full for words; then the pinnace drew near, and the mate called the men. All there but one! "Gregson!" . . . No Gregson! The bosun knew. He had seen what was Gregson lying still under the wreck of the topmost spars.

The captain and mate conferred long together. We had no sail in the gig, but the larger boat was fully equipped. "It's the only chance, mister," said Burke at last. "No food—no water! We can't hold out for long. Get sail on your boat and stand an hour or two to the east'ard. Ye may fall in with a ship; she w'was right in th' track whin she s-struck. We can but lie to in th' gig an' pray that a ship comes by."

"Aye, aye, sir." They stepped the mast and hoisted sail. "Good-bye all: God bless ye, captain," they said as the canvas swelled. "Keep heart!" For a time we heard their voices shouting us God-speed—then silence came!

Daybreak!

Thank God the bitter night was past. Out of the east the long-looked-for light grew on us, as we lay to sea-anchor, lurching unsteadily in the teeth of wind and driving rain. At the first grey break we scanned the now misty horizon. There was no sign of the pinnace; no God-sent sail in all the dreary round!

We crouched on the bottom boards of the little gig and gave way to gloomy thoughts. What else could be when we were alone and adrift on the broad Pacific, without food or water, in a tiny gig already perilously deep with the burden of eight of us? What a difference to the gay day when we manned the same little boat and set out in pride to the contest! Here was the same spare oar that we held up to the judges—the long oar that Jones was now swaying over the stern, keeping her head to the wind and sea! Out there in the tumbling water the sea-anchor held its place; the ten fathoms of good hemp "painter" was straining at the bows!

The same boat! The same gear! The same crew, but how different! A crew of bent heads and wearied limbs! Listless-eyed, despairing! A ghastly crew, with black care riding in the heaving boat with us!

Poor old Burke had hardly spoken since his last order to the mate to sail the pinnace to the east in search of help. When anything was put to him, he would say, "Aye, aye, b'ye," and take no further heed. He was utterly crushed by the disaster that had come so suddenly on the heels of his "good luck." He sat staring stonily ahead, deaf to our hopes and fears.

Water we had in plenty as the day wore on. The rain-soaked clothes of us were sufficient for the time, but soon hunger came and added a physical pain to the torture of our doubt. Again and again we stood up on the reeling thwarts and looked wildly around the sea-line. No pinnace— no ship—nothing! Nothing, only sea and sky, and circling sea-birds that came to mock at our misery with their plaintive cries.

A bitter night! A no less cruel day! Dark came on us again, chill and windy, and the salt spray cutting at us like a whiplash.

Boo-m-m!

Big Jones stood up in the sternsheets, swaying unsteadily. "D'ye hear anything there?. . . Like a gun?"

A gun? Gun?. . . Nothing new! . . . We had been hearing guns, seeing sails—in our minds—all the day! All day . . . guns . . . and sail! Boom-m-m-m!

"Gun! Oh God . . . a gun! Capt'n, a gun, d'ye hear! Hay—Hay-H. Out oars, there! A gun!" Hoarse in excitement Jones shook the old man and called at his ear. "Aye, aye, b'ye. Aye, aye," said the broken old man, seeming without understanding.

Jones ceased trying to rouse him, and, running out the steering oar, called on us to haul the sea-anchor aboard. We lay to our oars, listening for a further gunfire.

Whooo-o. . . . Boom-m-m.

A rocket! They were looking for us then! The pinnace must have been picked up! A cheer—what a cheer!—came brokenly from our lips; and we lashed furiously at the oars, steering to where a glare in the mist had come with the last report.

Roused by the thrash of our oars, the old man sat up. "Whatt now, b'ye? Whatt now?"

"Ship firin' rockets, sir," said Jones. "Rockets . . . no mistake." As he spoke, another coloured streamer went flaming through the eastern sky. "Give way, there! We'll miss her if she's running south! Give way, all!" The glare of the rocket put heart into our broken old skipper. "Steady now, b'yes," he said, with something of his old enthusiasm.

We laboured steadily at the oars, but our strength was gone. The sea too, that we had thought moderate when lying to sea-anchor, came at us broadside on and set our light boat to a furious dance. Wave crests broke and lashed aboard, the reeling boat was soon awash, and the spare men had to bale frantically to keep her afloat. But terror of the ship running south from us nerved our wearied arms, and we kept doggedly swinging the oars. Soon we made out the vessel's sidelight—the gleam of her starboard light, that showed that she was hauled to the wind, not running south as we had feared. They could not see on such a night, we had nothing to make a signal, but the faint green flame gave us heart in our distress.

The old man, himself again, was now steering, giving us Big Jones to bear at the oars. As we drew on we made out the loom of the vessel's sails—a big ship under topsails only, and sailing slowly to the west. We pulled down wind to cross her course, shouting together as we rowed. Would they never hear? . . . Again! . . . Again!

Suddenly there came a hail from the ship, a roar of orders, rattle of blocks and gear, the yards swung round and she layed up in the wind, while the ghostly glare of a blue light lit up the sea around.

A crowd of men were gathered at the waist, now shouting and cheering as we laboured painfully into the circle of vivid light. Among them a big man (huge he looked in that uncanny glare) roared encouragement in hoarse gutturals.

Old Schenke? The *Hedwig Rickmers*?

Aye—Schenke! But a different Schenke to the big, blustering, overbearing "Square-head" we had known in 'Frisco. Schenke as kind as a brother—a brother of the sea indeed. Big, fat, honest Schenke, passing his huge arm through that of our broken old skipper, leading him aft to his own bed, and silencing his faltering story by words of cheer. *"Ach, du lieber Gott!* It is all right, no? All right, Cabtin, now you come on board. Ah know all 'bout it! . . . Ah pick de oder boat up in de morning, und dey tells me. You come af mit me, Cabtin. . . . Goot, no?"

* * *

"Ninety-six days, Schenke, and here we are at the mouth of the Channel!" Old Burke had a note of regret in the saying. "Ninety-six days! Sure, this ship o' yours can sail. With a bit o' luck, now, ye'll be in Falmouth under the hundred."

"So. If de vind holds goot. Oh, de *Hedwig Rickmers* is a goot sheep, no? But if Ah dond't get de crew of de poor lettle *Hilda* to work mein sheep, Ah dond't t'ink ve comes home so quick as hundert days, no?'"

"God bless us, man. Shure, it's the least they cud do, now. An' you kaaping' us in food an' drink an' clothes, bedad—all the time."

"Vat Ah do, Cabtin. Ah leaf you starfe, no?"

"Oh. Some men would have put into the Falklands and landed——"

"Und spoil a goot bassage, eh? Ach nein. More better to go on. You know dese men Ah get in 'Frisco is no goot. Dem 'hoodlums,' they dond't know de sailorman vork. But your beoble is all recht, eh! Gott! If Ah dond't haf dem here, it is small sail ve can carry on de sheep."

"Och, now, ye just say that, Schenke, ye just say that! But it's glad I am if we're any use t' ye."

"Hundert days to Falmouth, eh?" Schenke grinned as he said it. "Vat 'bout dot bett now, Cabtin?"

"Oh that," said Burke queerly. "You win, of course. I'm not quite broke yet, Captain Schenke. I'll pay the twenty dollars all right."

"No, no. De bett is not von. No? De bett vass—'who is de first on shore come,' *Heim*? Goot. Ven de sheep comes to Falmouth ve goes on shore, you und me, together. Like dis, eh?" He seized Burke by the arm and made a motion that they two should thus step out together.

Burke, shamefacedly, said: "Aye, aye, b'ye."

"Ah dond't care about de bett," continued the big German. "De bett is noting, but, look here, Cabtin—Ah tell you Ah look to vin dot Merchants' Cup. *Gott!* Ah vass *verrickt* ven your boys come in first. Ach so!

Und now de Cup iss at de bottom of de Pacific." He sighed regretfully. "*Gott*! I van't t' be de first Sherman to vin dot Cup too!"

The mate of the *Rickmers* came on the poop and said something to his captain. Schenke turned to the old man in some wonderment. . . . "Vat dis is, eh? My mate tell me dot your boys is want to speak mit me. Vat it is, Cabtin? No troubles I hope?"

Burke looked as surprised as the other. "Send them up, Heinrich," he said. We, the crew of the *Hilda*'s gig, filed on to the poop, looking as hot and uncomfortable as proper sailorfolk should do when they come on a deputation. Jones headed us, and he carried a parcel under his arm.

"Captain Schenke," he said. "We are all here—the crew of the *Hilda*'s gig, that you picked up when—when—we were in a bad way. All here but poor Gregson."

The big lad's voice broke as he spoke of his lost watchmate. "An, if he was here he would want t' thank ye too for the way you've done by us. I can't say any more, Captain Schenke—but we want you to take a small present from us—the crew of the *Hilda*'s gig." He held out the parcel.

Only half understanding the lad's broken words, Schenke took the parcel and opened it. "*Ach Gott Lieber Gott*," he said, and turned to show the gift to old Burke. Tears stood in the big "squarehead's" eyes; stood, and rolled unchecked down his fat cheeks. Tears of pleasure! Tears of pity! Stretched between his hands was a weather-beaten flag, its white emblem stained and begrimed by sea-water!

A tattered square of blue silk—the flag of the Merchants' Cup!

Eighteen

Seventy Days in an Open Boat

Guy Pearce Jones

On the night of Wednesday, August 21, 1940, they had luncheon sausage and meatballs for tea on the *Anglo-Saxon*. The ship was making way steadily in a southwesterly direction and had left the Azores five hundred miles behind.

Widdicombe thought very little of the menu. It would be his wheel at eight o'clock, Paddy's lookout and Tapscott's standby. So he passed up his tea and slept instead.

Shortly before going forward, he rolled a half dozen cigarettes to take with him to the wheelhouse. It was against orders, but the *Anglo* steered like a yacht in heavy weather and like an automaton in calm. Two long hours of virtually no movement, with the binnacle paralyzing the optic nerves like a hypnotist's mirror, made it very hard for a man to keep his eyes open. Cigarettes helped.

Four explosions, so close together that they seemed one, shook the ship from stern to stem. The men stared at one another, the dreaded question in their eyes: mine or torpedo?

As he reached deck, a blinding glare struck Tapscott like a blow in the eyes. It lighted up the fireman's back against the curtain of blackness in front of them. The fireman plunged forward. Tapscott had only one idea: he must not lose sight of that back. He sprang after the fireman, struggled for balance, and the whole world seemed to blow up behind him in one terrific roar and shocking blast. He felt himself fly through the air and crash into something hard; then he knew nothing.

Tapscott came to on the starboard side of the deck twenty feet from the companionway. He was flat on his face, his nose jammed against the deckhouse bulkhead. He tried to move, but his muscles refused. He felt no pain—nothing at all. He wondered if he were dead. He lay there for what seemed a long time. Actually, it was only a few seconds.

It was 8:20 when the first salvo from the raider plowed into the *Anglo-Saxon*'s poop, demolishing the gun and killing everyone in the starboard fo'castle. Widdicombe knew the time to the tick, for he had just glanced at the wheelhouse clock and noted with satisfaction that twenty minutes of his trick was behind him.

A tearing crash and explosions shook the ship. Widdicombe ran out of the wheelhouse to the port end of the bridge. He could see nothing—only blackness and the water alongside. He ran back across the bridge to the starboard end and peered over the weather cloth. About a quarter of a mile away a dark shape was racing obliquely toward them, gun flashes stabbing from her as she came.

A raider!

Widdicombe tore back to the wheel and put it hard aport.

The hail of lead and steel that was pouring into the ship aft moved forward. It cut through the *Anglo*'s upper works with machine-like precision. It dropped to deck level and raked her fore and aft.

A breastwork of concrete building blocks protected the wheelhouse. Bullets were hammering into it. Shell fragments and shrapnel tore the ship all along the starboard side. Incendiary bullets crisscrossed into her in burning lines.

The Third Mate, whom Widdicombe had been unable to find, ran in from the bridge.

"Put her hard aport," he yelled.

"She's hard aport now," Widdicombe yelled back.

The raider was within a hundred yards now. Widdicombe could see her clearly. She was firing with everything she had. A red glow lighted up the *Anglo*'s poop. The starboard lifeboat was burning; the jolly boat on that side was smashed.

Widdicombe looked directly below. A body was slumped against the bulwark, just outside the Captain's quarters. It was Captain Flynn's. A machine-gun burst had caught him full in the chest as he was dumping the ship's papers overboard.

The storm of fire was moving forward again. Widdicombe ran back to the wheel. The First Mate was coming up the port ladder two steps at a time, followed by the Chief, Sparks.

"Antennas are all shot away and sets smashed, sir," Chief Sparks said. "No hope of an S.O.S. now."

"Right," said the Mate. "Hold her where she is," he ordered Widdicombe. "I'll be back in a minute."

The Mate returned to the bridge. "The Captain's gone," he said. "Bear a hand. I'm going to get the port boat away."

Widdicombe left the wheel and followed the Mate. They ran to the port jolly boat, the Mate fumbling for his knife as he ran. He got the knife opened and sawed at the ropes of the gripes. It seemed incredible to Widdicombe that they could be so tough and resistant.

Widdicombe stood by the after-fall. As the Mate's knife bit into the last strands of the gripe the boat went down with a run. The after-fall fouled round Widdicombe's body, pulling his hand into the block and jamming it there. The whipping rope stripped his trousers from his hips and seared his arm as it went. The boat hung by the stern.

Working frantically and in great pain from his jammed hand, Widdicombe managed to clear the block. The Mate leveled the boat and they lowered away. As it passed the deck below, two men leaped into it.

In a boat drill it was Tapscott's routine job to fend off. Automatically, after falling into the dropping jolly boat, he found the boat hook and held it off the *Anglo*'s side. The boat settled smoothly into the sea, rising and falling with the brisk swell. It was also his job to unhook the forward fall. He did this and then realized that there was no one to free the after-fall. He clambered over the hapless Penny, cast off, and went forward again to stand by the boat rope.

The Chief Mate was first down the lifeline. He slid so rapidly that the skin of his fingers and palms was badly burned. Widdicombe was next. As he swung off the ship, the Second, Sparks, came running up. He had on a hat and carried a sweater and an attaché case. "Wait for me," he begged.

A moment later Sparks came down the lifeline, just in time. Tapscott took only one turn of the boat rope—which is fastened fore and aft from ship to boat. Having reckoned without the fact that the ship was going full speed ahead, the rope ripped through his hands, burning and lacerating them when he tried to hold fast. The jolly boat went rapidly astern.

It was Widdicombe's job, too, to fend off. He and Tapscott attacked it with the energy of fear. The *Anglo*'s propeller, churning powerfully, was drawing them into the stern. The easterly swell made it difficult to keep away from the ship. One touch of the whirling blades and all would be over.

As they swept by amidships, the men in the jolly boat saw the smashed lifeboat over them and faces peering down. Two of the men overhead dropped into the boat. Tapscott and Widdicombe, frantically plying boat hook and oar, did not even see them. They gave two last mighty shoves against the ship's side and the boat swept around the stern, clearing the propeller by inches.

The *Anglo*'s poop was ablaze now and the raider was closing in to finish her. The men in the jolly boat crouched like hunted animals and scarcely breathed. The swell was carrying them right across the raider's bows. They waited for a burst from her guns.

It seemed impossible that they would not be discovered. But the raider was so intent on the kill, she hung on the Anglo's blazing trail. Hunter and hunted rushed ahead. The jolly boat drifted away on the swell. The boat was taking water badly but the men in her dared not move. When they were a good half mile astern they put out their oars, headed the boat into the sea and pulled.

They settled into their swing and made the boat leap. They were congratulating themselves on making good progress in their escape when, dead ahead, the moon rose, flooding the sea with brightness and throwing them into sharp relief. The men froze on their oars and swore.

Water had been coming over the side. After heading the boat to the swell there should not have been as much of it as there was. It was now slopping about their calves.

"That drainage plug must be loose," Tapscott said. No one paid any attention.

Widdicombe and Tapscott seized bailers and attacked the rising water. Pilcher and Morgan relieved them at the oars. But bail as fast as they could, the water did not lower as it should. The Mate plunged his arm into it and groped along the bottom of the boat. He found the drainage plug. Tapscott had been right; the plug was half out of the hole and water coming in around it. He drove it in with his axe and they were able to get the water down to the level of their ankles.

Between the boat and the *Anglo*, lights suddenly appeared bobbing up and down on the crests of the waves. "The life rafts!" exclaimed the men in the boat. The Mate tried to signal them with his torch; but, fearing the raider would see them, he stopped. They put the boat about and rowed toward the rafts. As they neared them, the raider swung its guns, and streams of incendiary bullets poured into them. On the rise of the next swell, the men in the boat saw, the bobbing lights were gone. The life rafts and the men clinging to them had been obliterated.

At nine o'clock by the Mate's wristwatch an explosion rent the Anglo's poop, followed, five minutes later, by another. "The magazines," said the men in the boat. Their ship's bow was rising now. When almost perpendicular, it snapped back with a jerk and she went down by the stern. There was a great hissing as the water reached her fires, and a cloud of steam shot up from her. It took less than a minute. The waves closed over her. Only a pall of black steam overhead remained to mark where she had been. The men in the boat watched silently. Even Morgan, the garrulous, had nothing to say. It's a terrible thing, Tapscott thought, to see your ship go down.

The raider, her work of destruction over, headed off into the east.

* * *

Tapscott and Widdicombe rigged the drogue and paid out the line. It brought the bows of the jolly boat into the sea but was not large enough to enable them to dispense with the oars altogether.

"Right," said the Mate, when he saw how matters stood. "We'll have to stand watches. Tapscott and Widdicombe will take the first. Keep her head into it or we're liable to swamp."

Changing places, Pilcher groaned.

"What's the matter?" the Mate asked.

"My foot. I think it's blown off."

"It can't be as bad as that. How could you have rowed? I'll take a look at it."

The Mate flashed his torch. The bottom of the boat was awash with bloody water and was full of the smell of blood. Tapscott's seat was smeared with it, from the shrapnel wounds in his back.

"You've been hit, right enough," said the Mate, straightening up, "but you still have a foot. I'll dress it for you in the morning. . . . Everybody try to get some sleep now."

No one could sleep; they were too cold, wet and miserable. The swell pounded the boat and the spray rained down in clouds. At midnight the Mate and Hawkes relieved the two A.B.'s. Tapscott and Widdicombe lay down where Hawkes and Denny had been, but could sleep no better than the rest. In the end they all gave it up and talked in lowered voices.

The sun rose at six o'clock on seven cold, sad, wet and thoroughly miserable men. As the long, level rays lighted up the surface of the sea and a golden light flooded the sky, with one accord all who could stood up and scanned the sea. In all that tossing circular plane they could see nothing, nothing save empty miles of water, nothing on the surface of the ocean and nothing in the sky. They were completely and singularly alone.

"Well," said the Mate, briskly, when he had satisfied himself that there was nothing human to be seen in the whole circle of the sea, "the sooner we start, the sooner we'll be there."

"Be where?" the men asked.

"The Leeward Islands," the Mate said. "There's no use going east. Wind and current are against us. Finding the Azores would be like looking for a needle in a haystack."

"Yes," said Widdicombe, "but supposing we miss the Leeward Islands?"

"Even if we sail right through them," the Mate said, "there is the Caribbean and the whole coast of North and South America."

Pilcher's wounded leg had stiffened so that it stood out straight before him. In daylight they saw that he was much more badly wounded than he had led them to believe. Shrapnel or a dum-dum bullet had torn through the length of his left foot, reducing it to a pulp of tissue and shattered bone. The men marveled that he could have rowed as he did in the first hours of their flight from the ship. "I didn't feel it," he said deprecatingly. They moved him forward into the bow-sheets, fashioning a sling from his scarf to keep his stiff leg from moving with the motion of the boat.

The gunner, Penny, had a badly torn hip, where shrapnel had caught him as he ran across the *Anglo-Saxon*'s deck. The bullet that had gone through his right forearm as he lay with Tapscott in the shelter of the bridge had left a hole, comparatively clean. Both wounds being on his right side, they arranged the gunner forward on his left, bolstering him as best they could against the roll.

Morgan, the second cook, had a jagged tear just above his right ankle, where shrapnel or flying metal of some variety had caught him as he scrambled to the boat. His whole foot was bruised and swollen. He had a badly contused hand as well.

Tapscott had the Mate check him up for wounds. One of his front teeth, which had been broken off, exposing the nerve, when he was hurled through the air by the explosion of a shell, was now paining him considerably, but blood had stopped seeping down his back. The Mate found that a small piece of shrapnel had buried itself in Tapscott's left buttock, another in the muscles of his back and another in the palm of his left hand. A long splinter, evidently spent, had gone through his shorts into his right groin. It had burned him painfully but had not penetrated his flesh.

The wounded men disposed of, the able men got in the sea anchor, stepped the stubby mast they found in the boat and set sail. They headed the boat due west to make west–southwest, a note on the compass advising them that it was out several points. A good breeze was blowing from the east. They were on their way to shelter and safety, they felt, doing all of four knots.

They took stock of the boat and supplies. The boat, eighteen feet long and six wide, was stoutly built and in sound condition. Seats ran fore and aft on either side inboard, converging at the bow. These were rounded off at the stern to join a roomy seat across it, underneath which was a locker. Two stout thwarts completed the seating arrangements.

The boat was propelled by a dipping lug, a powerful sail in the hands of experts. It was large enough to carry the boat along at decent speed but not so large as to be dangerous in squally weather. It is the favorite rig of fishermen and trawlers on the English coast. It has the disadvantage, however, of having to be lowered every time the boat is tacked, no maneuver for amateurs. This is the only way the spar from which it is suspended can be passed to the other side of the mast. Hence the name.

For equipment there was a sea anchor, a boat hook, six pairs of oars, a steering oar, an axe, the Mate's knife, two bailers, a rope painter, the boat's canvas cover, badly worn and torn, and a few ends of rope. In the locker under the stern seat was a boat's compass, a colza oil lamp with a bucket large enough to hold it—this for signaling in Morse—a dozen red flares in watertight canisters, a dozen matches in a watertight container and a medical kit.

Further to equip it as a lifeboat, three airtight tanks had been built into it—one in the bow, and two on either side in the middle—and a lifeline within easy reach from the surface of the sea fixed along both of the boat's sides.

Food supplies consisted of three tins of boiled mutton, weighing six pounds each; eleven tins of condensed milk and thirty-two pounds of ship's biscuit in an airtight metal tank fixed forward under the first thwart. Behind this was the water breaker, a keg in a cradle, with a wide bung in the top, and two long-handled, narrow-mouthed dippers for scooping up its contents.

Tapscott had, in addition to his underwear, khaki shorts, a singlet, heavy knitted Air Force socks and heavy shoes. Widdicombe had underwear, a pair of tattered trousers and a cotton shirt. Penny wore the khaki shirt and trousers of the Marine undress uniform. Morgan had denim cook's trousers and a singlet.

The sun was now high enough to shine down hotly. The men took off their soaked clothing and spread it dry.

Routine having been established, the Mate opened the medical kit. It consisted of a bottle of iodine, two rolls of bandage, a packet of medical lint and a pair of scissors. "Just the thing for taking to a picnic in the New Forest!" he said, on seeing the size of it. There were several rubber finger cots in the kit. He put these on his fingers that had been burned sliding down the lifeline from the ship; they were now raw and painful.

Sparks was treated first. With the aid of the Third Mate and Tapscott, the Mate bathed Sparks' mangled foot in seawater and cleaned it as best he could. On close examination it seemed crushed beyond any hope of saving; the shattered bone protruded through a bloody pulp.

"Man," said the Mate, shaking his head, "how could you have pulled a boat with a foot like that?"

"It wasn't anything," Sparks said. "I tell you I didn't feel it."

"If we were in port I am afraid the doctor would have it off," the Mate said.

Sparks was silent. Fever was already coursing through his depleted veins; he had lost a great deal of blood. "I know we must be careful with the water," he said, "but do you think that I might have a drop? Just enough to wet my throat."

They gave him a drink. It was then that the Mate discovered that the water breaker had lost some of its contents; it was little over half full—about four gallons.

After sterilizing and binding up the wounds of the injured men and settling them as comfortably as they could, the able men bailed the boat, made ready the lifeboat lamp and set up as orderly a routine as possible in quarters so cramped.

The Mate returned to his log and wrote:

Thursday, August 22nd, 1940. Wind N.E. 3, slight sea, slightly confused easterly swell. Course by compass W. All's well. Medical treatment given.

The men had their first rations that evening at six: a ship's biscuit apiece. No one asked for water. As night fell the wounded and the men off watch made themselves as comfortable as they could under the tattered boat cover. It kept off some of the spray.

When they came to light the small colza oil binnacle lamp—the compass was housed in a hoodlike affair of metal with the lamp fixed inside—they could not get it to ignite. They tried to light the large lamp with no better success.

"Too bad," said the Mate. "We'll have to steer by the stars."

They changed watches regularly during the night. The wind held and the boat ran before it, taking water now and then, which was promptly bailed out by the man on standby.

Everyone got some sleep. Those who could not sleep lay quietly. Now and again, one of the wounded men groaned in spite of himself. Underneath the canvas, someone snored.

Tapscott shifted about on his hard plank bed, trying to settle his hip bone so as to give him some ease. The boat was sailing handily. Seas struck it sharply and ran hissing by. Occasionally it rose on the crest of a wave and dropped with a booming shock that vibrated its timbers and sent a cloud of spray raining down.

The day passed uneventfully. Everyone's spirits were good, but conversation lagged. They had worn out their slim stock of conundrums. The sun was warm but not in the least oppressive. The wounded and the men off watch slept a good deal. It was not unlike their routine aboard ship, where sleep occupied most of their time off duty.

Physical functions practically ceased; their bodies had no waste to eliminate.

At six in the evening they had their biscuit and half dipper of water, changed the watch and settled down for the night. They knew by this time every bolt head, rib, knot and protuberance of the boat's planking. They had learned, too, how to dodge spray so as to get the minimum wetting. Under the tattered boat cover they were out of the wind and most of the wet; the night was appreciably warmer; they enjoyed good hours of sleep.

Tapscott and Widdicombe had finished their watch at ten o'clock and had turned in when they were awakened by a stir within the boat. It was eleven o'clock by Widdicombe's wristwatch, which was still functioning.

"What is it?" they asked, sleepily.

"Ship," said the Mate, fumbling hastily in the locker under the stern seat.

"Where?" they demanded, wide awake now and straining their eyes into the surrounding darkness.

The whole boat awoke and sat up eagerly trying to see.

"Aft," said Hawkes, who had the standby.

The Mate was unscrewing the top of the container that held the flares. As their eyes adjusted themselves to the light they saw a ship in outline heading north–northeast. She was blacked out; but they could see her clearly.

The Mate extracted a flare, struck it on the milled top of the container and held it, spluttering and hissing, over his head. The boat and the sea for yards around were lighted up with a dazzling red glare. They sat silent while the flare burnt itself out. When the last bead of the chemical composing it had melted, flamed, charred and gone out, the night closed in, blacker than before. The Mate flung the smoking stick into the sea. They brought the boat into the wind and waited.

When they could again see, the ship had turned and was coming back to them in a wide circle, as if afraid of a trap. They started to cheer. The Mate silenced them curtly. As she swung around they could see what he feared. Her build was German, or, at least, very like a German.

"I don't like the looks of her," said the Mate.

The stranger was now almost opposite them.

"Lower the sail," the Mate ordered. They dropped the dipping lug and lay to.

"If she's British she will put a searchlight on us," Widdicombe said.

"Stow it," the Mate said, "and lie low."

They crouched down in the boat and waited while the stranger hung off to the starboard.

"A Jerry, or I'm a Chinaman," said the Mate. "Look at the way she acts."

They waited there, their hearts stepping up to a higher beat. After some minutes of cruising at slow speed the stranger turned, gathered speed and made off to the north–northeast.

The next morning, after dealing out the early ration, the Mate cut another notch of his toll of days in the gunwale and wrote up his log for the previous one.

August 23rd. Friday. Wind E.N.E. 3, slight sea, slightly confused easterly swell, partly cloudy. Half a dipper of water per man 6 a.m. Also half a biscuit with a little condensed milk. Sighted a vessel showing no lights at 11 p.m. showed sea flare, she cruised around but was of the opinion she was raider as she was heading N.N.E. We were about a hundred miles from our original position. Kept quiet let her go off.

Saturday passed without incident. While everyone was cheerful enough, there were long periods of silence. The wind had hauled around to the northwest; the sky was cloudy. The Mate peered anxiously into the bung of the water breaker and noted the lowering level.

Sunday dawned cloudy with a slight swell on the sea. The wind had dropped during the night until the boat had lost all way. A light wind sprang up from the north–northwest. The watch trimmed sail hopefully and the boat moved forward once more. Everyone was cheerful over his half biscuit and pittance of water. Their bodies were so dehydrated it was impossible to swallow the hard biscuit without wetting it first. The Mate's eyes narrowed every time he looked inside the breaker. He scanned the sky anxiously for rain clouds, but nowhere in the idly moving mass was one remotely resembling the cloud he sought.

Pilcher and Morgan were suffering increasing pain in their wounds. Their lacerated feet had swollen during the night and were swelling more. It was necessary to loosen their bandages. When this had been done, an odor permeated the boat, an odor no man ever forgets once he has smelled it—the horrible stench of gangrene. The Mate examined the wounds intently, but said nothing. He returned to the stern sheets with another weight on his already heavily burdened sense of responsibility.

The men dozed as they drifted, or, sitting up, scanned the empty reaches of the sea. In all that space there was nothing but sky and water to be seen; not a bird, not a patch of seaweed, not even a drifting log.

The sun was sinking in red glory, betokening fair weather to come. The men made ready for the night. The Mate wrote up his log for the day. Had it not been for the disquieting smell of gangrene, which, once in the boat, infected the air, they might have been on a pleasure cruise, so well were they feeling after their hearty meal.

In the morning, Monday, August 26, their fifth day in the open boat, a bos'n bird appeared to the men of the *Anglo-Saxon*. It planed over them in a leisurely way, wheeled, came back and flew low over them again.

"Mean-looking blighter, he was," Widdicombe said. "Had that look in his eye: 'Just wait. I'll get you yet.'" He stared back at the bird malevolently. "Not me you won't get," he said.

The wind, which had been falling, now dropped away to an occasional fitful gust and the boat drifted or lay becalmed under the burning sun. To men without hats the direct rays were torture. Those who took shelter under the canvas boat cover found themselves in an oven.

They were very thirsty now. Their pores, denied any liquid to evaporate, closed up; their skin scorched and crisped; salivation ceased. The morning half dipper of water, gulped with such eagerness, was like a drop on a blotter.

They prayed for squalls and a decent wind.

"We should sight a vessel any time now," the Mate said; and they accepted this as inspired knowledge.

After breakfast, the Mate and the Third dressed the injured men's wounds. Pilcher's mangled foot had swollen to twice its natural size. Without the bandage it was a green and black gangrenous horror. The boat was almost untenable. They gave him the best treatment they could devise with the scant knowledge and means at their disposal. They got his leg over the gunwale. Tapscott bailed salt water over the wound for an hour. The Mate then bound it up with the last linen bandage. Pilcher never flinched, although in exquisite pain. He apologized for giving them so much trouble. "He has guts that one," Widdicombe said to Tapscott. "A proper man if I ever saw one."

Morgan's smashed ankle was swelling badly, too. They gave him the same treatment.

The bullet hole in the gunlayer's right forearm was still clean. They bathed it with a little of their precious fresh water, saturated it with iodine and bound it up again. The wound in his thigh, however, was not doing at all well. And they were at the end of their medical supplies.

As the sun reached meridian they writhed and twisted on the hot planking of the boat like fish on a griddle. Their mouths and throats were so parched that talking was painful. The able men bailed seawater over the wounded, and when they had them thoroughly doused, went over the side themselves, being careful to keep their faces out of water. Naked, they noted with surprise how much weight they had lost.

"At any rate," the Third said to Tapscott, "you're much better-looking without that fat belly."

They made good progress during the day. Whenever they spoke, which was seldom, they heard the changed notes of their voices distorted by thirst. Cracked and swollen lips, tightened skin and salient cheek bones gave them new expressions. They noticed that they were not so sure of their footing as before.

During the night the boat did a steady four knots on the port tack heading southwest true. The sun, the Mate noted, had set at 6:42.

The trade wind held all the next day. In the morning a bos'n bird and a common gull flew over them. "There's that bloody bird again," Widdicombe said, shaking his fist at it. "What's he want hanging about like this?" The men laughed. But Widdicombe was serious. "He had mean eyes," he said. "It made me fair narked."

The sun beat down tropically. The Mate, the Third, Tapscott and Widdicombe dipped their bodies overside again, holding on to the lifelines that festooned the side of the boat. Their bodies took up the water through their pores, leaving their skin white with salt. Saliva returned to their mouths. They all felt better.

The reek of gangrene was terrible now. Penny's torn hip was going the way of Pilcher's foot. No matter how brisk the breeze, it seemed impossible to get that devastating stench out of their nostrils. Sparks was apologetic. It was the only sign of suffering he allowed himself. Morgan's wound was gangrening, too. Everyone assured Sparks that it was nothing that could be helped, but on awakening to it, after the unconsciousness of sleep, their stomachs turned over.

In the morning Sparks' foot "went dead." He no longer felt pain in it. He no longer felt anything there. It was surcease of a sort but bought at the expense of his whole body. He was failing and they caught in his eyes the expression of a man who is looking beyond life.

The wind was strong and the sea boisterous. They bowled along handily, making fine time and shipping buckets of water. No one cared about the water. They were soaking wet, yet all they talked of were liquids. They talked of peaches, pears, oranges, the juice of pineapples. In the end, they talked of beer: light beer, dark beer, beer in glasses, mugs and tankards; English, Danish, French and German beer; beer foaming from inexhaustible taps.

They made ready for a wet night cheerfully. This, they told themselves, in voices croaking with thirst, was the last lap.

In the morning the wind dropped and the swell lessened. After the first bite no one cared to go on with his quarter of a ship's biscuit. It was impossible to swallow. The half dipper of water gave them a few moments' release from the grip at their throats; then it was gone, a splatter of rain in the desert of their desiccated tissues.

Pilcher lay very quietly now under the shelter of the boat cover. As the sun climbed the sky and the temperature mounted, the men bailed water over one another. Morgan, the cook, seemed better; he could move his stiffened leg. Penny seemed better, too; so much so, he took a short trick at the tiller.

Everyone remarked that they had stopped sweating, even at noon. The dryness of their throats crawled up their tongues, which, thick and discolored, filled their mouths like a gag of hot felt. They grinned at each other now and then as if sharing a painful, sardonic joke.

In the evening, the wind fell away completely. The boat pitched, tossed and drifted on the swell. It was hard to rest, much less to sleep, thrown about on a narrow seat. They were awake until late at night.

Just as, fatigued to the point of sleep, some of them achieved an uneasy doze, Pilcher's voice rose from the bow in a long and plangent wail. An inhuman quality in it, a severance from all direction of the intellect, brought them awake with a start.

"What is it, Sparks?" the Mate asked.

There was no answer.

The Mate went forward to him. He was lying face upward, staring unseeingly into the sky.

"What's the matter?" the Mate repeated.

The glazed eyes showed no flicker of recognition. A voice, a detached caricature of Pilcher's, apostrophized an unseen, abstract personage in bitter, obscene invective. The voice, matter and manner were utterly unlike the normal, gentle Sparks! Then he laughed—hysterical bursts of unmotivated laughter, which stopped as abruptly and unreasonably as they had started.

"Off his head," the Mate said, "and that's a fact."

Sparks was singing now, a street ballad of the most scabrous nature, a song they would not have thought him capable of. He sang it to the end, then started again. He sang it over and over.

"For God's sake, stop him!" Morgan, who lay next him, begged the Mate.

The Mate tried again to quiet the delirious operator, but to no use. Sparks, insulated by hysteria, gave no sign of hearing anything. All they could do was wait uneasily for the seizure to wear itself out.

After a bit Sparks subsided into low moans and a running mutter of talk. The men settled themselves again. They could not sleep, no matter how they twisted and turned; expectancy of what they did not want to hear kept them tense. It came very soon, heralded by that long-drawn, rising, animal cry—singing, invective, bursts of maniacal laughter.

Toward the morning Sparks drifted off into sleep or semi-coma.

At sunup the next morning the Mate called the able men to him.

"We've got to do something about Sparks' foot," he said.

All of the previous day Sparks had alternated between delirium and comatose sleep. He had been off his head most of the night and was now exhausted but lucid.

"What can we do?" they asked.

"Amputate it."

"With what?" Tapscott asked.

"The axe. If we don't, he's going to die."

Tapscott was shocked. The axe was rusty, dirty and dull. They had no antiseptic. He was certain, too—but he did not mention this to the others—that Sparks was going to die anyway. Why torture him with a clumsy operation that would in all likelihood end in him bleeding to death?

The men looked uncertainly at one another.

"I know it's taking a long chance," the Mate said, "but that foot is poisoning him. It will have to come off."

Hawkes and Widdicombe agreed that the Mate was right, but Tapscott still demurred.

"Very well," said the Mate, "we'll put it up to Sparks himself."

They went forward to Pilcher, who lay in the bow-sheets, weak but uncomplaining.

"Sparks, old chap," the Mate said, "your foot is in bad shape."

"I know it," Sparks said, feebly.

"We think it will have to come off. The sooner the better. Do you want us to do it?"

"Yes. Please. Anything," Pilcher said, his face contorted with pain.

The Mate fetched the axe. They removed the boat cover from Pilcher. Widdicombe and Hawkes stood by to hold him. But when it came to the actual business of lopping off Sparks' foot with the axe, even the resolute Mate quailed. "I can't do it," he said. "He'll just have to take the chance."

Everyone breathed a sigh of relief.

The Mate rearranged the boat cover over Pilcher so as to keep the sun off his face. "Carry on, old boy," he said. "We're not going to do it now. We're certain to be picked up soon and a proper doctor will make it right for you."

Pilcher smiled weakly and closed his eyes.

Saturday night was grim and miserable. They were so thirsty they could not sleep. In the morning the Mate wrote in his log:

During Saturday night crew felt very thirsty; boiled mutton could not be digested and some felt sick. Doubled water ration that night.

This was a triumph of understatement. When the Mate used the word "sick" he did not mean ill, as Americans do, but, literally, nauseated. Nor did he say that of the "some" who were sick he was the sickest. Nausea and cramps seized him early in the night, causing him to retch agonizedly. Had the others not been so tormented themselves, they would have been more concerned.

When the morning dawned, they saw to their consternation that he, the symbol of discipline and fountainhead of knowledge, had suffered some sort of internal collapse. His face was livid and lined with pain. All of his strength seemed to have gone from him. His abraded fingers were suppurating badly, and his flesh, even where burned by the sun, had a lifeless, claylike appearance.

Pilcher, too, was very low. He was so weak he could hardly speak. He was lucid, though. When they took him his morning ration of water he turned his head from the dipper and told them to give it to someone who needed it more than he.

They went about the business of the boat in a vague and silent fashion. The sun was already hot enough to be uncomfortable. It reflected back from the sea, giving them a taste of the burning misery to come.

At eight o'clock, Morgan, from his place next to Pilcher in the bow-sheets said, suddenly: "I say, I think Sparks has gone."

The Mate, the Third, Tapscott and Widdicombe went forward. Morgan had spoken the truth. Sparks had, indeed, gone, as silently and unobtrusively as he had done everything in life. They looked at one another incredulously. So soon! It couldn't be possible. But when they looked at Sparks' fallen jaw and saw the subtle changes that had already taken place in the contours of his face, they knew that it could not be otherwise. They stood impotently by, overwhelmed by the awful finality of death.

In curt, low orders the Mate arranged all that was left to do. Tapscott and the Third lifted the body over the gunwale and lowered it gently into the sea. They had nothing to wrap it in and nothing with which to weigh it.

The wind was south–southwest; the boat was making several knots. The body drifted away on the swell, off into the immensity of the Atlantic. They watched it until they could see it no more.

Very little was said in the boat the rest of that day.

* * *

When the sun came up the next morning, their twelfth in the open boat, there was no expectant stir, no sense of a new day. Pilcher's death was heavy upon them and the Mate's condition gave them new cause for alarm. He could scarcely crawl to the breaker to issue the morning half dipper of water. His face was ghastly, his will and vitality almost gone.

Several of them refused the ration of dry ship's biscuit. Those who took it gave up trying to eat; they were unable to chew.

Penny, the gunner, lay quiet in the bow. He was visibly much weaker. Morgan, who had lain beside him, almost as quietly, since the ordeal by thirst began, now chattered and sang.

Widdicombe, too, seemed bordering on hysteria, alternately violently optimistic, confident that they would come through all right in the end, or sunk in apathetic pessimism. He suffered more than the others from the heat. He attributed this to a sunstroke he had experienced on a voyage through the Arabian Sea. Frequently he would be overcome while taking his turn at the tiller and would have to get Tapscott to relieve him until the faintness had passed.

Tapscott, engrossed in his own misery, brooded over this. He doubted Widdicombe's peculiar susceptibility as he doubted many of Widdicombe's stories. He had never liked Widdicombe and he felt that his watchmate was exploiting him now, trading on his good nature to get out of onerous duty. This growing resentment blazed forth in open quarrel that morning, which might have ended tragically had the Third not intervened. Discipline, obviously weakening, was restored for the time.

Discipline now was sustained only by a symbol, the Mate. So long as he lived and retained his reason he held command. But as the sun rose higher and the heat increased, violent nausea and spasms of pain racked him. He lay in the bottom of the boat, retching horribly. After one of these spells he would lie exhausted, in a state of semi-coma. His arms were a mass of blisters and scales. In spite of all this, he took a turn at the tiller whenever he could.

To Morgan, also, they gave a trick at the rudder. It seemed to allay his growing dementia, and weak as they now were, everyone had to help. But Morgan gave them more trouble than aid. He seemed incapable of grasping the first principles of sailing a boat. He yawed and blundered, shipping water and nearly capsizing them several times.

They made little progress that day. No one had the energy to dip himself overside. They simply lay and suffered, protecting themselves from the worst of the sun with the boat cover.

That evening the Mate made the last entry he was able to write in the log. He reported their condition with his habitual understatement. As ill and despairing as he was, he remained the officer. He included a suggestion for bettering lifeboat equipment based upon his all-too-actual experience. In a hand that could just trace the letters he wrote:

Sept. 2nd. Monday. 6:15 a.m. issued half a dipper of water per man and same in evening with a little condensed milk diluted with it. Crew now feeling rather low, unable to masticate hard biscuit owing to low ration of water.

Suggestion for lifeboat stocks. At the very least two breakers of water for each boat, tins of fruit such as peaches, apricots, pears, fruit juices and lime juices, baked beans, etc. Our stores consisted of—

One tank filled with dry biscuit.

11 tins condensed milk

3 tins each 6 pounds boiled mutton.

One breaker of water, half filled.

These were the last words of any kind the Mate ever wrote. He was ill all during the night and so weak in the morning he could not get to his feet. They made a bed for him in the thwarts, using boards and life belts. He was too ill now to command.

Whatever was in store for them, they must keep going west. It was their only hope. They trimmed sail and kept the boat on her course.

When the Third went to the water breaker Widdicombe proposed issuing a whole dipperful instead of the usual ration. Tapscott objected; he was as thirsty as anyone but he felt they should maintain the ration the Mate had decided upon up to the very end. And that would be all too soon, they knew; the water was very near the bottom of the breaker now.

The Third hesitated. The Mate was too ill to intervene. Penny and Morgan were nullities so far as making decisions was concerned; Penny was nearly as weak as the Mate and Morgan had to be handled like a feebleminded child.

"The water won't last much longer anyway," Widdicombe insisted, "so why not have a decent drink now?"

In the end Hawkes and Tapscott gave in. There was that in Widdicombe's voice and manner which caused them to fear that he would attempt to take the water by force. They did not want a fight to complete their distress. A fight was the last thing in the world they wanted at that moment.

That day they made better time than on the day before. They took their turns at the tiller, bailed the boat and sought shelter from the sun under the boat cover, trying not to think of the thirst that was consuming them.

They were dull with apprehension, but, whatever their thoughts were, they did not express them. Each man was wholly preoccupied with his chief concern, his own life. That is, all but Morgan, who kept up a babble of senseless talk and singing until one of them, driven into a passion of irritation, would silence him temporarily with a curse.

Tapscott thought of nothing in particular; he was sunk into a heat-drenched stupor, a sort of anesthesia. He was not despondent, though. Things were pretty bad, he realized, but they need not inevitably end in disaster. He was nineteen years old, and life, even in such circumstances as these, was essentially a hopeful affair.

Widdicombe, mercurial in his calmest moments, alternated between energetic optimism and angry despair. This he relieved at times by cursing, long and fervently, an abstract fate that could not be reached with his fists and feet.

The older men, one poisoned with gangrenous wounds, the other, ravaged by some internal disorder in addition to the thirst and starvation they were all suffering, said nothing. Neither could have had much hope, unless that of a miraculous landfall or rescue ship, but neither of them would admit it. Both must have known that rescue for them would be too late now.

The night was chill and wet. Their teeth chattered as they huddled together for warmth. The Mate kept them awake with the violence of his vomiting, the cook with hysterical ramblings and complaint. Toward morning the cook became quieter.

The sun rose on a calmer sea. The wind fell off and the boat made slow progress under a sky that promised no escape from another day of unbearable heat and thirst. They had the morning ration of water. It was plain that after the next drink there would be no more. They were at the very end of their tether.

The noon sun reduced them to such an agony of thirst they voted to drink the last of the water then and there. It was measured out, the three able men jealously watching that an exact division was made.

Tapscott sat on the starboard seat. Widdicombe and the Third were at the breaker. Morgan was lying in the bow-sheets and the Mate on his bed of boards amidships. Penny had just taken the tiller. He said nothing, but there was nothing unusual in that. Normally taciturn, Penny had practically quit speech in the last few days.

The boat yawed suddenly. Tapscott turned to see what the helmsman was up to. But there was no helmsman to be seen. Where Penny had sat there was an empty seat and the rudder handle swinging purposelessly. Tapscott sprang to his feet and searched the sea. Then he saw Penny. He was floating away rapidly, face downward, his arms outstretched before him. "Like doing the dead man's float," Tapscott said. He was making no attempt to swim. He had not gone over for that.

Even if they had been able to bring the boat about in time—which was impossible—it is doubtful anyone would have moved to go after the gunner. He was doing what he deemed best. Unwritten law gave a man in these circumstances, provided he was or appeared to be in his right mind, the inalienable right to choose his own way out of his suffering. The gunner had "gone over the side."

It was hard to take in, but, they had to admit, this very likely was the end. It confronted them suddenly with a horrible actuality. There was nothing more for them, saving a miracle, than prolonged torture, madness and inevitable death. They had known that it was a possibility but had kept it so buried in their minds, resolutely determined not to admit it, it seemed a new and unfair hardship, considering what they had already endured.

Better end it now and spare themselves further suffering. That would be the sensible course, they agreed. But a physical lassitude weighed down their bodies and paralyzed their wills. No one wanted the initiative in so final a move. In the end they did nothing but mechanically sail the boat, and there was no more talk of going over the side.

Night brought escape from the heat, at least. They sprawled or lay in misery, all power gone from their limbs. When they moved to relieve one another at the tiller the physical action lagged appreciably behind the act of will. It seemed too great an effort to get themselves from the bow to the stern of the boat.

The wind was slight and fitful. The boat drifted or sailed listlessly. At times it seemed to float stationary, its only motion a slight heaving with the swell.

The Mate was scarcely breathing. Morgan had quieted down. There was no sound just before dawn, save the sighs of the sleeping men and the murmur of the sea.

Morgan sat up suddenly, pushed away the canvas that was muffling him, stared into the void and said peremptorily: "I want my mother."

* * *

Day dawned to the five men of the *Anglo-Saxon*—their thirteenth in the open boat—as the morning of execution to a condemned man.

The sun came up from the sea; the breeze freshened; the boat picked up speed. But the unkempt, hollow-eyed, thirst-tortured creatures in it hardly stirred.

From habit they edged toward the water breaker. Then, remembering, they sat down again. They were heavy with apprehension. Each regarded the other warily, awaiting the gesture or word they all knew would come.

As if to put the seal of finality on disaster, the rudder, which lacked a lower pintle when they left the ship, was carried away by a heavy swell. The Mate was steering at the time. He watched it disappear as if it were his last hope for life, dropped the useless tiller, went to his bed of life belts and boards, and lay down with closed eyes.

Tapscott and Widdicombe got out the steering oar and shipped it in place of the lost rudder. The Third, who had been watching the Mate narrowly, stretched himself out beside him. The two lay this way for a long time, while the sun, mounting higher, burned into their eyelids and whipped up the demon inside, which was relentlessly squeezing the juice of life from them.

They sailed for long hours this way, hearing nothing, seeing nothing. Then the Mate opened his eyes, raised himself on his elbow and said from swollen and discolored lips, "I'm going over. Who's coming with me?"

"No, no! Don't! You can't! Don't leave me! You won't, will you?" Morgan cried from the bow-sheets.

"Shut up!" Widdicombe ordered him, menacing him with a gesture. Morgan was still.

Tapscott, Widdicombe and the Third stared at one another, each waiting for the other to speak.

"I'll go," the Third said, finally; "only you'll have to help me, you know."

The Mate nodded assent. He turned his eyes to Morgan.

"No, no, no!" Morgan cried. "I can't. I can't. I don't want to die."

The Mate turned next to Tapscott. Tapscott thought a minute before answering. "No," he said. He felt that he must explain. "I mean to say," he added, "it isn't like I was dying; and you can't tell; we may be picked up yet."

The Mate turned to Widdicombe. Widdicombe shook his head violently. "No fear," he said, showing the whites of his eyes like a nervous horse.

The Mate lay back with eyes closed.

"That settles that," said the Third. "I might as well have some sleep myself before we go." He lay down beside the Mate again.

The dread of what they must see overwhelmed the other three. Morgan lay back in the bow-sheets staring at the self-condemned men. It was the formal end; those in command were laying down their arms.

Maybe, Tapscott thought, they'll change their minds. He could not imagine anyone making such a dreadful decision and then going calmly to sleep.

Neither Tapscott nor Widdicombe would have dreamed of remonstrating or trying to interfere. Not only were the two officers senior to them and entitled to make decisions without question, but they had decided the most private question in life. Interference would have been presumptuousness of the most outrageous character. Moreover, they were too dazed and miserable themselves to formulate more than their own desire to hold on until the last. Tapscott did not even think that far; blind instinct had spoken for him, an instinct he had never thought to question.

There was something awesome in this vigil. They sailed for an hour or more in this fashion.

Somewhere near ten o'clock Hawkes sat up and said: "Ready?"

The Mate opened his eyes and then got to his feet with surprising speed and sureness, considering how weak he had been. He picked up the axe and placed it on a thwart near the port gunwale.

"Just a minute," Hawkes said, almost gaily. "I'm going to have something to eat and drink."

He dipped a can of water from the sea and gulped it down greedily. He filled the can again and drank that off. Then he softened a biscuit in seawater and ate it.

The Mate drew off his signet ring and handed it to Widdicombe. "Give it—my mother—if you get through," he gasped.

They shook hands all around.

It was then that Hawkes had a moment of bitterness. "To think," he said, "that I put in four years of training—to come to this."

The Mate took off his coat. Hawkes removed his, too.

"Give it to me," Widdicombe said.

"No," the Third said, "I can't. I promised it to Bob if anything happened to me."

"Then give me your trousers," Widdicombe said. "You won't need them."

"Can't do that either," the Third said, grinning through cracked and blistered lips; "I might meet a mermaid where I'm going."

Tapscott's eyes were blinded with unexpected tears when the Mate took his hand.

"Keep going west," the Mate said, scarcely able to force the words through his stiffened lips. His voice fell off into a throaty croak; he gestured loosely with his left hand and struggled to achieve what he wanted to say. "No more south."

The Mate and the Third then stood up in the thwarts near the port gunwale. They shook hands. Tapscott turned his head away and held his breath. He heard a crunching thud, a great splash, and, when he dared look again, the Mate and the Third were floating yards away, clasped, apparently, in each other's arms.

The Third's hair, the lightest of yellows normally, had bleached to white in the boat. Long and untrimmed, it floated out on the sea like a patch of bright sea-growth, a brilliant note in the monotone of blue water. The men in the boat could see it for a long time. Getting smaller and smaller, they found and lost it in the swell until it, too, became a part of the heaving whole.

The passing of the Mate and the Third Engineer had a curious effect upon Widdicombe. It vitalized him into hope and action. He took to himself command.

Morgan and Tapscott were sunk in despair. Widdicombe, on the contrary, was full of energy. He trimmed sail, set the course and rallied the others to new effort. They had nothing to maintain life—neither food nor water—but Widdicombe was sure that somehow, perhaps miraculously, they would come through.

They moved now like figures in a slow-motion film. They had to rest continuously to husband what little strength they had left. They divided the day into hour watches, which they kept more or less accurately, although they had to shorten or lengthen them often according to their physical state.

The mere thought of food gave Tapscott a painful contraction of the stomach, which seized him like a cramp. He would have to lie down while suffering it and wait until it had passed before he could do a thing.

Widdicombe suffered more and more from the sun. It gave him fits of vertigo that rendered him useless for whatever he was doing at the time. They tried Morgan on the steering oar and it seemed to steady him for a time. But, always, in the end, they had to snatch it away from him to avoid disaster. He could not stay on the wind or learn the rudiments of handling the boat in the swell. They took bucketfuls of water which had to be laboriously bailed out. And they could scarcely stand on their legs.

The second day after the Mate and the Third went over was very like the first, save that the breeze lessened and the heat increased. They bailed seawater over each other and dipped their heads over the side when the sun was at its worst. Tapscott noticed impersonally, as if it was not his body he was considering, that his skin was cracking and scaling, and that he was covered with boil-like eruptions. The piece of shrapnel in his left palm was working to the surface. It made a painful lump there which prevented him from grasping anything firmly. As consumed as he was by his thirst, he drank no more seawater. The Third's last actions before going over made him somehow chary of that.

Morgan, on the other hand, drank seawater in greater quantities. It did no good to warn him not to or to stop him in the act. At night he could dip it up when he chose.

That night they lay hove to again, but got very little sleep. Morgan kept them awake with his noise and lamentations. The next day he was worse. They put him in the bow-sheet under the boat cover, wedged in well with life belts, and ordered him to lie still. Tapscott had no sooner made his way aft again than Morgan was out of his shelter and working his way aft. Or, worse, he would pull himself up on the gunwale and try to walk, at imminent peril of falling overboard.

The breeze was lighter and the humidity intense. Widdicombe sat a great deal of the time with his head in his hands. His burst of optimism and energy had departed as suddenly as it had arrived. He would shut out Morgan's ravings with the palms of his hands, or bear them until his frayed nerves recoiled. Then he would stagger up and rave and curse like a madman himself.

In the heat of the day, while staring at nothing, Widdicombe got up suddenly, went to the water breaker, took the keg from its cradle and hurled it into the sea.

"That's a damn silly thing to do," Tapscott said to him. "What did you want to do that for?"

"Why is it sitting there with nothing in it?" Widdicombe demanded furiously.

"Now what will we put the water in when it rains?" Tapscott asked.

"We aren't going to get any water. It's never going to rain. What do I care, anyway?" Widdicombe shrieked.

Tapscott mentally shrugged his shoulders. He wondered if Widdicombe, too, were going completely off his head.

They lay to that night and tried to get some sleep. The second cook was in and out of the bow-sheets all the time. When not raving he sank into coma and at such moments they dozed.

The next day was like the one before. Widdicombe was now sunk in deepest dejection. Morgan laughed, cried, shouted and sang, practically continuously. He varied this with demands for his mother. This, particularly, roused Widdicombe to fury.

"Shut up!" he yelled at Morgan, unable to support more. "Do you think *you're* the only one that wants his mother?"

The insane man was quiet for several minutes. He gave no sign of understanding a word of what had been said to him, but something in the tone of Widdicombe's voice penetrated to that distant spot to which his intelligence had withdrawn.

In the afternoon Morgan seemed to be clearing up mentally. He was quiet and spoke almost normally. Encouraged, Tapscott and Widdicombe put him at the steering oar and tried to get some of the sleep they had lost the night before. They had just dozed off when the boat swung wildly and took a big sea, drenching Widdicombe to the skin. Widdicombe leaped up. Morgan was on his feet, too, the boat left to the vagaries of the swell. Widdicombe threw himself on Morgan. In the struggle that followed the cook went overboard. Tapscott, always a heavy sleeper, awoke in time to see the cook disappear. Flinging himself over the gunwale he seized him by the hair, and, after a desperate effort, managed to get him back into the boat.

They drifted and sailed aimlessly until sundown and then went through the familiar and weary ritual of lowering the sail, putting out the sea anchor, and lying to. They pulled the boat cover over them and longed for sleep—anything to get away from the slow fire that was consuming them inwardly. But Morgan's dementia would not let them rest. What was torture by day became nightmare at night. The peak came toward early morning; then he subsided into coma, broken now and then by muttering and groaning.

The sun rose in a sky bright and clear. With scorched and bloodshot eyes they searched the heavens for signs of rain. The air was heavier and more humid than it had been, but there were only a few thin, high clouds and a steady breeze. With leaden hands they got in the sea anchor and set sail again west, resigning themselves to another day of suffering.

Morgan lay in the bow quietly, his eyes closed. Tapscott nodded at the steering oar; Widdicombe was stretched out on the port seat, the side that shipped the least water. They would go on this way until midday, when the sun, they knew, would be unendurable and they would have to bail water over one another.

Morgan pushed away the boat cover and got up. His expression was normal and his voice firm, clear and without the detached quality of insanity.

"I think," he said, as casually as if they had all been at home in a Newport house, "I'll go down the street for a drink." He climbed up on the seat, walked aft rapidly, and before they could rouse themselves, or even realize what was afoot, stepped over the side. He went down like a shot. When his body reappeared it was being carried away by the swell. He made no more movement, no outcry. They stood staring stupidly after him. Then they sat down and stared at each other. Of the seven men of the *Anglo-Saxon* they were all that was left.

* * *

They had no idea of their whereabouts, or even in what direction they were drifting. They had failed to put out the sea anchor in their orgy of sleep; but, luckily, the sea was smooth and windless. They got the anchor over and stretched out on their seats.

Sometime toward morning, Tapscott, who had dozed off into a catnap, was roused by a terrific peal of thunder. Lightning flashed on the horizon, throwing the clouds into livid relief. The surface of the sea showed a ghastly green for a moment; then all was dark again. The rumbling and muttering of thunder drew nearer. Balls of fire, a variety of lightning he had never seen before, flamed in the heavens. More clouds closed over them and a darkness so close it was impossible to see on the boat.

Tapscott peered uneasily ahead, wondering if squalls were bearing down on them. With a terrifying crash a tongue of lightning flickered down from the cloud mass and hissed into the sea. A moment later there was a splatter of drops on the boat cover.

They rolled themselves out from under the cover and stretched it across the thwarts. Rain! Now that it was here they disbelieved the fact of it; they had looked for it so long in vain. The drops became a shower and a puddle formed in the hollow of the canvas. They could not wait, but dipped their tins into it, scraping the tin edges frantically into the grain of the canvas in their eagerness to drink. At the first swallow, they spat the water out. It was saltier than seawater. The boat cover was so impregnated with salt from the gallons of spray that had fallen on it, the water was spoiled. Regretfully they drained the canvas overside and sat waiting for it to fill again.

They had collected a few mouthfuls when the shower ceased as suddenly as it had started. They threw themselves down in disgust, but did not despair. Surely there would be more rain. The very air smelt of it.

Toward dawn the rain came, a cloudburst of it. The skies opened and it sluiced down in steady, heavy sheets. It plastered their long hair to their skulls and their rags to their bodies, washing the salt out of them and filling the boat with almost fresh water.

With the cover spread again they soon had a good puddle in the middle of it. This time it was fresh. The first rain had washed the boat cover clean. They poured water down their throats by the canful, spilling it out of the corners of their mouths, down their chins and chests, with joyful, gluttonous, animal noises. Tapscott had drunk three canfuls when his stomach, constricted and unconditioned by the long drought, revolted and sent it all up again. After that he was cautious and took the water by sips. No man ever held an old wine in his mouth, savoring its bouquet, as Tapscott did each mouthful of rainwater. He let it roll voluptuously round the root of his tongue, and, tilting back his head with infinite caution, let it trickle down his damaged throat. Never before had he known such pleasure in drink.

Widdicombe, who had not drunk as much as Tapscott at first, had no trouble keeping his water down.

Using the boat hook for a tool, Tapscott pried out the bow airtight tank from its fastening. He cut a hole in the soft copper with the Mate's knife, making a bung in the top of it. They rigged a series of creases and runoffs and let the water drain off the canvas into the tank. Tapscott fashioned a plug for it from the loom of an oar. They caught about six gallons of water.

Their thirst quenched, Tapscott and Widdicombe were aware of hunger as a sensation apart from their general misery. It was the first recognizable hunger they had felt in days. They soaked sea biscuit in water and ate it.

By early afternoon the storm had passed and the rain ceased. A light breeze sprang up. Life flowed back into them. They were very weak, but the tide, definitely, had changed. Widdicombe was jubilant.

"I knew we'd make it," he declared. "I knew it the moment we got back into the boat. If we couldn't go then, it stands to reason that we're going to be O.K."

On the morning of September 18, they were becalmed. Worse, they had reached the bottom of the water tank again.

For two days they drifted this way. It did not seem so bad as before. They had learned a technique of suffering. They were unable to eat dry biscuit, and the horrible dryness was throttling them again. But they held on hopefully, confident that they would get rain.

It came early on the morning of the twentieth. It was a good rain. They rigged the cover and runoffs. They drank copiously. Tapscott, having learned his lesson, took his water slowly. He had no trouble keeping it down.

While the tank was filling they soaked six biscuits each in rainwater. When they were soft they ate them. Their supply was getting low, but they had been without food for two days. The rain did not last as long as they would have liked. Still, they had enough water for several days.

"By the time that's gone we'll have reached land," Widdicombe said confidently.

The biscuit tank was nearly empty now, so they decided to limit themselves to one biscuit a day. But they drank all the water they wanted.

Their stomachs swelled ludicrously with the food and water. They had sores on their hips from lying on the hard thwarts. In spite of life belts as cushions these sores grew worse. They noted with some surprise that their bowels had started functioning again. They had a decent night's rest, which helped their morale enormously.

Widdicombe got out the log and wrote:

Sept. 20th. Rain again for four days. Getting very weak but trusting in God to pull us through.
  W. R. Widdicombe
  R. Tapscott.

They had no thought of sailing that night. If they were so near land, why hurry? They had no suspicion that miles and miles of Atlantic lay between them and the nearest land.

"Keep your eye peeled, Bobby," Widdicombe said. "We may see it any minute now."

Toward sunset on September 24, they were sure that they had made landfall, a low, blue-black mass on the horizon. But as they drew nearer it shifted. Even then they were hopeful; a landmass gives the appearance of shifting sometimes with a boat's change of position. But when they had sailed a while longer they saw that it was cloud.

Several times they had arguments over this: a cloud mass that looked like land. It always revealed itself finally as mirage. They began to wonder what had gone wrong. They had held a course generally west. Even allowing for the aimless drifting they had done at times, wind and current could not have carried them anywhere but in the direction they wanted to go.

The next morning they upended the tank and dribbled the last of the water into their cans. They fumbled in the biscuit tank, but it yielded only broken bits and crumbs. They were without food and water.

Widdicombe fell into the deepest despondency. Fate had done him in the eye again! Water, he felt, they would get; but where would they get food?

No use giving up now, Tapscott said. They could take it. It couldn't go on this way much longer. He was hungry, it was true, very hungry, and both of them looked like scarecrows, but being without food was easy compared to being without water.

Widdicombe made a final entry in the log:

Sept. 24th. All water and biscuits gone. Still hoping to make land.
   W. R. W.

During the muggy, stifling night the tension cracked and rain fell again. They had all they wanted to drink and collected a good tankful. The rain soaked them through and through, blanching and wrinkling their hands. It gathered in the boat and they were forced to bail.

When the sun rose, they stripped off their rags and put everything out to dry. The breeze lifted and they set sail. Had it not been for that gnawing vacuity in their middles they would have been gay.

Widdicombe was purposeful again.

"We'll have to sail nights," he said. "Something's wrong with our figuring. We should have made land by now."

They decided, too, that they had best ration the water. They agreed that three dipperfuls a day was about right.

They made good progress that day. When darkness came they did not lie to, but held their course. They took regular watches.

The boat sailed steadily in the starlit night, carrying them westward. As Tapscott sat in the stern sheets, holding the oar, giving an occasional glance at Arcturus or brilliant Jupiter and his satellite in their traverse of the sky, he thought of beef—thick, red, juicy, dripping slices of beef. He dozed off and then awoke with a guilty start. He must stay awake now. It helped to think of beef. What went best with it, a pint of bitter or a tankard of Irish ale?

The five weeks that followed their reprieve from death—the last lap—were like a long, bad dream to Tapscott and Widdicombe. They remembered certain days, the one when a flying fish flew into the boat, for example, but there were stretches when one day followed another in such an unvarying pattern of hunger, sun and sea that they ran together, an indistinguishable blur in the continuity of their suffering.

There were periods of great clarity, under the stress of extreme danger, and long periods of stupor from heat and starvation. All of this time the rhythm of physical existence was slowing down. The subtle debility of slow starvation mounted in unmarked stages until it reached and sapped the very faculty of memory. Their last week in the boat was almost a blank to them.

The day and night following their decision to sail at night were distinguished by nothing but hunger. They were new to a total lack of food. The water they drank revealed as a special craving what had before been an indistinguishable element in their general sufferings.

They cast about the boat for something to eat, anything. There was nothing, nothing they could even think of chewing.

Tapscott had an idea.

"Why not seaweed?" he asked. They had passed clumps of it in considerable number and he remembered the succulent variety of it sold and eaten in Wales under the name of rock laver, which went very well, indeed, particularly when fried with the breakfast bacon.

They kept sharp watch for the next seaweed they should encounter. They were not long in finding it. But far from resembling the long, black, edible laver, which, soft as it is, had, at home, to be boiled in huge pots before being sold for further cooking, this seaweed was tough, rubbery and salty. It was covered with small, hard bulbs resembling berries.

They snared a large bunch of it with the boat hook and stowed it in the bow-sheets. It gave them hours of chewing, but very little nourishment. It was impossible to reduce it to a pulp they could swallow without prolonged mastication. The salt in it, even when they had rinsed it in rainwater, gave them intolerable thirst.

The seaweed was not very successful as food; but it was better than nothing.

Tapscott could not sleep. Sometime later he thought he heard a thud against the sail, another on the boat cover over him, and a desperate flapping in the boat. They had seen flying fish in schools but none had come very near them. He was sure, now, that one of them had barged into the sail and was somewhere in the boat. He fumbled for the Mate's torch, which still gave forth a sickly glow, and looked for his quarry. He could find no trace of it. He decided, regretfully, that it had flopped overboard. He went back under the boat cover to sleep.

The next morning, remembering the fish, Tapscott made a thorough search for it. To his incredulous joy he found it, wedged between two battens at the bottom of the boat. He got out Sparks' razor and cut it in two. He took the head half. Widdicombe had the other. They ate it all, every scrap, eyes, bones and fins.

What maddened them was that there were fish all about now. Small ones swam alongside them for hours. They saw dorsal fins. One large fish, which they were convinced was a shark, came up boldly alongside. Widdicombe thrust at it with an oar.

Widdicombe dabbled his fingers overside. The larger fish came to this dangerous bait. He stabbed at them with the Mate's knife. He spent fruitless hours at this and was never able to do more than wound one of them. They became an obsession with him. When they reappeared near the boat he cursed and swore at them, defying them to come within fair stroke of his knife.

The next morning they saw an impressive sight. A whole gam of whales appeared off to the starboard, ten or twelve of them. They cruised along serenely, apparently unaware of the boat, or, if aware of it, completely undisturbed. Tapscott and Widdicombe watched them with respect and apprehension. Suppose one of these huge creatures should take it into his head to charge them? They said nothing until the whole shoal had safely distanced the boat. Then Widdicombe apologized for having doubted Tapscott's story of the other night. He was overcome with belated awe that they had got off as easily as they had when they grounded on one.

Neither Tapscott nor Widdicombe knew enough about whales to identify the species. All that they could report was that they were long-headed, about twenty feet long over all, and that they did not spout.

The next day was without incident. Their initial supply of seaweed had dried out and they saw none that day. They chewed the hardened, leathery strings for hours; but, chew as they might, hunger was now an active torture.

"I wonder what became of that little fish," Tapscott said.

Several days before they grounded on the whale they had seen a small, pinkish object come over the gunwale in a sea.

"It must be somewhere about," Tapscott said, groping on his knees.

He searched the boat systematically, coiling rope, folding the boat cover, stacking the sodden life belts, stowing the sea anchor and generally tidying up. He bailed the boat laboriously, finally finding what he was looking for. The fish was small, soft and rotten; but they ate it, gladly.

"An odd taste it had," Tapscott said, "rather coppery."

After eating the spoiled fish they feared they had poisoned themselves. They waited for the symptoms to appear. Tapscott did feel some qualms in his stomach, but they passed. Widdicombe was unaffected.

Now they were encountering great patches of seaweed. They got a large supply of it aboard and were delighted to find a tiny variety of crab in its meshes. There were, also, some small shellfish, rather like the winkles they got at the seaside resorts at home, which one ate with the aid of a pin. They winnowed out a large number of these, but it took handfuls to make a decent mouthful.

In searching Sparks' attaché case for an implement with which to extract the meat from their shellfish, Tapscott found a safety pin, which gave him another idea. He bent it into a hook, fixed it to a length of spun yarn he found in the boat, sacrificed half a crab for bait and put it over the side. One of the big fish struck it almost immediately. He jerked the line, hooked his fish and let out a howl of joy.

"Play him!" Widdicombe yelled, wild with excitement at the prospect of a real meal. "Play him up to the boat and I'll get him with the knife!"

But the metal of the pin was too soft. The fish gave an outraged leap, plunged, and the hook straightened out, releasing him. Furious, they bent the pin back into shape and tried again. They had another strike. Again they lost the fish; this time he wriggled off the barbless point. They tried over and over. They had no trouble getting strikes. But, always, the hook was too soft and their fish got away. Disgusted, they gave it up in the end.

Seaweed was their sole source of food. The patches of it were growing larger in area daily. They scorned the small detached clumps now, waiting until they sighted a floating field of it. Sometimes there were leads through it, like open water in an ice floe. When they found one of these they steered the boat into it. Moving slowly, they picked crabs, winkles and anything that seemed edible out of it.

The seaweed patches they were now encountering contained a small shrimp as well as crabs and winkles. These were a most welcome addition to their diet—if such it could be called—but, like the crabs, were too small to supply much substance. Hours of work were required to furnish a meal. Unsatisfied after such labor, and after dark, they picked over their supply of seaweed in the boat for the most tender morsels and chewed on that.

The weather held generally fair until October 8. Except for their side excursions into kelp fields and the extra mileage on tacks they made good progress westward. Late that day, however, squalls blew up, and rain. They lay to, spreading the boat cover and the sail over them in an attempt to keep dry. They passed a damp but, by comparison with some others during the trip, comfortable night.

Day broke overcast and drizzling. They decided to stay as they were until the weather lifted. They had already collected all the water they needed. There was nothing to do but sleep. Widdicombe was still snoring when Tapscott wakened. Rain was drumming on the canvas over them. Tapscott lifted the canvas, sat up and looked about. The sky was leaden and the visibility poor. There was some swell on the sea. He checked the sea anchor and found it in order. He was about to pull the canvas over him again when his glance, roving aft, fixed on a sight that riveted his eyes. Not more than a half mile away, off the port bow, bearing southerly, was a large steamer.

"Roy! Roy!" he yelled, shaking Widdicombe savagely. "A ship!"

Widdicombe sat up sleepily. He did not take it in for a minute. Tapscott was already heaving in the sea anchor. Together they got it in, grabbed oars and rowed frantically toward her.

Pulling as hard as they could, the steamer was distancing them. They rowed until they could go on no longer. They stood up in the boat, waving their arms and shouting. The steamer showed no signs of seeing them. They swung their oars, semaphore fashion, and Tapscott, finding the Mate's whistle, stood up in the bow and blew it until he was breathless. The liner steamed steadily ahead.

They picked up their oars again and tried to row to it. They were so weak and winded they achieved only a few strokes. Tapscott seized the whistle and blew it as hard and as long as he could. The liner kept her course. Then, turning sharply, she went off east.

Tapscott and Widdicombe collapsed on their oars, completely spent. Their hearts were beating as if to burst; their lungs heaved and they gulped down air in sobbing spasms. When they had recovered enough to handle the boat, they put it about and hoisted sail. Sick with disappointment and fatigue, they resumed their course west.

* * *

For four days after losing the passenger liner the weather was unsettled. The wind shifted about the compass uneasily. Rain squalls struck Tapscott and Widdicombe from unexpected quarters and the sea lost its reliable northeasterly swell. Confused winds and choppy cross waves buffeted them about, making steering difficult and exhausting. They shipped a great deal of water and had to bail at all hours. They noticed when bailing how much weaker they had become. They tired quickly and had to take more frequent rests.

On the morning of October 27, just after having started for the day, Widdicombe, whose turn it was, complained of dizziness and called to Tapscott to relieve him. Tapscott studied Widdicombe intently and decided he was speaking the truth. He took the oar and Widdicombe stretched himself out on one of the side seats. In a few minutes he was snoring.

After three hours at the helm, or thereabouts, as near as he could judge it, Tapscott called for relief. Widdicombe stirred, opened his eyes and heard Tapscott repeat his request. He rolled over again and paid no attention to it.

"All right, then," Tapscott said, shipping the oar. "Let the bloody boat sail itself. I'm through."

He left the stern sheets and made his way to the middle of the boat, where he sat down, braced against a thwart. Widdicombe, alongside him, watched him with narrowed eyes. He sat up slowly, his jaw jutted out. Tapscott had seen that expression before—the night Widdicombe had slugged Elliott. He was certain Widdicombe was about to hit. Before he could, Tapscott, gathering up all his remaining strength, punched Widdicombe flush on the jaw. Widdicombe went down on the seat, but was up a moment later flinging himself upon Tapscott. They rolled about in the bottom of the boat weakly pummeling each other. Suddenly, Widdicombe quit.

"I'm too weak," he said.

Tapscott pulled himself up on a thwart and waited. He did not relax his guard until Widdicombe had crawled aft and manned the steering oar.

They sailed and drifted for a long time without moving or speaking. Tapscott was feeling sorry now. He had not wanted to hit Widdicombe, but he was certain that it had been self-defense. Still, they were absolutely dependent upon each other. And when Tapscott thought of what they had been through and of the chaps who were gone, he was sorrier still.

"I'm sorry, Roy," he said, finally. "I'm sorry I hit you. It's crazy for us to fight."

Widdicombe grunted but said nothing. His lips were puffed and discolored from Tapscott's blow.

All that day, part of the night and until midnight the next day, they made fair progress. They said little or nothing, communicating when necessary by signs and grunts. It was an effort to talk and in spite of Tapscott's apology, feeling between them was strained.

To Tapscott it seemed as if they were moving in a child's dream of impotence, when he wants desperately to run, jump, hit or cry out and his body refuses to obey the command of his brain.

At midnight the wind fell away completely. They let the boat drift with the current and slept, not troubling even to lower the sail. Sometime between then and dawn Tapscott thought he heard a fish flapping in the boat. The night was dark, the Mate's torch had gone dead and he was too weak to look for it. He decided to wait until daylight.

With the first light of dawn Tapscott was in the bottom of the boat, looking for the fish. He found it; what he called a gar. Actually it was a Bahamian houndfish, a long needlelike creature, almost transparent, which is considered inedible by the natives. Houndfish lie near the surface of the sea and when frightened leap clear out of it. An enemy had chased it into the boat.

"I've found it," Tapscott said.

Widdicombe said nothing.

"I've found the fish," Tapscott repeated, looking up to see why Widdicombe received this important news so apathetically. Widdicombe was staring straight ahead, his eyes straining from the sockets. "Look," he said, pointing.

Tapscott, holding the fish firmly, raised himself on a thwart to see. Dead ahead lay a long line of lowland and beach, stretching north and south, apparently, as far as they could see.

They had been deceived so many times before, they did not dare to believe it.

"Land?" said Tapscott.

Widdicombe nodded.

"You're sure we're not seeing things?"

Widdicombe shook his head. The breeze was rising and the boat picked up way. As they drew nearer, they could see a line of reef with the sea breaking over it. Those rocks and the spray they threw up were no mirage.

"It *is* land, Bobby," Widdicombe said, sounding as if he were about to cry. "It must be the Leeward Islands."

Tapscott stared and stared at the reef and the beach beyond. There was no doubt in his mind now. Had the night lasted longer or the wind held the night through, they would have sailed right into it. Suddenly a wrenching pain racked his bowels. Still staring at the land, he grabbed for the bucket that served them as latrine.

* * *

A line of reef and broken water separated them from the land. They could see bush in back of the beach, and, in places, clumps of higher bush or trees, but no sign of human habitation.

"Land or no land," Tapscott said, "I'm going to eat this fish."

He cut it in half with Sparks' razor and together they ate it, staring at the land.

When they had finished the fish, they stood up unsteadily and studied the water ahead.

"I think I see a place to get through," Tapscott said, indicating a patch of smooth water. They got in the sea anchor, set sail and steered for it. It was a channel right enough, but extremely narrow. Tapscott lay in the bow and directed their course.

They threaded their way through patches of shoal and sharp, rocky heads. It was easy to see them; the water was the clearest Tapscott had ever seen. It was the most brilliantly colored he had ever seen, too, ranging from ultramarine in the depths to aquamarine in the shallows. Below him, around rocky drops and coral-encrusted banks, brilliantly colored fish swam and long, purple sea fans waved with the movement of the water. Here and there were patches of grass, and white sand between the rocks. He saw reticulated coral mounds like gargantuan brains. White trees of the same substance grew in this sea garden.

Twenty minutes later the bow of their boat grounded on the beach. Tapscott clambered over the bow, surprised at his strength. Widdicombe followed with the boat hook. He had some idea of mooring the boat to it, driven in the beach. The tide was ebbing, but they did not know it at the time. They were on land!

The sun beat down on the white sand. In the lee of the bush there was shade. Like drunken men they staggered up the beach to it and collapsed. The effort of walking was too much for them.

They lay in the shade until Widdicombe was feeling better. He proposed that they start northward up the beach. He felt that they might find human beings there. Tapscott got to his feet with infinite effort. Now that they were actually on land, all his remaining strength seemed to have gone from him. Laboriously they started up the beach. Thirty or forty feet of walking and Tapscott collapsed. The job he had been carrying on by sheer willpower and momentum was through.

How long Tapscott and Widdicombe lay on the edge of the bush they could not say. They were roused by the sound of blows somewhere in the bush in back of them. Someone was cutting a way through. Then they heard voices. They saw no one, but they knew someone was here, someone who retreated precipitously.

"They speak English whoever they are," Widdicombe said. "I heard them say 'fetch.'"

Tapscott was beyond answering.

"We'd better get back to the boat. They'll be coming back and might miss us," Widdicombe said. "Can you make it?"

Tapscott indicated that he would try.

They staggered and crawled back to a point on the edge of the bush opposite the boat. The tide had left it stranded. They lay there for a long time; then they heard voices again. The voices spoke English, an English such as they had never heard before, but, indubitably, English. A moment later, a man, a woman and several more men emerged from the bush and stood over them.

"Who you are?" one of them asked.

"English," Widdicombe said. "Off the *Anglo-Saxon*. Our ship was sunk by a German raider. We got away in the boat."

"When dat?" the man asked, dubiously.

"August twenty-first," Widdicombe said. "We were sixty-five days in the boat. We're starving."

A chorus of exclamations went up.

"Save me, Lawd!"

"Hey, man?"

"Today's the twenty-fifth, isn't it?" Widdicombe said.

"Today's the thirtieth," one of the men said.

"Then it's seventy days."

The group seemed suspicious. Nobody stirred.

"Look," said Widdicombe, taking Sparks' wallet from his pocket and handing it to the nearest man. "Papers."

The man took the wallet and extracted the log sheets and other papers from it. He studied them a long time and handed them to the others to look at.

Tapscott only half heard all this. He lay with closed eyes.

The spokesman was satisfied with what he read. He said something to the others. Neither Tapscott nor Widdicombe understood. The group broke into fresh exclamations of wonder and commiseration.

The seamen knew they were safe.

NINETEEN

Loss of the *Pequod*

Herman Melville

FIRST DAY

THAT NIGHT, IN THE MID-WATCH, WHEN THE OLD MAN—AS HIS WONT at intervals—stepped forth from the scuttle in which he leaned, and went to his pivot-hole, he suddenly thrust out his face fiercely, snuffing up the sea air as a sagacious ship's dog will, in drawing nigh to some barbarous isle. He declared that a whale must be near. Soon that peculiar odor, sometimes to a great distance given forth by the living Sperm Whale, was palpable to all the watch; nor was any mariner surprised when, after inspecting the compass, and then the dog-vane, and then ascertaining the precise bearing of the odor as nearly as possible, Ahab rapidly ordered the ship's course to be slightly altered, and the sail to be shortened.

The acute policy dictating these movements was sufficiently vindicated at daybreak by the sight of a long sleek on the sea directly and lengthwise ahead, smooth as oil, and resembling in the pleated watery wrinkles bordering it, the polished metallic-like marks of some swift tiderip, at the mouth of a deep, rapid stream.

"Man the mastheads! Call all hands!"

Thundering with the butts of three clubbed handspikes on the forecastle deck, Daggoo roused the sleepers with such judgment claps that they seemed to exhale from the scuttle, so instantaneously did they appear with their clothes in their hands.

"What d'ye see?" cried Ahab, flattening his face to the sky.

"Nothing, nothing, sir!" was the sound hailing down in reply.

"T'gallantsails! stunsails alow and aloft, and on both sides!"

All sail being set, he now cast loose the lifeline, reserved for swaying him to the mainroyal masthead; and in a few moments they were hoisting him thither, when, while but two-thirds of the way aloft, and while peering ahead through the horizontal vacancy between the maintopsail and topgallantsail, he raised a gull-like cry in the air, "There she blows!—there she blows! A hump like a snow hill! It is Moby Dick!"

Fired by the cry which seemed simultaneously taken up by the three lookouts, the men on deck rushed to the rigging to behold the famous whale they had so long been pursuing. Ahab had now gained his final perch, some feet above the other lookouts, Tashtego standing just beneath him on the cap of the topgallant-mast, so that the Indian's head was almost on a level with Ahab's heel. From this height the whale was now seen some mile or so ahead, at every roll of the sea revealing his high sparking hump, and regularly jetting his silent spout into the air. To the credulous mariners it seemed the same silent spout they had so long ago beheld in the moonlit Atlantic and Indian Oceans.

"And did none of ye see it before?" cried Ahab, hailing the perched men all around him.

"I saw him almost that same instant, sir, that Captain Ahab did, and I cried out," said Tashtego.

"Not the same instant; not the same—no, the doubloon is mine, Fate reserved the doubloon for me. I only; none of ye could have raised the White Whale first. There she blows! There she blows!—there she blows! There again!—there again!" he cried, in long-drawn, lingering, methodic tones, attuned to the gradual prolongings of the whale's visible jets. "He's going to found! In stunsails! Down top-gallant-sails! stand by three boats. Mr. Starbuck, remember, stay on board, and keep the ship. Helm there! Luff, luff a point! So; steady, man, steady! There go flukes! No, no; only black water! All ready the boats there? Stand by, stand by! Lower me, Mr. Starbuck; lower, lower,—quick, quicker!" and he slid through the air to the deck.

"He is heading straight to leeward, sir," cried Stubb; "right away from us; cannot have seen the ship yet."

"Be dumb, man! Stand by the braces! Hard down the helm!—brace up! Shiver her!—shiver her! So; well that! Boats, boats!"

Soon all the boats but Starbuck's were dropped; all the boatsails set—all the paddles plying; with rippling swiftness, shooting to leeward; and Ahab heading the onset. A pale, death glimmer lit up Fedallah's sunken eyes; a hideous motion gnawed his mouth.

Like noiseless nautilus shells, their light prows sped through the sea; but only slowly they neared the foe. As they neared him, the ocean grew still more smooth; seemed drawing a carpet over its waves; seemed a noon-meadow, so serenely it spread. At length the breathless hunter came so nigh his seemingly unsuspecting prey, that his entire dazzling hump was distinctly visible, sliding along the sea as if an isolated thing, and continually set in a revolving ring of finest, fleecy, greenish foam. He saw the vast involved wrinkles of the slightly projecting head beyond. Before it, far out on the soft Turkish—rugged waters, went the glistening white shadow from his broad, milky forehead, a musical ripping playfully accompanying the shade; and behind, the blue waters interchangeably flowed over into the moving valley of his steady wake; and on either hand bright bubbles arose and danced by his side. But these were broken again by the light toes of hundreds of gay fowl softly feathering the sea, alternate with their fitful flight; and like to some flagstaff rising from the painted hull of an argosy, the tall but shattered pole of a recent lance projected from the White Whale's back; and at intervals one of the cloud of soft-toed fowls hovering, and to and fro skimming like a canopy over the fish, silently perched and rocked on this pole, the long tail feathers streaming like pennons.

A gentle joyousness—a mighty mildness of repose in swiftness, invested the gliding whale. Not the white bull Jupiter swimming away with ravished Europa clinging to his graceful horns; his lovely, leering eyes sideways intent upon the maid; with smooth bewitching fleetness, rippling straight for the nuptial bower in Crete; not Jove did surpass the glorified White Whale as he so divinely swam.

On each soft side—coincident with the parted swell, that but once laving him, then flowed so wide away—on each bright side, the whale shed off enticings. No wonder there had been some among the hunters who namelessly transported and allured by all this serenity, had ventured to assail it; but had fatally found that quietude but the vesture of tornadoes. Yet calm, enticing calm, oh, whale! thou glidest on, to all who for the first time eye thee, no matter how many in that same way thou may'st have bejuggled and destroyed before.

And thus, through the serene tranquilities of the tropical sea, among waves whose hand clappings were suspended by exceeding rapture, Moby Dick moved on, still withholding from sight the full terrors of his submerged trunk, entirely hiding the wretched hideousness of his jaw. But soon the fore part of him slowly rose from the water; for an instant his whole marbleized body formed a high arch, like Virginia's Natural Bride, and warningly waving his bannered flukes in the air; the giant god revealed himself, sounded, and went out of sight. Hoveringly halting, and dipping on the wing, the white seafowls longingly lingered over the agitated pool that he left.

With oars apeak, and paddles down, the sheets of their sails adrift, the three boats now stilly floated, awaiting Moby Dick's reappearance.

"An hour," said Ahab, standing rooted in his boat's stern, and he gazed beyond the whale's place, towards the dim blue spaces and wide wooing vacancies to leeward. It was only an instant; for again his eyes seemed whirling round in his head as he swept the watery circle. The breeze now freshened; the sea began to swell.

"The birds!—the birds!" cried Tashtego.

In long Indian file, as when herons take wing, the white birds were now all flying towards Ahab's boat; and when within a few yards began fluttering over the water there, wheeling round and round, with joyous, expectant cries. Their vision was keener than man's; Ahab could discover no sign in the sea. But suddenly as he peered down and down into its depths, he profoundly saw a white living spot no bigger than a white weasel, with wonderful celerity uprising, and magnifying as it rose, till it turned, and then there were plainly revealed two long crooked rows of white, glistening teeth, floating up from the undiscoverable bottom. It was Moby Dick's open mouth and scrolled jaw; his vast, shadowed bulk still half blending with the blue of the sea. The glittering mouth yawned beneath the boat like an open-doored marble tomb; and giving one sidelong sweep with his steering oar, Ahab whirled the craft aside from this tremendous apparition. Then, calling upon Fedallah to change places with him, he went forward to the bows, and seizing Perth's harpoon, commanded his crew to grasp their oars and stand by to stern.

Now, by reason of this timely spinning round the boat upon its axis, its bow, by anticipation, was made to face the whale's head while yet under water. But as if perceiving this stratagem. Moby Dick, with that malicious intelligence ascribed to him, sidelingly transplanted himself, as it were, in an instant, shooting his plaited head lengthwise beneath the boat.

Through and through; through every plank and each rib, it thrilled for an instant, the whale obliquely lying on his back, in the manner of a biting shark, slowly and feelingly taking its bows full within his mouth, so that the long, narrow, scrolled lower jaw curled high up into the open air, and one of the teeth caught in a rowlock. The bluish pearl white of the inside of the jaw was within six inches of Ahab's head, and reached higher than that. In this attitude the White Whale now shook the slight cedar as a mildly cruel cat her mouse. With unastonished eyes Fedallah gazed and crossed his arms; but the tiger-yellow crew were tumbling over each other's heads to gain the uttermost stern.

And now, while both elastic gunwales were springing in and out, as the whale dallied with the doomed craft in this devilish way; and from his body being submerged beneath the boat, he could not be darted at from the bows, for the bows were almost inside of him, as it were, and while the other boats involuntarily paused, as before a quick crisis impossible to withstand, then it was that monomaniac Ahab, furious with this tantalizing vicinity of his foe, which placed him all alive and helpless in the very jaws he hated; frenzied with all this, he seized the long bone with his naked hands, and wildly strove to wrench it from its gripe. As now he thus vainly strove, the jaw slipped from him; the frail gunwales bent in, collapsed, and snapped, as both jaws, like an enormous shears, sliding further aft, bit the craft completely in twain, and locked themselves fast again in the sea, midway between the two floating wrecks. These floated aside, the broken ends drooping, the crew at the sternwreck clinging to the gunwales, and striving to hold fast to the oars to lash them across.

At that preluding moment, ere the boat was yet snapped, Ahab, the first to perceive the whale's intent, by the crafty upraising of his head, a movement that loosed his hold for the time; at that moment his hand had made one final effort to push the boat out of the bite. But only slipping further into the whale's mouth, and tilting over sideways as it slipped, the boat had shaken off his hold on the jaw; spilled him out of it, as he leaned to the push; and so he fell flat-faced upon the sea.

Ripplingly withdrawing from his prey, Moby Dick now lay at a little distance, vertically thrusting his oblong white head up and down in the billows; and at the same time slowly revolving his whole spindled body; so that when his vast wrinkled forehead rose—some twenty or more feet out of the water—the now rising swells, with all their confluent waves, dazzling broke against it; vindictively tossing their shivered spray still higher into the air. So, in a gale, the but half baffled channel billows only recoil from the base of the Eddystone, triumphantly to overleap its summit with their scud.

But soon resuming his horizontal attitude, Moby Dick swam swiftly round and round the wrecked crew; sideways churning the water in his vengeful wake, as if lashing himself up to still another and more deadly assault. The sight of the splintered boat seemed to madden him, as the blood of grapes and mulberries cast before Antiochus's elephants in the book of Maccabees. Meanwhile Ahab half smothered in the foam of the whale's insolent tail, and too much of a cripple to swim,—though he could still keep afloat, even in the heart of such a whirlpool as that; helpless Ahab's head was seen, like a tossed bubble which the least chance shock might burst. From the boat's fragmentary stern, Fedallah incuriously and mildly eyed him; the clinging crew, at the other drifting end, could not succor him; more than enough was it for them to look to themselves. For so revolvingly appalling was the White Whale's aspect, and so planetarily swift the ever-contracting circles he made, that he seemed horizontally swooping upon them. And though the other boats, unharmed, still hovered hard by, still they dared not pull into the eddy to strike, lest that should be the signal for the instant destruction of the jeopardized castaways, Ahab and all; nor in that case could they themselves hope to escape. With straining eyes then, they remained on the outer edge of the direful zone, whose center had now become the old man's head.

Meantime, from the beginning all this had been descried from the ship's mastheads; and squaring her yards, she had borne down upon the scene; and was now so nigh, that Ahab in the water hailed her;—"Sail on the"—but that moment a breaking sea dashed on him from Moby Dick, and whelmed him for the time. But struggling out of it again, and chancing to rise on a towering crest, he shouted,—"Sail on the whale!—Drive him off!"

The *Pequod's* prows were pointed; and breaking up the charmed circle, she effectually parted the White Whale from his victim. As he sullenly swam off, the boats flew to the rescue.

Dragged into Stubb's boat with bloodshot, blinded eyes, the white brine caking in his wrinkles; the long tension of Ahab's bodily strength did crack, and helplessly he yielded to his body's doom: for a time, lying all crushed in the bottom of Stubb's boat, like one trodden under foot of herds of elephants. Far inland, nameless wails came from him, as desolate sounds from out ravines.

But this intensity of his physical prostration did but so much the more abbreviate it. In an instant's compass, great hearts sometimes condense to one deep pang, the sum-total of those shallow pains kindly diffused through feebler men's whole lives. And so, such hearts, though summary in each one suffering; still, if the gods decree it, in their lifetime aggregate a whole age of woe, wholly made up of instantaneous intensities; for even in their pointless centers, those noble natures contain the entire circumferences of inferior souls.

"The harpoon," said Ahab, halfway rising, and draggingly leaning on one bended arm—"is it safe?"

"Aye, sir, for it was not darted; this is it," said Stubb, showing it.

"Lay it before me;—any missing men?"

"One, two, three, four, five;—there were five oars, sir, and here are five men."

"That's good.—Help me, man; I wish to stand. So, so, I see him! there! there! going to leeward still; what a leaping spout!—Hands off from me! The eternal sap runs up in Ahab's bones again! Set the sail; out oars; the helm!"

It is often the case that when a boat is stove, its crew, being picked up by another boat, help to work that second boat; and the chase is thus continued with what is called double-banked oars. It was thus now. But the added power of the boat did not equal the added power of the whale, but he seemed to have treble-banked his every fin; swimming with a velocity which plainly showed, that if now, under these circumstances, pushed on, the chase would prove an indefinitely prolonged, if not a hopeless one; nor could any crew endure for so long a period, such an unintermitted, intense straining at the oar; a thing barely tolerable only in some one brief vicissitude. The ship itself, then, as it sometimes happens, offered the most promising intermediate means of overtaking the chase. Accordingly the boats now made for her, and were soon swayed up to their cranes—the two parts of the wrecked boat having been previously secured by her—and then hoisting everything to her side, and stacking her canvas high up, and sideways outstretching it with stunsails, like the double-jointed wings of an albatross; the *Pequod* bore down in the leeward wake of Moby Dick. At the well-known, methodic intervals, the whale's glittering spout was regularly announced from the manned mastheads; and when he would be reported as just gone down, Ahab would take the time, and then pacing the deck, binnacle watch in hand, so soon as the last second of the allotted hours expired, his voice was heard.—"Whose is the doubloon now? D'ye see him?" and if the reply was, "No, sir!" straightway he commanded them to lift him to his perch. In this way the day wore on; Ahab, now aloft and motionless; anon, unrestingly pacing the planks.

As he was thus walking, uttering no sound, except to hail the men aloft, or to bid them hoist a sail still higher, or to spread one to a still greater breadth—thus to and fro pacing, beneath his slouched hat, at every turn he passed his own wrecked boat, which had been dropped upon the quarterdeck, and lay there reversed; broken bow to shattered stern. At last he paused before it; and as in an already over-clouded sky fresh troops of clouds will sometimes sail across, so over the old man's face there now stole some such added gloom as this.

Stubb saw him pause; and perhaps intending, not vainly, though to evince his own unabated fortitude, and thus keep up a valiant place in his captain's mind, he advanced, and eyeing the wreck exclaimed—"the thistle the ass refused; it pricked his mouth too keenly, sir; ha! ha!"

"What soulless thing is this that laughs before a wreck? Man, man! did I not know thee brave as fearless fire (and as mechanical) I could swear thou wert poltroon. Groan nor laugh should be heard before a wreck."

"Aye, sir," said Starbuck, drawing near, " 'tis a solemn sight; an omen, and an ill one."

"Omen? omen?—the dictionary! If the gods think to speak outright to man, they will honorably speak outright; not shake their heads, and give an old wife's darkling hint.—Begone! Ye two are the opposite poles of one thing; Starbuck is Stubb revered, and Stubb is Starbuck; and ye two are all mankind; and Ahab stands alone among the millions of the peopled earth, nor gods nor men his neighbors! Cold, cold—I shiver!—How now? Aloft there! D'ye see him? Sing out for every spout, though he spout ten times a second!"

The day was nearly done; only the hem of his golden robe was rustling. Soon, it was almost dark, but the lookout men still remained unset.

"Can't see the spout now, sir; too dark"—cried a voice from the air.

"How heading when last seen?"

"As before, sir,—straight to leeward."

"Good! he will travel slower now 'tis night. Down royals and topgallant stunsails, Mr. Starbuck. We must not run over him before morning; he's making a passage now, and may heaveto a while. Helm there! keep her full before the wind!—Aloft! come down!—Mr. Stubb, send a fresh hand to the foremast head and see it manned till morning."—Then advancing towards the doubloon in the mainmast—"Men, this gold is mine, for I earned it; but I shall let it abide here till the White Whale is dead; and then, whosoever of ye first raises him, upon the day he shall be killed, this gold is that man's; and if on that day I shall again raise him, then ten times its sum shall be divided among all of ye! Away now!—the deck is thine, sir."

And so saying, he placed himself halfway within the scuttle, and slouching his hat, stood there till dawn, except when at intervals rousing himself to see how the night wore on.

Second Day

At daybreak, the three mastheads were punctually manned afresh.

"D'ye see him?" cried Ahab after allowing a little space for the light to spread.

"See nothing, sir."

"Turn up all hands and make sail! he travels faster than I thought for;—the top-gallant-sails!—aye, they should have been kept on her all night. But no matter—'tis but resting for the rush."

Here be it said, that this pertinacious pursuit of one particular whale, continued through day into night, and through night into day, is a thing by no means unprecedented in the South Sea fisher. For such is the wonderful skill, prescience of experience, and invincible confidence acquired by some great natural geniuses among the Nantucket commanders, that from the simple observation of a whale when last described, they will, under certain given circumstances, pretty accurately foretell both the direction in which he will continue to swim for a time, while out of sight, as well as his probable rate of progression during that period. And, in these cases, somewhat as a pilot, when about losing sight of a coast, whose general trending he well knows, and which he desires shortly to return to again, but at some further point; like as this pilot stands by his compass, and takes the precise bearing of the cape at present visible, in order the more certainly to hit aright the remote, unseen headland, eventually to be visited: so does the fisherman, at his compass, with the whale; for after being chased, and diligently marked, through several hours of daylight, then, when night obscures the fish, the creature's future wake through darkness is almost as established to the sagacious mind of the hunter, as the pilot's coast is to him. So that to this hunter's wondrous skill, the proverbial evanescence of a thing writ in water, a wake, is to all desired purposes well-nigh as reliable as the steadfast land. And as the mighty iron Leviathan of the modern railway is so familiarly known in its every pace, that, with watches in their hands, men time his rate as doctors that of a baby's pulse; and lightly say of it, "the up train or the down train will reach such or such a spot, at such or such an hour," even so, almost, there are occasions when these Nantucketers time that other Leviathan of the deep, according to the observed humor of his speed; and say to themselves, "So many hours hence this whale will have gone two hundred miles, will have about reached this or that degree of latitude or longitude." But to render this acuteness at all successful in the end, the wind and the sea must be the whaleman's allies; for of what present avail to the becalmed or windbound mariner is the skill that assures him he is exactly ninety-three leagues and a quarter from his port? Inferable from these statements are many collateral subtle matters touching the chase of whales.

The ship tore on; leaving such a furrow in the sea as when a cannon-ball, missent, becomes a ploughshare and turns up the level field.

"By salt and hemp!" cried Stubb, "but this swift motion of the deck creeps up one's legs and tingles at the heart. This ship and I are two brave fellows!—Ha! ha! Some one take me up, and launch me, spine-wise, on the sea,—for by live-oaks! My spine's a keel. Ha, ha! we go the gait that leaves no dust behind!"

"There she blows—she blows!—she blows!—right ahead!" was now the masthead cry.

"Aye, aye!" cried Stubb; "I knew it—ye can't escape—blow on and split your spout, O whale! the mad fiend himself is after ye! Blow your trump—blister your lungs!—Ahab will dam off your blood, as a miller shuts his water-gate upon the stream!"

And Stubb did but speak out for well-nigh all that crew. The frenzies of the chase had by this time worked them bubblingly up, like old wine worked anew. Whatever pale fears and forebodings some of them might have felt before; these were not only now kept out of sight through the growing awe of Ahab, but they were broken up, and on all sides routed, as timid prairie hares that scatter before the bounding bison. The hand of Fate had snatched all their souls; and by the stirring perils of the previous day; the rack of the past night's suspense; the fixed, unfearing, blind, reckless way in which their wild craft went plunging towards its flying mark; by all these things, their hearts were bowled along. The wind that made great bellies of their sails, and rushed the vessel on by arms invisible as irresistible; this seemed the symbol of that unseen agency which so enslaved them to the race.

They were one man, not thirty. For as the one ship that held them all; though it was put together of all contrasting things—oak, and maple, and pine wood; iron, and pitch, and hemp—yet all these ran into each other in the one concrete hull, which shot on its way, both balanced and directed by the long central keel; even so, all the individualities of the crew. This man's valor, that man's fear; guilt and guiltiness, all varieties were welded into oneness, and were all directed to that fatal goal which Ahab their one lord and keel did point to.

The rigging lived. The mastheads, like the tops of tall palms, were outspreadingly tufted with arms and legs. Clinging to a spar with one hand, some reached forth the other with impatient wavings; others, shading their eyes from the vivid sunlight, sat far out on the rocking yards; all the spars in full bearing of mortals, ready and ripe for their fate. Ah! how they still strove through that infinite blueness to seek out the thing that might destroy them!

"Why sing ye not out for him, if ye see him?" cried Ahab, when, after the lapse of some minutes since the first cry, no more had been heard. "Sway me up, men; ye have been deceived; not Moby Dick casts one odd jet that way, and then disappears."

It was even so; in their headlong eagerness, the men had mistaken some other thing for the whale-spout, as the event itself soon proved; for hardly had Ahab reached his perch; hardly was the rope belayed to its pin on deck, when he struck the keynote to an orchestra, that made the air vibrate as with the combined discharges of rifles. The triumphant halloo of thirty buckskin lungs was heard, as—much nearer to the ship than the place of the imaginary jet, less than a mile ahead—Moby Dick bodily burst into view! For not by any calm and indolent spoutings; not by the peaceable gush of that mystic fountain in his head, did the White Whale now reveal his vicinity; but by the far more wondrous phenomenon of breaching. Rising with his utmost velocity from the further depths, the Sperm Whale thus booms his entire bulk into the pure element of air, and piling up a mountain of dazzling foam, shows his place to the distance of seven miles and more. In those moments, the torn, enraged waves he shakes off seem his mane; in some cases this breaching is his act of defiance.

"There she breaches! there she breaches!" was the cry, as in his immeasurable bravadoes the White Whale tossed himself salmon-like to Heaven. So suddenly seen in the blue plain of the sea, and relieved against the still bluer margin of the sky, the spray that he raised, for the moment, intolerably glittered and glared like a glacier; and stood there gradually fading and fading away from its first sparkling intensity, to the dim mistiness of an advancing shower in a vale.

"Aye, breach your last to the sun, Moby Dick!" cried Ahab, "thy hour and thy harpoon are at hand!—Down! down all of ye, but one man at the fore. The boats!—stand by!"

Unmindful of the tedious rope-ladders of the shrouds, the men, like shooting starts, slid to the deck by the isolated backstays and halyards; while Ahab, less dartingly, but still rapidly, was dropped from his perch.

"Lower away," he cried, so soon as he had reached his boat—a spare one, rigged the afternoon previous. "Mr. Starbuck, the ship is thine—keep away from the boats but keep near them. Lower, all!"

As if to strike a quick terror into them, by this time being the first assailant himself, Moby Dick had turned, and was now coming for the three crews. Ahab's boat was central; and cheering his men, he told them he would take the whale head-and-head,—that is pull straight up to his forehead,—a not uncommon thing; for when within a certain limit, such a course excludes the coming onset from the whale's sidelong vision. But ere that close limit was gained, and while yet all three boats were plain as the ship's three masts to his eye; the White Whale churning himself into furious speed, almost in an instant as it were, rushing among the boats with open jaws, and a lashing tail, offered appalling battle on every side; and heedless of the irons darted at him from every boat, seemed only intent on annihilating each separate plank of which those boats were made. But skilfully maneuvered, incessantly wheeling like trained chargers in the field; the boats for a while eluded him, though, at times, but by a plank's breadth; while all the time, Ahab's unearthly slogan tore every other cry but his to shreds.

But at last in his untraceable evolutions, the White Whale so crossed and re-crossed, and in a thousand ways entangled the slack of the three lines now fast to him, that they foreshortened, and, of themselves, warped the devoted boats towards the planted irons in him; though now for a moment the whale drew aside a little, as if to rally for a more tremendous charge. Seizing that opportunity, Ahab first paid out more line: and then was rapidly hauling and jerking in upon it again—hoping that way to disencumber it of some snarls—when lo!—a sight more savage than the embattled teeth of sharks!

Caught and twisted—corkscrewed in the mazes of the line—loose harpoons and lances, with all their bristling barbs and points, came flashing and dripping up to the chocks in the bows of Ahab's boat. Only one thing could be done. Seizing the boat-knife, he critically reached within—through—and then, without—the rays of steel; dragged in the line beyond, passed it, inboard, to the bowsman, and then, twice sundering the rope near the chocks—dropped the intercepted fagot of steel into the sea; and was all fast again. That instant, the White Whale made a sudden rush among the remaining tangles of the other lines; by so doing, irresistibly dragged the more involved boats of Stubb and Flask towards his flukes; dashed them together like two rolling husks on a surf-beaten beach, and then, diving down into the sea, disappeared in a boiling maelstrom, in which, for a space, the odorous cedar chips of the wrecks danced round and round, like the grated nutmeg in a swiftly stirred bowl of punch.

While the two crews were yet circling in the waters, reaching out after the revolving linetubs, oars, and other floating furniture, while aslope little Flask bobbed up and down like an empty vial, twitching his legs upwards to escape the dreaded jaws of sharks; and Stubb was lustily singing out for someone to ladle him up; and while the old man's line—now parting—admitted of his pulling into the creamy pool to rescue whom he could;—in that wild simultaneousness of a thousand concreted perils,—Ahab's yet unstricken boat seemed drawn up toward Heaven by invisible wires,—as arrow-like, shooting perpendicularly from the sea, the White Whale dashed his broad forehead against its bottom, and sent it, turning over and over, into the air; till it fell again—gunwale downwards—and Ahab and his men struggled out from under it, like seals from a seaside cave.

The first uprising momentum of the whale—modifying its direction as he struck the surface—involuntarily launched him along it, to a little distance from the center of the destruction he had made; and with his back to it, he now lay for a moment slowly feeling with his flukes from side to side; and whenever a stray oar, bit of plank, the least chip or crumb of the boats touched his skin, his tail swiftly drew back and came sideways, smiting the sea. But soon, as if satisfied that his work for that time was done, he pushed his plaited forehead through the ocean, and trailing after him the intertangled lines, continued his leeward way at a traveler's method pace.

As before, the attentive ship having described the whole fight, again came bearing down to the rescue, and dropping a boat, picked up the floating mariners, tubs, oars, and whatever else could be caught at, and safely landed them on her decks. Some sprained shoulders, wrists, and ankles; livid contusions; wrenched harpoons and lances: inextricable intricacies of rope; shattered oars and planks; all these were there; but no fatal or even serious ill seemed to have befallen anyone. As with Fedallah the day before, so Ahab was now found grimly clinging to his boat's broken half, which afforded a comparatively easy float; nor did it so exhaust him as the previous day's mishap.

But when he was helped to the deck, all eyes were fastened upon him; as instead of standing by himself he still halfhung upon the shoulder of Starbuck, who had thus far been the foremost to assist him. His ivory leg had been snapped off, leaving but one short sharp splinter.

"Aye, aye, Starbuck, 'tis sweet to lean sometimes, be the leaner who he will; and would old Ahab had leaned oftener than he has."

"The ferrule has not stood, sir," said the carpenter, now coming up; "I put good work into that leg."

"But no bones broken, sir, I hope," said Stubb with true concern.

"Aye! and all splintered to pieces, Stubb—d'ye see it.—But even with a broken bone, old Ahab is untouched; and I account no living bone of mine one jot more me, than this dead one that's lost. Nor White Whale, nor man, nor fiend, can so much as graze old Ahab in his own proper and inaccessible being. Can any lead touch younger floor, any mast scrape yonder roof?—Aloft there! which way?"

"Dead to leeward, sir."

"Up helm, then; pile on the sail again, shipkeepers! down the rest of the spare boats and rig them—Mr. Starbuck, away, and muster the boat's crews."

"Let me first help thee towards the bulwarks, sir."

"Oh, oh, oh! how this splinter gores me now! Accursed fate! that the unconquerable captain in the soul should have such a craven mate!"

"Sir?"

"My body, man, not thee. Give me something for a cane—there, that shivered lance will do. Muster the men. Surely I have not seen him yet. By heaven, it cannot be!—missing?—quick! call them all."

The old man's hinted thought was true. Upon mustering the company, the Parsee was not there.

"The Parsee!" cried Stubb—"he must have been caught in———"

"The black vomit wrench thee!—run all of ye above, alow, cabin, forecastle—find him—not gone—not gone!"

But quickly they returned to him with the tidings that the Parsee was nowhere to be found.

"Aye, sir," said Stubb—"caught among the tangles of your line—I thought I saw him dragging under."

"*My* line? *my* line? Gone?—gone? What means that little word?— What deathknell rings in it, that old Ahab shakes as if he were the belfry. The harpoon, too!—toss over the litter there,—d'ye see it?—the forged iron, men, the White Whale's—no, no, no,—blistered fool! this hand did dart it!—'tis in the fish—Aloft there! Keep him nailed—Quick!—all hands to the rigging of the boats—collect the oars—harpooneers!—the irons, the irons!—hoist the royals higher—a pull on the sheets!—helm there! steady, steady for your life! I'll ten times girdle the unmeasured globe; yea and dive straight through it, but I'll slay him yet!"

"Great God! but for one single instant show thyself," cried Starbuck; "never never wilt thou capture him, old man.—In Jesus' name no more of this, that's worse than devil's madness. Two days chased; twice stove to splinters; thy very leg once more snatched from under thee; thy evil shadow gone—all good angels mobbing thee with warnings:—what more wouldst thou have?—Shall we keep chasing this murderous fish till he swamps the last man? Shall we be dragged by him to the bottom of the sea? Shall we be towed by him to the infernal world? Oh, oh!—Impiety and blasphemy to hunt him more!"

"Starbuck, of late I've felt strangely moved to thee; ever since that hour we both saw—though know'st what, in one another's eyes. But in this matter of the whale, be the front of thy face to me as the palm of this hand—a lipless, unfeatured blank. Ahab is for ever Ahab, man. This whole act's immutably decreed. 'Twas rehearsed by thee and me a billion years before this ocean rolled. Fool! I am the Fates' lieutenant; I act under orders. Look thou, underling! that thou obeyest mine.—Stand round me, men. Ye see an old man cut down to the stump; leaning on a shivered lance; propped up on a lonely foot. 'Tis Ahab—his body's part; but Ahab's soul's a centipede, that moves upon a hundred legs. I feel strained, half stranded, as ropes that tow dismasted frigates in a gale; and I may look so. But ere I break, ye'll hear me crack; and till ye hear that, know that Ahab's hawser tows his purpose yet. Believe ye, men, in the things called omens? Then laugh aloud, and cry encore! For ere they drown, drowning things will twice rise to the surface; then rise again, to sink for evermore. So with Moby Dick—two days he's floated—tomorrow will be the third. Aye, man, he'll rise once more—but only to spout his last! D'ye feel brave men, brave?"

"As fearless fire," cried Stubb.

"And as mechanical," muttered Ahab. Then as the men went forward, he muttered on:—"The things called omens! And yesterday I talked the same to Starbuck there, concerning my broken boat. Oh! how valiantly I seek to drive out of others' hearts what's clinched so fast in mine!—The Parsee—the Parsee!—gone, gone? And he was to go before:—but still was to be seen again ere I could perish—How's that?—There's a riddle now might baffle all the lawyers backed by the ghosts of the whole line of judges:—like a hawk's beak it pecks my brain. *I'll, I'll* solve it, though!"

When dusk descended, the whale was still in sight to leeward.

So once more the sail was shortened, and everything passed nearly as on the previous night; only, the sound of hammers, and the hum of the grindstone was heard till nearly daylight, as the men toiled by lanterns in the complete and careful rigging of the spare boats and sharpened their fresh weapons for the morrow. Meantime, of the broken keel of Ahab's wrecked craft the carpenter made him another leg; while still as on the night before, slouched Ahab stood fixed within his scuttle; his hid helio-trope glance anticipatingly gone backward on its dial; set due eastward for the earliest sun.

THIRD DAY

The morning of the third day dawned fair and fresh, and once more the solitary nightman at the foremasthead was relieved by crowds of the day-light lookouts, who dotted every mast and almost every spar.

"D'ye see him?" cried Ahab; but the whale was not yet in sight.

"In his infallible wake, though; but follow that wake, that's all. Helm there; steady as thou goest, and hast been going. What a lovely day again! were it a new-made world, and made for a summerhouse to the angels, and this morning the first of its throwing open to them, a fairer day could not dawn upon that world. Here's food for thought, had Ahab time to think, but Ahab never thinks; he only feels, feels, feels, *that's* tingling enough for mortal man! to think's audacity. God only has that right and privilege. Thinking is, or ought to be a coolness and a calmness; and our poor hearts throb, and our poor brains beat too much for that. And yet, I've sometimes thought my brain was very calm—frozen calm, this old skull cracks so, like a glass in which the contents turn to ice, and shiver it. And still this hair is growing now; this moment growing, and heat must breed it; but no, it's like that sort of common grass that will grow anywhere, between the earthly clefts of Greenland ice or in Vesuvius lava. How the wild winds blow it; they whip it about me as the torn shreds of split sails lash the tossed ship they cling to. A vile wind that has no doubt blown ere this through prison corridors and cells, and wards of hospitals, and ventilated them, and now comes blowing hither as inno-cent as fleeces. Out upon it!—it's tainted. Were I the wind, I'd blow no more on such a wicked, miserable world. I'd crawl somewhere to a cave, and slink there. And yet, 'tis a noble and heroic thing, the wind! Who ever conquered it? In every fight it has the last and bitterest blow. Run tilting at it, and you but run through it. Ha! a coward wind that strikes stark naked men, but will not stand to receive a single blow. Even Ahab is a braver thing—a nobler thing than that. Would now the wind but had a body; but all the things that most exasperate and outrage mortal man, all these things are bodiless, but only bodiless as objects, not as agents. There's a most special, a most cunning, oh a most malicious difference! And yet, I say again, and swear it now, that there's something all glorious and gracious in the wind. These warm Trade Winds, at least, that in the clear heavens blow straight on, in strong and steadfast, vigorous mildness; and veer not from their mark, however the baser current of the sea may turn and tack, and the mightiest Mississippis of the land shift and swerve about, uncertain where to go at last. And by the eternal Poles; these same Trades that so directly blow my good ship on; these Trades, or something like them—something so unchangeable, and full as strong, blow my keeled soul along! To it! Aloft there! What d'ye see?"

"Nothing, sir."

"Nothing! and noon at hand! The doubloon goes a-begging! See the sun! Aye, aye it must be so. I've oversailed him. How, got the start? Aye, he's chasing me now; not I, *him*—that's bad; I might have known it, too. Fool! The lines—the harpoons he's towing. Aye, aye, I have run him by last night. About! about! Come down, all of ye, but the regular lookouts! Man the braces!"

Steering as she had done, the wind had been somewhat on the *Pequod*'s quarter, so that now being pointed in the reverse direction, the braced ship sailed hard upon the breeze as she rechurned the cream in her own white wake.

"Against the wind he now steers for the open jaw," murmured Starbuck to himself, as he coiled the new-hauled main-brace upon the rail. "God keep us, but already my bones feel damp within me, and from the inside wet my flesh. I misdoubt me that I disobey my God in obeying him!"

"Stand by to sway me up!" cried Ahab, advancing to the hempen basket. "We should meet him soon."

"Aye, aye, sir," and straightway Starbuck did Ahab's bidding, and once more Ahab swung on high.

A whole hour now passed; gold-beaten out to ages. Time itself now held long breaths with keen suspense. But at last, some three points off the weatherbow, Ahab descried the spout again, and instantly from the three mastheads three shrieks went up as if the tongues of fire had voiced it.

"Forehead to forehead I meet thee, this third time, Moby Dick! On deck there!—brace sharper up; crowd her into the wind's eye. He's too far off to lower yet, Mr. Starbuck. The sails shake! Stand over the helmsman with a topmaul! So, so; he travels fast, and I must down. But let me have one more good round look aloft here at the sea; there's time for that. An old, old sight, and yet somehow so young; aye, and not changed a wink since I first saw it, a boy, from the sandhills of Nantucket! The same!—the same!—the same to Noah as to me. There's a soft shower to leeward. Such lovely leewardings! They must lead somewhere—to something else than common land, more palmy than the palms. Leeward! the White Whale goes that way; look to windward, then; the better if the bitter quarter. But good-bye, good-bye, old masthead! What's this?—green? aye, tiny mosses in these warped cracks. No such green weather stains on Ahab's head! There's the difference now between man's old age and matter's. But aye, old mast, we both grow old together; sound in our hulls, though, are we not, my ship? Aye, minus a leg, that's all. By heaven! this dead wood has the better of my live flesh every way, I can't compare with it; and I've known some ships made of dead trees outlast the lives of man made of the most vital stuff of vital fathers. What's that he said? he should still go before me, my pilot; and yet to be seen again? But where? Shall I have eyes at the bottom of the sea, supposing I descend those endless stairs? and all night I've been sailing from him, wherever he did sink to. Aye, aye, like many more thou told'st direful truth as touching thyself, O Parsee; but, Ahab, there thy shot fell short. Good-bye, masthead—keep a good eye upon the whale, the while I'm gone. We'll talk tomorrow, nay, tonight, when the White Whale lies down there, tied by head and tail." He gave the word; and still gazing round him, was steadily lowered through the cloven blue air to the deck.

In due time the boats were lowered; but as standing in his shallop's stern, Ahab just hovered upon the point of the descent, he waved to the mate,—who held one of the tackleropes on deck—and bade him pause.

"Starbuck!"

"Sir?"

"For the third time my soul's ship starts upon this voyage, Starbuck."

"Aye, sir, thou wilt have it so."

"Some ships sail from their ports, and ever afterwards are missing, Starbuck!"

"Truth, sir: saddest truth."

"Some men die at ebb tide; some at low water; some at the full of the flood;—and I feel now like a billow that's all one crested comb, Starbuck. I am old;—shake hands with me, man."

Their hands met; their eyes fastened; Starbuck's tears the glue.

"Oh, my captain, my captain!—noble heart—go not—go not!—see, it's a brave man that weeps; how great the agony of the persuasion then!"

"Lower away!"—cried Ahab, tossing the mate's arm from him. "Stand by the crew?"

In an instant the boat was pulling round close under the stern.

"The sharks! the sharks!" cried a voice from the low cabin window there; "O master, my master, come back!"

But Ahab heard nothing; for his own voice was highlifted then; and the boat leaped on.

Yet the voice spake true; for scarce had he pushed from the ship, when numbers of sharks, seemingly rising from out dark waters beneath the hull, maliciously snapped at the blades of the oars, every time they dipped in the water; and in this way accompanied the boat with their bites. It is a thing not uncommonly happening to the whaleboats in those swarming seas; the sharks at times apparently following them in the same prescient way that vultures hover over the banners of marching regiments in the east. But these were the first sharks that had been observed by the *Pequod* since the White Whale had been first described; and whether it was that Ahab's crew were all such tiger-yellow barbarians, and therefore their flesh more musky to the senses of the sharks—a matter sometimes well known to affect them,—however it was, they seemed to follow that one boat without molesting the others.

"Heart of wrought steel!" murmured Starbuck, gazing over the side, and following with his eyes the receding boat—"canst thou yet ring boldly to that sight?—lowering thy keel among ravening sharks, and followed by them, open-mouthed, to the chase; and this the critical third day?—For when three days flow together in one continuous intense pursuit; be sure the first is the morning, the second the noon, and the third the evening and the end of that thing—be that end what it may. Oh! my God! what is this that shoots through me, and leaves me so deadly calm, yet expectant,—fixed at the top of a shudder! Future things swim before me, as in empty outlines and skeletons; all the past is somehow grown dim. Mary, girl! thou fadest in pale glories behind me; boy! I seem to see but thy eyes grown wondrous blue. Strangest problems of life seem clearing; but clouds sweep between—Is my journey's end coming? My legs feel faint; like his who has footed it all day. Feel thy heart,—beats it yet?—Stir thyself, Starbuck!—stave it off—move, move! speak aloud!—Masthead there! See ye my boy's hand on the hill?—Crazed;—aloft there!—keep thy keenest eye upon the boats:—mark well the whale—Ho! again!—drive off that hawk! see! he pecks—he tears the vane"—pointing to the red flag flying at the maintruck—"Ha! he soars away with it!—Where's the old man now? sees't thou that sight, oh Ahab!—shudder, shudder!"

The boats had not gone very far, when by a signal from the mastheads—a downward pointed arm, Ahab knew that the whale had sounded; but intending to be near him at the next rising, he held on his way a little sideways from the vessel; the becharmed crew maintaining the profoundest silence, as the head-beat waves hammered and hammered against the opposing bow.

"Drive, drive in your nails, oh ye waves! To their uttermost heads drive them in! ye but strike a thing without a lid; and no coffin and no hearse can be mine:—and hemp only can kill me! Ha! ha!"

Suddenly the waters around them slowly swelled in broad circles; then quickly upheaved, as if sideways sliding from a submerged berg of ice, swiftly rising to the surface. A low rumbling sound was heard; a subterraneous hum; and then all held their breaths; as bedraggled with trailing ropes, and harpoons, and lances, a vast form shot lengthwise, but obliquely from the sea. Shrouded in a thin drooping veil of mist, it hovered for a moment in the rainbowed air; and then fell swamping back into the deep. Crushed thirty feet upwards, the waters flashed for an instant like heaps of fountains, then brokenly sank in a shower of flakes, leaving the circling surface creamed like new milk round the marble trunk of the whale.

"Give way!" cried Ahab to the oarsmen and the boats darted forward to the attack; but maddened by yesterday's fresh irons that corroded in him, Moby Dick seemed combinedly possessed by all the angels that fell from heaven. The wide tiers of welded tendons overspreading his broad white forehead, beneath the transparent skin, looked knitted together, as head on, he came churning his tail among the boats; and once more flailed them apart; spilling out the irons and lances from the two mates' boats, and dashing in one side of the upper part of their bows, but leaving Ahab's almost without a scar.

While Daggoo and Queequeg were stopping the strained planks; and as the whale swimming out from them, turned, and showed one entire flank as he shot by them again; at that moment a quick cry went up. Lashed round and round to the fish's back; pinioned in the turns upon turns in which during the past night, the whale had reeled the involutions of the lines around him, the half torn body of the Parsee was seen, his sable raiment frayed to shreds; his distended eyes turned full upon old Ahab.

The harpoon dropped from his hand.

"Befooled, befooled!"—drawing in a long lean breath—"aye, Parsee! I see thee again.—Aye, and thou goest before; and this, this then is the hearse that thou didst promise. But I hold thee to the last letter of thy word. Where is the second hearse? Away, mates, to the ship! Those boats are useless now; repair them if ye can in time, and return to me; if not Ahab is enough to die—

"Down men! the first thing that but offers to jump from this boat I stand in, that thing I harpoon. Ye are not other men, but my arms and my legs; and so obey me.—Where's the whale? gone down again?"

But he looked too nigh the boat; for as if bent upon escaping with the corpse he bore and as if the particular place of the last encounter had been but a stage in his leeward voyage, Moby Dick was now again steadily swimming forward; and had almost passed the ship,—which thus far had been sailing in the contrary direction to him, though for the present her headway had been stopped. He seemed swimming with his utmost velocity, and now only intent upon pursuing his own straight path in the sea.

"Oh! Ahab," cried Starbuck, "not too late is it, even now, the third day, to desist. See! Moby Dick seeks thee not. It is thou, thou, that madly seekest him!"

Setting sail to the rising wind, the lonely boat was swiftly impelled to leeward, by both oars and canvas. And at last when Ahab was sliding by the vessel, so near as plainly to distinguish Starbuck's face as he leaned over the rail, he hailed him to turn the vessel about, and follow him, not too swiftly, at a judicious interval. Glancing upwards, he saw Tashtego, Queequeg, and Daggoo, eagerly mounting to the three mastheads; while the oarsmen were rocking in the two staved boats which had just been hoisted to the side, and were busily at work in repairing them. One after the other, through the port-holes, as he sped, he also caught flying glimpses of Stubb and Flask, busying themselves on deck among bundles of new irons and lances. As he saw all this; as he heard the hammers in the broken boats; far other hammers seemed driving a nail into his heart. But he rallied. And now marking that the vane or flag was gone from the main masthead, he shouted to Tashtego, who had just gained that perch, to descend again for another flag, and a hammer and nails, and so nail it to the mast.

Whether fagged by the three days' running chase, and the resistance to his swimming in the knotted hamper he bore; or whether it was some latent deceitfulness and malice in him: whichever was true, the White Whale's way now began to abate, as it seemed, from the boat so rapidly nearing him once more; though indeed the whale's last start had not been so long a one as before. And still as Ahab glided over the waves the unpitying sharks accompanied him, and so pertinaciously stuck to the boat; and so continually bit at the plying oars, that the blades became jagged and crunched, and left small splinters in the sea, at almost every dip.

"Heed them not! those teeth but give new rowlocks to your oars. Pull on! 'tis the better rest, the sharks' jaw than the yielding water."

"But at every bite, sir, the thin blades grow smaller and smaller."

"They will last long enough! pull on!—But who can tell"—he muttered—"whether these sharks swim to feast of a whale or on Ahab?—But pull on! Aye, all alive, now—we near him. The helm! take the helm; let me pass,"—and so saying, two of the oarsmen helped him forward to the bows of the still flying boat.

At length as the craft was cast to one side, and ran ranging along with the White Whale's flank, he seemed strangely oblivious of its advance—as the whale sometimes will—and Ahab was fairly within the smoky mountain mist, which thrown off from the whale's spout, curled round his great, Monadnock hump. He was even thus close to him; when, with body arched back, and both arms lengthwise highlifted to the poise, he darted his fierce iron, and his far fiercer curse into the hated whale. As both steel and curse sank to the socket, as if sucked into a morass, Moby Dick sideways writhed; spasmodically rolled his nigh flank against the bow, and, without staving a hole in it, so suddenly canted the boat over, that had it not been for the elevated part of the gunwale to which he then clung, Ahab would once more have been tossed into the sea. As it was, three of the oarsmen—who foreknew not the precise instant of the dart, and were therefore unprepared for its effects—these were flung out; but so fell, that in an instant two of them clutched the gunwale again, and rising to its level on a combing wave, hurled themselves bodily inboard again; the third man helplessly drooping astern, but still afloat and swimming.

Almost simultaneously, with a mighty volition of ungraduated, instantaneous swiftness, the White Whale darted through the weltering sea. But when Ahab cried out to the steersman to take new turns with the line, and hold it so; and commanded the crew to turn round on their seats, and tow the boat up to the mark; the moment the treacherous line felt that double strain and tug, it snapped in the empty air!

"What breaks in me? Some sinew cracks!—'tis whole again; oars! oars! Burst in upon him!"

Hearing the tremendous rush of the sea-crashing boat, the whale wheeled round to present his blank forehead at bay; but in that evolution, catching sight of the nearing black hull of the ship; seemingly seeing in it the source of all his persecutions; bethinking it—it may be—a larger and nobler foe; of a sudden, he bore down upon its advancing prow, smiting his jaws amid fiery showers of foam.

Ahab staggered; his hand smote his forehead. "I grow blind; hands! Stretch out before me that I may yet grope my way. Is't nigh?"

"The whale! The ship!" cried the cringing oarsman.

"Oars! Oars! Slope downwards to thy depths, O sea, that ere it be for ever too late, Ahab may slide this last, last time upon his mark! I see: the ship! The ship! Dash on, my men! Will ye not save my ship?"

But as the oarsmen violently forced their boat through the sledge-hammering seas, the before whalesmitten bowends of two planks burst through, and in an instant almost, the temporarily disabled boat lay nearly level with the waves; its halfwading, splashing crew, trying hard to stop and gap and bail out the pouring water.

Meantime, for that one beholding instant, Tashtego's masthead hammer remained suspended in his hand; and the red flag, half-wrapping him as with a plaid, then streamed itself straight out from him, as his own forward-flowing heart; while Starbuck and Stubb, standing upon the bowsprit beneath, caught sight of the down-coming monster just as soon as he.

"The whale, the whale! Up helm, up helm! Oh, all ye sweet powers of air, now hug me close! Let not Starbuck die, if die he must, in a woman's fainting fit. Up helm, I say—ye fools, the jaw! The jaw! Is this the end of all my bursting prayers? all my lifelong fidelities? Oh, Ahab, Ahab, lo, thy work. Steady! helmsman, steady. Nay, nay! Up helm again! He turns to meet us! Oh, his unappeasable brow drives on towards one, whose duty tells him he cannot depart. My God, stand by me now!"

"Stand not by me, but stand under me, whoever you are that will now help Stubb; for Stubb, too, sticks here. I grin at thee, thou grinning whale! Whoever helped Stubb, or kept Stubb awake, but Stubb's own unwinking eye? And now poor Stubb goes to bed upon a mattress that is all too soft; would it were stuffed with brushwood! I grin at thee, thou grinning whale! Look ye, sun, moon and stars! I call ye assassins of as good a fellow as ever spouted up his ghost. For all that, I would yet ring glasses with ye, would ye but hand the cup! Oh, oh, oh, oh! thou grinning whale, but there'll be plenty of gulping soon! Why fly ye not, O Ahab? For me, off shoes and jacket to it, let Stubb die in his drawers! A most moldy and oversalted death, though;—cherries! cherries! cherries! Oh, Flask, for one red cherry ere we die!"

"Cherries? I only wish that we were where they grow. Oh, Stubb, I hope my poor mother's drawn my part-pay ere this; if not, few coppers will come to her now, for the voyage is up."

From the ship's bows, nearly all the seamen now hung inactive; hammers, bits of plank, lances, and harpoons, mechanically retained in their hands, just as they had darted from their various employments; all their enchanted eyes intent upon the whale, which from side to side strangely vibrating his predestinating head, sent a broad band of overspreading semicircular foam before him as he rushed. Retribution, swift vengeance, eternal malice were in his whole aspect, and spite of all that mortal man could do, the solid white buttress of his forehead smote the ship's starboard bow, till men and timbers reeled. Some fell flat upon their faces. Like dislodged trucks, the heads of the harpooneers aloft shook on their bulllike necks. Through the breach, they heard the waters pour, as mountain torrents down a flume.

"The ship! The hearse!—the second hearse!" cried Ahab from the boat; "its wood could only be American!"

Diving beneath the settling ship, the whale ran quivering along its keel; but turning under water, swiftly shot to the surface again, far off the other bow, but within a few yards of Ahab's boat, where, for a time, he lay quiescent.

"I turn my body from the sun. What ho, Tashtego! let me hear thy hammer. Oh! ye three unsurrendered spires of mine; thou uncracked keel; and only god-bullied hull; thou firm deck, and haughty helm, and Polepointed prow,—death-glorious ship! Must ye then perish, and without me? Am I cut off from the last fond pride of meanest shipwrecked captains? Oh, lonely death on lonely life! Oh, now I feel my topmost greatness lies in my topmost grief. Ho, ho! from all your furthest bounds, pour ye now in, ye bold billows of my whole foregone life, and top this one piled comber of my death! Towards thee I roll, thou all-destroying but unconquering whale; to the last I grapple with thee; from hell's heart I stab at thee; for hate's sake I spit my last breath at thee. Sink all coffins and all hearses to one common pool! and since neither can be mine let me then tow to pieces, while still chasing thee, though tied to thee, thou damned whale! *Thus*, I give up the spear!" The harpoon was darted; the stricken whale flew forward; with igniting velocity the line ran through the groove; ran foul. Ahab stopped to clear it; he did clear it; but the flying turn caught him round the neck, and voicelessly as Turkish mutes bowstring their victim, he was shot out of the boat, ere the crew knew he was gone. Next instant, the heavy eyesplice in the rope's final end flew out of the stark empty tub, knocked down an oarsman, and smiting the sea, disappeared in its depths.

For an instant, the tranced boat's crew stood still; then turned. "The ship? Great God, where is the ship?" Soon they through dim bewildering mediums saw her sidelong facing phantom, as in the gaseous Fata Morgana; only the uppermost masts out of water; while fixed by infatuation, or fidelity, or fate, to their once lofty perches, the pagan harpooneers still maintained their sinking lookouts on the sea. And now, concentric circles seized the lone boat itself, and all its crew, and each floating oar, and every lancepole, and spinning, animate and inanimate, all round and round in one vortex, carried the smallest chip of the *Pequod* out of sight.

But as the last whelmings intermixingly poured themselves over the sunken head of the Indian at the mainmast, leaving a few inches of the erect spar yet visible, together with long streaming yards of the flag, which calmly undulated, with ironical coincidings, over the destroying billows they almost touched;—at that instant, a red arm and a hammer hovered backwardly uplifted in the open air, in the act of nailing the flag faster and yet faster to the subsiding spar. A sky-hawk that tauntingly had followed the maintruck downwards from its natural home among the start, pecking at the flag, and incommoding Tashtego there, this bird now chanced to intercept its broad fluttering wing between the hammer and the wood; and simultaneously feeling that ethereal thrill, the sub-merged savage beneath, in his deathgasp, kept his hammer frozen there; and so the bird of heaven, with unearthly shrieks, and his imperial beak thrust upwards, and his whole captive form folded in the flag of Ahab, went down with his ship, which, like Satan, would not sink to hell till she had dragged a living part of heaven along with her, and helmeted herself with it.

Now small fowls flew screaming over the yet yawning gulf; a sullen white surf beat against its steep sides; then all collapsed, and the great shroud of the sea rolled on as it rolled five thousand years ago.

TWENTY

Blueskin, the Pirate

Merle Johnson

CAPE MAY AND CAPE HENLOPEN FORM, AS IT WERE, THE UPPER AND lower jaws of a gigantic mouth, which disgorges from its monstrous gullet the cloudy waters of the Delaware Bay into the heaving, sparkling blue-green of the Atlantic Ocean. From Cape Henlopen as the lower jaw there juts out a long, curving fang of high, smooth-rolling sand dunes, cutting sharp and clean against the still, blue sky above—silent, naked, utterly deserted, excepting for the squat, white-walled lighthouse standing upon the crest of the highest hill. Within this curving, shelter- ing hook of sand hills lie the smooth waters of Lewes Harbor, and, set a little back from the shore, the quaint old town, with its dingy wooden houses of clapboard and shingle, looks sleepily out through the masts of the shipping lying at anchor in the harbor, to the purple, clean-cut, level thread of the ocean horizon beyond.

Lewes is a queer, odd, old-fashioned little town, smelling fragrant of salt marsh and sea breeze. It is rarely visited by strangers. The people who live there are the progeny of people who have lived there for many generations, and it is the very place to nurse, and preserve, and care for old legends and traditions of bygone times, until they grow from bits of gossip and news into local history of considerable size. As in the busier world men talk of last year's elections, here these old bits, and scraps, and odds and ends of history are retailed to the listener who cares to listen— traditions of the War of 1812, when Beresford's fleet lay off the harbor threatening to bombard the town; tales of the Revolution and of Earl Howe's warships, tarrying for a while in the quiet harbor before they sailed up the river to shake old Philadelphia town with the thunders of their guns at Red Bank and Fort Mifflin.

With these substantial and sober threads of real history, other and more lurid colors are interwoven into the web of local lore—legends of the dark doings of famous pirates, of their mysterious, sinister comings and goings, of treasures buried in the sand dunes and pine barrens back of the cape and along the Atlantic beach to the southward.

Of such is the story of Blueskin, the pirate.

It was in the fall and the early winter of the year 1750, and again in the summer of the year following, that the famous pirate, Blueskin, became especially identified with Lewes as a part of its traditional history.

For some time—for three or four years—rumors and reports of Blueskin's doings in the West Indies and off the Carolinas had been brought in now and then by sea captains. There was no more cruel, bloody, desperate, devilish pirate than he in all those pirate-infested waters. All kinds of wild and bloody stories were current concerning him, but it never occurred to the good folk of Lewes that such stories were some time to be a part of their own history.

But one day a schooner came drifting into Lewes harbor—shattered, wounded, her forecastle splintered, her foremast shot half away, and three great tattered holes in her mainsail. The mate with one of the crew came ashore in the boat for help and a doctor. He reported that the captain and the cook were dead and there were three wounded men aboard. The story he told to the gathering crowd brought a very peculiar thrill to those who heard it. They had fallen in with Blueskin, he said, off Fenwick's Island (some twenty or thirty miles below the capes), and the pirates had come aboard of them; but, finding that the cargo of the schooner consisted only of cypress shingles and lumber, had soon quitted their prize. Perhaps Blueskin was disappointed at not finding a more valuable capture; perhaps the spirit of deviltry was hotter in him that morning than usual; anyhow, as the pirate craft bore away she fired three broadsides at short range into the helpless coaster. The captain had been killed at the first fire, the cook had died on the way up, three of the crew were wounded, and the vessel was leaking fast, betwixt wind and water.

Such was the mate's story. It spread like wildfire, and in half an hour all the town was in a ferment. Fenwick's Island was very near home; Blueskin might come sailing into the harbor at any minute and then—! In an hour Sheriff Jones had called together most of the able-bodied men of the town, muskets and rifles were taken down from the chimney places, and every preparation was made to defend the place against the pirates, should they come into the harbor and attempt to land.

But Blueskin did not come that day, nor did he come the next or the next. But on the afternoon of the third the news went suddenly flying over the town that the pirates were inside the capes. As the report spread the people came running—men, women, and children—to the green before the tavern, where a little knot of old seamen were gathered together, looking fixedly out toward the offing, talking in low voices. Two vessels, one bark-rigged, the other and smaller a sloop, were slowly creeping up the bay, a couple of miles or so away and just inside the cape. There appeared nothing remarkable about the two crafts, but the little crowd that continued gathering upon the green stood looking out across the bay at them none the less anxiously for that. They were sailing close-hauled to the wind, the sloop following in the wake of her consort as the pilot fish follows in the wake of the shark.

But the course they held did not lie toward the harbor, but rather bore away toward the Jersey shore, and by and by it began to be apparent that Blueskin did not intend visiting the town. Nevertheless, those who stood looking did not draw a free breath until, after watching the two pirates for more than an hour and a half, they saw them—then about six miles away—suddenly put about and sail with a free wind out to sea again.

"The bloody villains have gone!" said old Captain Wolfe, shutting his telescope with a click.

But Lewes was not yet quit of Blueskin. Two days later a half-breed from Indian River bay came up, bringing the news that the pirates had sailed into the inlet—some fifteen miles below Lewes—and had careened the bark to clean her.

Perhaps Blueskin did not care to stir up the country people against him, for the half-breed reported that the pirates were doing no harm, and that what they took from the farmers of Indian River and Rehoboth they paid for with good hard money.

It was while the excitement over the pirates was at its highest fever heat that Levi West came home again.

Even in the middle of the last century the grist mill, a couple of miles from Lewes, although it was at most but fifty or sixty years old, had all a look of weather-beaten age, for the cypress shingles, of which it was built, ripen in a few years of wind and weather to a silvery, hoary gray, and the white powdering of flour lent it a look as though the dust of ages had settled upon it, making the shadows within dim, soft, mysterious. A dozen willow trees shaded with dappling, shivering ripples of shadow the road before the mill door, and the mill itself, and the long, narrow, shingle-built, one-storied, hip-roofed dwelling house. At the time of the story the mill had descended in a direct line of succession to Hiram White, the grandson of old Ephraim White, who had built it, it was said, in 1701.

Hiram White was only twenty-seven years old, but he was already in local repute as a "character." As a boy he was thought to be half-witted or "natural," and, as is the case with such unfortunates in small country towns where everybody knows everybody, he was made a common sport and jest for the keener, crueler wits of the neighborhood. Now that he was grown to the ripeness of manhood he was still looked upon as being—to use a quaint expression—"slack," or "not jest right." He was heavy, awkward, ungainly and loose-jointed, and enormously, prodigiously strong. He had a lumpish, thick-featured face, with lips heavy and loosely hanging, that gave him an air of stupidity, half droll, half pathetic. His little eyes were set far apart and flat with his face, his eyebrows were nearly white and his hair was of a sandy, colorless kind. He was singularly taciturn, lisping thickly when he did talk, and stuttering and hesitating in his speech, as though his words moved faster than his mind could follow. It was the custom for local wags to urge, or badger, or tempt him to talk, for the sake of the ready laugh that always followed the few thick, stammering words and the stupid drooping of the jaw at the end of each short speech. Perhaps Squire Hall was the only one in Lewes Hundred who mis-doubted that Hiram was half-witted. He had had dealings with him and was wont to say that whoever bought Hiram White for a fool made a fool's bargain. Certainly, whether he had common wits or no, Hiram had managed his mill to pretty good purpose and was fairly well off in the world as prosperity went in southern Delaware and in those days. No doubt, had it come to the pinch, he might have bought some of his tormentors out three times over.

Hiram White had suffered quite a financial loss some six months before, through that very Blueskin who was now lurking in Indian River inlet. He had entered into a "venture" with Josiah Shippin, a Philadelphia merchant, to the tune of seven hundred pounds sterling. The money had been invested in a cargo of flour and corn meal which had been shipped to Jamaica by the bark *Nancy Lee*. The *Nancy Lee* had been captured by the pirates off Currituck Sound, the crew set adrift in the longboat, and the bark herself and all her cargo burned to the water's edge.

Five hundred of the seven hundred pounds invested in the unfortunate "venture" was money bequeathed by Hiram's father, seven years before, to Levi West.

Eleazer White had been twice married, the second time to the widow West. She had brought with her to her new home a good-looking, long-legged, black-eyed, black-haired ne'er-do-well of a son, a year or so younger than Hiram. He was a shrewd, quick-witted lad, idle, shiftless, willful, ill-trained perhaps, but as bright and keen as a pin. He was the very opposite to poor, dull Hiram. Eleazer White had never loved his son; he was ashamed of the poor, slack-witted oaf. Upon the other hand, he was very fond of Levi West, whom he always called "our Levi," and whom he treated in every way as though he were his own son. He tried to train the lad to work in the mill, and was patient beyond what the patience of most fathers would have been with his stepson's idleness and shiftlessness. "Never mind," he was used to say. "Levi 'll come all right. Levi's as bright as a button."

It was one of the greatest blows of the old miller's life when Levi ran away to sea. In his last sickness the old man's mind constantly turned to his lost stepson. "Mebby he'll come back again," said he, "and if he does I want you to be good to him, Hiram. I've done my duty by you and have left you the house and mill, but I want you to promise that if Levi comes back again you'll give him a home and a shelter under this roof if he wants one." And Hiram had promised to do as his father asked.

After Eleazer died it was found that he had bequeathed five hundred pounds to his "beloved stepson, Levi West," and had left Squire Hall as trustee.

Levi West had been gone nearly nine years and not a word had been heard from him; there could be little or no doubt that he was dead.

One day Hiram came into Squire Hall's office with a letter in his hand. It was the time of the old French war, and flour and corn meal were fetching fabulous prices in the British West Indies. The letter Hiram brought with him was from a Philadelphia merchant, Josiah Shippin, with whom he had had some dealings. Mr. Shippin proposed that Hiram should join him in sending a "venture" of flour and corn meal to Kingston, Jamaica. Hiram had slept upon the letter overnight and now he brought it to the old Squire. Squire Hall read the letter, shaking his head the while. "Too much risk, Hiram!" said he. "Mr Shippin wouldn't have asked you to go into this venture if he could have got anybody else to do so. My advice is that you let it alone. I reckon you've come to me for advice?" Hiram shook his head. "Ye haven't? What have ye come for, then?"

"Seven hundred pounds," said Hiram.

"Seven hundred pounds!" said Squire Hall. "I haven't got seven hundred pounds to lend you, Hiram."

"Five hundred been left to Levi—I got hundred—raise hundred more on mortgage," said Hiram.

"Tut, tut, Hiram," said Squire Hall, "that'll never do in the world. Suppose Levi West should come back again, what then? I'm responsible for that money. If you wanted to borrow it now for any reasonable venture, you should have it and welcome, but for such a wildcat scheme—"

"Levi never come back," said Hiram—"nine years gone—Levi's dead."

"Mebby he is," said Squire Hall, "but we don't know that."

"I'll give bond for security," said Hiram.

Squire Hall thought for a while in silence. "Very well, Hiram," said he by and by, "if you'll do that. Your father left the money, and I don't see that it's right for me to stay his son from using it. But if it is lost, Hiram, and if Levi should come back, it will go well to ruin ye."

So Hiram White invested seven hundred pounds in the Jamaica venture and every farthing of it was burned by Blueskin, off Currituck Sound.

Sally Martin was said to be the prettiest girl in Lewes Hundred, and when the rumor began to leak out that Hiram White was courting her the whole community took it as a monstrous joke. It was the common thing to greet Hiram himself with, "Hey, Hiram; how's Sally?" Hiram never made answer to such salutation, but went his way as heavily, as impassively, as dully as ever.

The joke was true. Twice a week, rain or shine, Hiram White never failed to scrape his feet upon Billy Martin's doorstep. Twice a week, on Sundays and Thursdays, he never failed to take his customary seat by the kitchen fire. He rarely said anything by way of talk; he nodded to the farmer, to his wife, to Sally and, when he chanced to be at home, to her brother, but he ventured nothing further. There he would sit from half past seven until nine o'clock, stolid, heavy, impassive, his dull eyes following now one of the family and now another, but always coming back again to Sally. It sometimes happened that she had other company—some of the young men of the neighborhood. The presence of such seemed to make no difference to Hiram; he bore whatever broad jokes might be cracked upon him, whatever grins, whatever giggling might follow those jokes, with the same patient impassiveness. There he would sit, silent, unresponsive; then, at the first stroke of nine o'clock, he would rise, shoulder his ungainly person into his overcoat, twist his head into his three-cornered hat, and with a "Good night, Sally, I be going now," would take his departure, shutting the door carefully to behind him.

Never, perhaps, was there a girl in the world had such a lover and such a courtship as Sally Martin.

It was one Thursday evening in the latter part of November, about a week after Blueskin's appearance off the capes, and while the one subject of talk was of the pirates being in Indian River inlet. The air was still and wintry; a sudden cold snap had set in and skins of ice had formed over puddles in the road; the smoke from the chimneys rose straight in the quiet air and voices sounded loud, as they do in frosty weather.

Hiram White sat by the dim light of a tallow dip, poring laboriously over some account books. It was not quite seven o'clock, and he never started for Billy Martin's before that hour. As he ran his finger slowly and hesitatingly down the column of figures, he heard the kitchen door beyond open and shut, the noise of footsteps crossing the floor and the scraping of a chair dragged forward to the hearth. Then came the sound of a basket of corncobs being emptied on the smoldering blaze and then the snapping and crackling of the reanimated fire. Hiram thought nothing of all this, excepting, in a dim sort of way, that it was Bob, the mill hand, or old Dinah, the housekeeper, and so went on with his calculations.

At last he closed the books with a snap and, smoothing down his hair, arose, took up the candle, and passed out of the room into the kitchen beyond.

A man was sitting in front of the corncob fire that flamed and blazed in the great, gaping, sooty fireplace. A rough overcoat was flung over the chair behind him and his hands were spread out to the roaring warmth. At the sound of the lifted latch and of Hiram's entrance he turned his head, and when Hiram saw his face he stood suddenly still as though turned to stone. The face, marvelously altered and changed as it was, was the face of his stepbrother, Levi West. He was not dead; he had come home again. For a time not a sound broke the dead, unbroken silence excepting the crackling of the blaze in the fireplace and the sharp ticking of the tall clock in the corner. The one face, dull and stolid, with the light of the candle shining upward over its lumpy features, looked fixedly, immovably, stonily at the other, sharp, shrewd, cunning—the red wavering light of the blaze shining upon the high cheek bones, cutting sharp on the nose and twinkling in the glassy turn of the black, ratlike eyes. Then suddenly that face cracked, broadened, spread to a grin. "I have come back again, Hi," said Levi, and at the sound of the words the speechless spell was broken.

Hiram answered never a word, but he walked to the fireplace, set the candle down upon the dusty mantelshelf among the boxes and bottles, and, drawing forward a chair upon the other side of the hearth, sat down.

His dull little eyes never moved from his stepbrother's face. There was no curiosity in his expression, no surprise, no wonder. The heavy under lip dropped a little farther open and there was more than usual of dull, expressionless stupidity upon the lumpish face; but that was all.

As was said, the face upon which he looked was strangely, marvelously changed from what it had been when he had last seen it nine years before, and, though it was still the face of Levi West, it was a very different Levi West than the shiftless ne'er-do-well who had run away to sea in the Brazilian brig that long time ago. That Levi West had been a rough, careless, happy-go-lucky fellow; thoughtless and selfish, but with nothing essentially evil or sinister in his nature. The Levi West that now sat in a rush-bottom chair at the other side of the fireplace had that stamped upon his front that might be both evil and sinister. His swart complexion was tanned to an Indian copper. On one side of his face was a curious discoloration in the skin and a long, crooked, cruel scar that ran diagonally across forehead and temple and cheek in a white, jagged seam. This discoloration was of a livid blue, about the tint of a tattoo mark. It made a patch the size of a man's hand, lying across the cheek and the side of the neck. Hiram could not keep his eyes from this mark and the white scar cutting across it.

There was an odd sort of incongruity in Levi's dress; a pair of heavy gold earrings and a dirty red handkerchief knotted loosely around his neck, beneath an open collar, displaying to its full length the lean, sinewy throat with its bony "Adam's apple," gave to his costume somewhat the smack of a sailor. He wore a coat that had once been of fine plum color—now stained and faded—too small for his lean length, and furbished with tarnished lace. Dirty cambric cuffs hung at his wrists and on his fingers were half a dozen and more rings, set with stones that shone, and glistened, and twinkled in the light of the fire. The hair at either temple was twisted into a Spanish curl, plastered flat to the cheek, and a plaited queue hung halfway down his back.

Hiram, speaking never a word, sat motionless, his dull little eyes traveling slowly up and down and around and around his stepbrother's person.

Levi did not seem to notice his scrutiny, leaning forward, now with his palms spread out to the grateful warmth, now rubbing them slowly together. But at last he suddenly whirled his chair around, rasping on the floor, and faced his stepbrother. He thrust his hand into his capacious coat pocket and brought out a pipe which he proceeded to fill from a skin of tobacco. "Well, Hi," said he, "d'ye see I've come back home again?"

"Thought you was dead," said Hiram, dully.

Levi laughed, then he drew a red-hot coal out of the fire, put it upon the bowl of the pipe and began puffing out clouds of pungent smoke. "Nay, nay," said he; "not dead—not dead by odds. But [puff] by the Eternal Holy, Hi, I played many a close game [puff] with old Davy Jones, for all that."

Hiram's look turned inquiringly toward the jagged scar and Levi caught the slow glance. "You're lookin' at this," said he, running his finger down the crooked seam. "That looks bad, but it wasn't so close as this"— laying his hand for a moment upon the livid stain. "A devil off Singapore gave me that cut when we fell foul of an opium junk in the China Sea four years ago last September. This," touching the disfiguring blue patch again, "was a closer miss, Hi. A Spanish captain fired a pistol at me down off Santa Catharina. He was so nigh that the powder went under the skin and it'll never come out again. —— his eyes—he had better have fired the pistol into his own head that morning. But never mind that. I reckon I'm changed, ain't I, Hi?"

He took his pipe out of his mouth and looked inquiringly at Hiram, who nodded.

Levi laughed. "Devil doubt it," said he, "but whether I'm changed or no, I'll take my affidavy that you are the same old half-witted Hi that you used to be. I remember dad used to say that you hadn't no more than enough wits to keep you out of the rain. And, talking of dad, Hi, I hearn tell he's been dead now these nine years gone. D'ye know what I've come home for?"

Hiram shook his head.

"I've come for that five hundred pounds that dad left me when he died, for I hearn tell of that, too."

Hiram sat quite still for a second or two and then he said, "I put that money out to venture and lost it all."

Levi's face fell and he took his pipe out of his mouth, regarding Hiram sharply and keenly. "What d'ye mean?" said he presently.

"I thought you was dead—and I put—seven hundred pounds— into *Nancy Lee*—and Blueskin burned her—off Currituck."

"Burned her off Currituck!" repeated Levi. Then suddenly a light seemed to break upon his comprehension. "Burned by Blueskin!" he repeated, and thereupon flung himself back in his chair and burst into a short, boisterous fit of laughter. "Well, by the Holy Eternal, Hi, if that isn't a piece of your tarnal luck. Burned by Blueskin, was it?" He paused for a moment, as though turning it over in his mind. Then he laughed again. "All the same," said he presently, "d'ye see, I can't suffer for Blueskin's doings. The money was willed to me, fair and true, and you have got to pay it, Hiram White, burn or sink, Blueskin or no Blueskin." Again he puffed for a moment or two in reflective silence. "All the same, Hi," said he, once more resuming the thread of talk, "I don't reckon to be too hard on you. You be only half-witted, anyway, and I sha'n't be too hard on you. I give you a month to raise that money, and while you're doing it I'll jest hang around here. I've been in trouble, Hi, d'ye see. I'm under a cloud and so I want to keep here, as quiet as may be. I'll tell ye how it came about: I had a set-to with a land pirate in Philadelphia, and somebody got hurt. That's the reason I'm here now, and don't you say anything about it. Do you understand?"

Twenty-One

The Adventures of Captain Horn

Frank Stockton

EARLY IN THE SPRING OF THE YEAR 1884 THE THREE-MASTED SCHOO-ner *Castor*, from San Francisco to Valparaiso, was struck by a tornado off the coast of Peru. The storm, which rose with frightful suddenness, was of short duration, but it left the *Castor* a helpless wreck. Her masts had snapped off and gone overboard, her rudder-post had been shattered by falling wreckage, and she was rolling in the trough of the sea, with her floating masts and spars thumping and bumping her sides.

The *Castor* was an American merchant-vessel, commanded by Captain Philip Horn, an experienced navigator of about thirty-five years of age. Besides a valuable cargo, she carried three passengers—two ladies and a boy. One of these, Mrs. William Cliff, a lady past middle age, was going to Valparaiso to settle some business affairs of her late husband, a New England merchant. The other lady was Miss Edna Markham, a school-teacher who had just passed her twenty-fifth year, although she looked older. She was on her way to Valparaiso to take an important position in an American seminary. Ralph, a boy of fifteen, was her brother, and she was taking him with her simply because she did not want to leave him alone in San Francisco. These two had no near relations, and the education of the brother depended upon the exertions of the sister. Valparaiso was not the place she would have selected for a boy's education, but there they could be together, and, under the circumstances, that was a point of prime importance.

But when the storm had passed, and the sky was clear, and the mad waves had subsided into a rolling swell, there seemed no reason to believe that any one on board the *Castor* would ever reach Valparaiso. The vessel had been badly strained by the wrenching of the masts, her sides had been battered by the floating wreckage, and she was taking in water rapidly. Fortunately, no one had been injured by the storm, and although the captain found it would be a useless waste of time and labor to attempt to work the pumps, he was convinced, after a careful examination, that the ship would float some hours, and that there would, therefore, be time for those on board to make an effort to save not only their lives, but some of their property.

All the boats had been blown from their davits, but one of them was floating, apparently uninjured, a short distance to leeward, one of the heavy blocks by which it had been suspended having caught in the cordage of the topmast, so that it was securely moored. Another boat, a small one, was seen, bottom upward, about an eighth of a mile to leeward. Two seamen, each pushing an oar before him, swam out to the nearest boat, and having got on board of her, and freed her from her entanglements, they rowed out to the capsized boat, and towed it to the schooner. When this boat had been righted and bailed out, it was found to be in good condition.

The sea had become almost quiet, and there was time enough to do everything orderly and properly, and in less than three hours after the vessel had been struck, the two boats, containing all the crew and the passengers, besides a goodly quantity of provisions and water, and such valuables, clothing, rugs, and wraps as room could be found for, were pulling away from the wreck.

The captain, who, with his passengers, was in the larger boat, was aware that he was off the coast of Peru, but that was all he certainly knew of his position. The storm had struck the ship in the morning, before he had taken his daily observation, and his room, which was on deck, had been carried away, as well as every nautical instrument on board. He did not believe that the storm had taken him far out of his course, but of this he could not be sure. All that he knew with certainty was that to the eastward lay the land, and eastward, therefore, they pulled, a little compass attached to the captain's watch-guard being their only guide.

For the rest of that day and that night, and the next day and the next night, the two boats moved eastward, the people on board suffering but little inconvenience, except from the labor of continuous rowing, at which everybody, excepting the two ladies, took part, even Ralph Markham being willing to show how much of a man he could be with an oar in his hand.

The weather was fine, and the sea was almost smooth, and as the captain had rigged up in his boat a tent-like covering of canvas for the ladies, they were, as they repeatedly declared, far more comfortable than they had any right to expect. They were both women of resource and courage. Mrs. Cliff, tall, thin in face, with her gray hair brushed plainly over her temples, was a woman of strong frame, who would have been perfectly willing to take an oar, had it been necessary. To Miss Markham this boat trip would have been a positive pleasure, had it not been for the unfortunate circumstances which made it necessary.

On the morning of the third day land was sighted, but it was afternoon before they reached it. Here they found themselves on a portion of the coast where the foot-hills of the great mountains stretch themselves almost down to the edge of the ocean. To all appearances, the shore was barren and uninhabited.

The two boats rowed along the coast a mile or two to the southward, but could find no good landing-place, but reaching a spot less encumbered with rocks than any other portion of the coast they had seen, Captain Horn determined to try to beach his boat there. The landing was accomplished in safety, although with some difficulty, and that night was passed in a little encampment in the shelter of some rocks scarcely a hundred yards from the sea.

The next morning Captain Horn took counsel with his mates, and considered the situation. They were on an uninhabited portion of the coast, and it was not believed that there was any town or settlement near enough to be reached by walking over such wild country, especially with ladies in the party. It was, therefore, determined to seek succor by means of the sea. They might be near one of the towns or villages along the coast of Peru, and, in any case, a boat manned by the best oarsmen of the party, and loaded as lightly as possible, might hope, in the course of a day or two, to reach some port from which a vessel might be sent out to take off the remainder of the party.

But first Captain Horn ordered a thorough investigation to be made of the surrounding country, and in an hour or two a place was found which he believed would answer very well for a camping-ground until assistance should arrive. This was on a little plateau about a quarter of a mile back from the ocean, and surrounded on three sides by precipices, and on the side toward the sea the ground sloped gradually downward. To this camping-ground all of the provisions and goods were carried, excepting what would be needed by the boating party.

When this work had been accomplished, Captain Horn appointed his first mate to command the expedition, deciding to remain himself in the camp. When volunteers were called for, it astonished the captain to see how many of the sailors desired to go.

The larger boat pulled six oars, and seven men, besides the mate Rynders, were selected to go in her. As soon as she could be made ready she was launched and started southward on her voyage of discovery, the mate having first taken such good observation of the landmarks that he felt sure he would have no difficulty in finding the spot where he left his companions. The people in the little camp on the bluff now consisted of Captain Horn, the two ladies, the boy Ralph, three sailors,—one an Englishman, and the other two Americans from Cape Cod,—and a jet-black native African, known as Maka.

Captain Horn had not cared to keep many men with him in the camp, because there they would have little to do, and all the strong arms that could be spared would be needed in the boat. The three sailors he had retained were men of intelligence, on whom he believed he could rely in case of emergency, and Maka was kept because he was a cook. He had been one of the cargo of a slave-ship which had been captured by a British cruiser several years before, when on its way to Cuba, and the unfortunate men had been landed in British Guiana. It was impossible to return them to Africa, because none of them could speak English, or in any way give an idea as to what tribes they belonged, and if they should be landed anywhere in Africa except among their friends, they would be immediately re-enslaved. For some years they lived in Guiana, in a little colony by themselves, and then, a few of them having learned some English, they made their way to Panama, where they obtained employment as laborers on the great canal. Maka, who was possessed of better intelligence than most of his fellows, improved a good deal in his English, and learned to cook very well, and having wandered to San Francisco, had been employed for two or three voyages by Captain Horn. Maka was a faithful and willing servant, and if he had been able to express himself more intelligibly, his merits might have been better appreciated.

Englishman Davis, each armed with a gun, set out on a tour of investigation, hoping to be able to ascend the rocky hills at the back of the camp, and find some elevated point commanding a view over the ocean. After a good deal of hard climbing they reached such a point, but the captain found that the main object was really out of his reach. He could now plainly see that a high rocky point to the southward, which stretched some distance out to sea, would cut off all view of the approach of rescuers coming from that direction, until they were within a mile or two of his landing-place. Back from the sea the hills grew higher, until they blended into the lofty stretches of the Andes, this being one of the few points where the hilly country extends to the ocean.

The coast to the north curved a little oceanward, so that a much more extended view could be had in that direction, but as far as he could see by means of a little pocket-glass which the boy Ralph had lent him, the captain could discover no signs of habitation, and in this direction the land seemed to be a flat desert. When he returned to camp, about noon, he had made up his mind that the proper thing to do was to make himself and his companions as comfortable as possible and patiently await the return of his mate with succor.

Captain Horn was very well satisfied with his present place of encampment. Although rain is unknown in this western portion of Peru, which is, therefore, in general desolate and barren, there are parts of the country that are irrigated by streams which flow from the snow-capped peaks of the Andes, and one of these fertile spots the captain seemed to have happened upon. On the plateau there grew a few bushes, while the face of the rock in places was entirely covered by hanging vines. This fertility greatly puzzled Captain Horn, for nowhere was to be seen any stream of water, or signs of there ever having been any. But they had with them water enough to last for several days, and provisions for a much longer time, and the captain felt little concern on this account.

As for lodgings, there were none excepting the small tent which he had put up for the ladies, but a few nights in the open air in that dry climate would not hurt the male portion of the party.

In the course of the afternoon, the two American sailors came to Captain Horn and asked permission to go to look for game. The captain had small hopes of their finding anything suitable for food, but feeling sure that if they should be successful, every one would be glad of a little fresh meat, he gave his permission, at the same time requesting the men to do their best in the way of observation, if they should get up high enough to survey the country, and discover some signs of habitation, if such existed in that barren region. It would be a great relief to the captain to feel that there was some spot of refuge to which, by land or water, his party might make its way in case the water and provisions gave out before the return of the mate.

As to the men who went off in the boat, the captain expected to see but a few of them again. One or two might return with the mate, in such vessel as he should obtain in which to come for them, but the most of them, if they reached a seaport, would scatter, after the manner of seamen.

The two sailors departed, promising, if they could not bring back fish or fowl, to return before dark, with a report of the lay of the land.

It was very well that Maka did not have to depend on these hunters for the evening meal, for night came without them, and the next morning they had not returned. The captain was very much troubled. The men must be lost, or they had met with some accident. There could be no other reason for their continued absence. They had each a gun, and plenty of powder and shot, but they had taken only provisions enough for a single meal.

Davis offered to go up the hills to look for the missing men. He had lived for some years in the bush in Australia, and he thought that there was a good chance of his discovering their tracks. But the captain shook his head.

"You are just as likely to get lost, or to fall over a rock, as anybody else," he said, "and it is better to have two men lost than three. But there is one thing that you can do. You can go down to the beach, and make your way southward as far as possible. There you can find your way back, and if you take a gun, and fire it every now and then, you may attract the attention of Shirley and Burke, if they are on the hills above, and perhaps they may even be able to see you as you walk along. If they are alive, they will probably see or hear you, and fire in answer. It is a very strange thing that we have not heard a shot from them."

Ralph begged to accompany the Englishman, for he was getting very restless, and longed for a ramble and scramble. But neither the captain nor his sister would consent to this, and Davis started off alone.

"If you can round the point down there," said the captain to him, "do it, for you may see a town or houses not far away on the other side. But don't take any risks. At all events, make your calculations so that you will be back here before dark."

The captain and Ralph assisted the two ladies to a ledge of rock near the camp from which they could watch the Englishman on his way. They saw him reach the beach, and after going on a short distance he fired his gun, after which he pressed forward, now and then stopping to fire again. Even from their inconsiderable elevation they could see him until he must have been more than a mile away, and he soon after vanished from their view.

As on the previous day darkness came without the two American sailors, so now it came without the Englishman, and in the morning he had not returned. Of course, every mind was filled with anxiety in regard to the three sailors, but Captain Horn's soul was racked with apprehensions of which he did not speak. The conviction forced itself upon him that the men had been killed by wild beasts. He could imagine no other reason why Davis should not have returned. He had been ordered not to leave the beach, and, therefore, could not lose his way. He was a wary, careful man, used to exploring rough country, and he was not likely to take any chances of disabling himself by a fall while on such an expedition.

Although he knew that the great jaguar was found in Peru, as well as the puma and black bear, the captain had not supposed it likely that any of these creatures frequented the barren western slopes of the mountains, but he now reflected that there were lions in the deserts of Africa, and that the beasts of prey in South America might also be found in its deserts.

A great responsibility now rested upon Captain Horn. He was the only man left in camp who could be depended upon as a defender,—for Maka was known to be a coward, and Ralph was only a boy,—and it was with a shrinking of the heart that he asked himself what would be the consequences if a couple of jaguars or other ferocious beasts were to appear upon that unprotected plateau in the night, or even in the daytime. He had two guns, but he was only one man. These thoughts were not cheerful, but the captain's face showed no signs of alarm, or even unusual anxiety, and, with a smile on his handsome brown countenance, he bade the ladies good morning as if he were saluting them upon a quarter-deck.

"I have been thinking all night about those three men," said Miss Markham, "and I have imagined something which may have happened. Isn't it possible that they may have discovered at a distance some inland settlement which could not be seen by the party in the boat, and that they thought it their duty to push their way to it, and so get assistance for us? In that case, you know, they would probably be a long time coming back."

"That is possible," said the captain, glad to hear a hopeful supposition, but in his heart he had no faith in it whatever. If Davis had seen a village, or even a house, he would have come back to report it, and if the others had found human habitation, they would have had ample time to return, either by land or by sea.

The restless Ralph, who had chafed a good deal because he had not been allowed to leave the plateau in search of adventure, now found a vent for his surplus energy, for the captain appointed him fire-maker. The camp fuel was not abundant, consisting of nothing but some dead branches and twigs from the few bushes in the neighborhood. These

Ralph collected with great energy, and Maka had nothing to complain of in regard to fuel for his cooking.

Toward the end of that afternoon, Ralph prepared to make a fire for the supper, and he determined to change the position of the fireplace and bring it nearer the rocks, where he thought it would burn better. It did burn better—so well, indeed, that some of the dry leaves of the vines that there covered the face of the rocks took fire. Ralph watched with interest the dry leaves blaze and the green ones splutter, and then he thought it would be a pity to scorch those vines, which were among the few green things about them, and he tried to put out the fire. But this he could not do, and, when he called Maka, he was not able to help him. The fire had worked its way back of the green vines, and seemed to have found good fuel, for it was soon crackling away at a great rate, attracting the rest of the party.

"Can't we put it out?" cried Miss Markham. "It is a pity to ruin those beautiful vines."

The captain smiled and shook his head. "We cannot waste our valuable water on that conflagration," said he. "There is probably a great mass of dead vines behind the green outside. How it crackles and roars! That dead stuff must be several feet thick. All we can do is to let it burn. It cannot hurt us. It cannot reach your tent, for there are no vines over there."

The fire continued to roar and blaze, and to leap up the face of the rock.

"It is wonderful," said Mrs. Cliff, "to think how those vines must have been growing and dying, and new ones growing and dying, year after year, nobody knows how many ages."

"What is most wonderful to me," said the captain, "is that the vines ever grew there at all, or that these bushes should be here. Nothing can grow in this region, unless it is watered by a stream from the mountains, and there is no stream here."

Miss Markham was about to offer a supposition to the effect that perhaps the precipitous wall of rock which surrounded the little plateau, and shielded it from the eastern sun, might have had a good effect upon the vegetation, when suddenly Ralph, who had a ship's biscuit on the end of a sharp stick, and was toasting it in the embers of a portion of the burnt vines, sprang back with a shout.

"Look out!" he cried. "The whole thing's coming down!" And, sure enough, in a moment a large portion of the vines, which had been clinging to the rock, fell upon the ground in a burning mass. A cloud of smoke and dust arose, and when it had cleared away the captain and his party saw upon the perpendicular side of the rock, which was now revealed to them as if a veil had been torn away from in front of it, an enormous face cut out of the solid stone.

The great face stared down upon the little party gathered beneath it. Its chin was about eight feet above the ground, and its stony countenance extended at least that distance up the cliff. Its features were in low relief, but clear and distinct, and a smoke-blackened patch beneath one of its eyes gave it a sinister appearance. From its wide-stretching mouth a bit of half-burnt vine hung, trembling in the heated air, and this element of motion produced the impression on several of the party that the creature was about to open its lips.

Mrs. Cliff gave a little scream,—she could not help it,—and Maka sank down on his knees, his back to the rock, and covered his face with his hands. Ralph was the first to speak.

"There have been heathen around here," he said. "That's a regular idol."

"You are right," said the captain. "That is a bit of old-time work. That face was cut by the original natives."

The two ladies were so interested, and even excited, that they seized each other by the hands. Here before their faces was a piece of sculpture doubtless done by the people of ancient Peru, that people who were discovered by Pizarro; and this great idol, or whatever it was, had perhaps never before been seen by civilized eyes. It was wonderful, and in the conjecture and exclamation of the next half-hour everything else was forgotten, even the three sailors.

Because the captain was the captain, it was natural that every one should look to him for some suggestion as to why this great stone face should have been carved here on this lonely and desolate rock. But he shook his head.

"I have no ideas about it," he said, "except that it must have been some sort of a landmark. It looks out toward the sea, and perhaps the ancient inhabitants put it there so that people in ships, coming near enough to the coast, should know where they were. Perhaps it was intended to act as a lighthouse to warn seamen off a dangerous coast. But I must say that I do not see how it could do that, for they would have had to come pretty close to the shore to see it, unless they had better glasses than we have."

The sun was now near the horizon, and Maka was lifted to his feet by the captain, and ordered to stop groaning in African, and go to work to get supper on the glowing embers of the vines. He obeyed, of course, but never did he turn his face upward to that gaunt countenance, which grinned and winked and frowned whenever a bit of twig blazed up, or the coals were stirred by the trembling man.

After supper and until the light had nearly faded from the western sky, the two ladies sat and watched that vast face upon the rocks, its features growing more and more solemn as the light decreased.

"I wish I had a long-handled broom," said Mrs. Cliff, "for if the dust and smoke and ashes of burnt leaves were brushed from off its nose and eyebrows, I believe it would have a rather gracious expression."

As for the captain, he went walking about on the outlying portion of the plateau, listening and watching. But it was not stone faces he was thinking of. That night he did not sleep at all, but sat until day-break, with a loaded gun across his knees, and another one lying on the ground beside him.

When Miss Markham emerged from the rude tent the next morning, and came out into the bright light of day, the first thing she saw was her brother Ralph, who looked as if he had been sweeping a chimney or cleaning out an ash-hole.

"What on earth has happened to you!" she cried. "How did you get yourself so covered with dirt and ashes?"

"I got up ever so long ago," he replied, "and as the captain is asleep over there, and there was nobody to talk to, I thought I would go and try to find the back of his head"—pointing to the stone face above them. "But he hasn't any. He is a sham."

"What do you mean?" asked his sister.

"You see, Edna," said the boy, "I thought I would try if I could find any more faces, and so I got a bit of stone, and scratched away some of the burnt vines that had not fallen, and there I found an open place in the rock on this side of the face. Step this way, and you can see it. It's like a narrow doorway. I went and looked into it, and saw that it led back of the big face, and I went in to see what was there."

"You should never have done that, Ralph," cried his sister. "There might have been snakes in that place, or precipices, or nobody knows what. What could you expect to see in the dark?"

"It wasn't so dark as you might think," said he. "After my eyes got used to the place I could see very well. But there was nothing to see—just walls on each side. There was more of the passageway ahead of me, but I began to think of snakes myself, and as I did not have a club or anything to kill them with, I concluded I wouldn't go any farther. It isn't so very dirty in there. Most of this I got on myself scraping down the burnt vines. Here comes the captain. He doesn't generally oversleep himself like this. If he will go with me, we will explore that crack."

When Captain Horn heard of the passage into the rock, he was much more interested than Ralph had expected him to be, and, without loss of time, he lighted a lantern and, with the boy behind him, set out to investigate it. But before entering the cleft, the captain stationed Maka at a place where he could view all the approaches to the plateau, and told him if he saw any snakes or other dangerous things approaching, to run to the opening and call him. Now, snakes were among the few things that Maka was not afraid of, and so long as he thought these were the enemies to be watched, he would make a most efficient sentinel.

When Captain Horn had cautiously advanced a couple of yards into the interior of the rock, he stopped, raised his lantern, and looked about him. The passage was about two feet wide, the floor somewhat lower than the ground outside, and the roof but a few feet above his head. It was plainly the work of man, and not a natural crevice in the rocks. Then the captain put the lantern behind him, and stared into the gloom ahead of them. As Ralph had said, it was not so dark as might have been expected. In fact, about twenty feet forward there was a dim light on the right-hand wall.

The captain, still followed by Ralph, now moved on until they came to this lighted place, and found it was an open doorway. Both heads together, they peeped in, and saw it was an opening like a doorway into a chamber about fifteen feet square and with very high walls. They scarcely needed the lantern to examine it, for a jagged opening in the roof let in a good deal of light.

Passing into this chamber, keeping a good watch out for pitfalls as he moved on, and forgetting, in his excitement, that he might go so far that he could not hear Maka, should he call, the captain saw to the right another open doorway, on the other side of which was another chamber, about the size of the one they had first entered. One side of this was a good deal broken away, and through a fracture three or four feet wide the light entered freely, as if from the open air. But when the two explorers peered through the ragged aperture, they did not look into the open air, but into another chamber, very much larger than the others, with high, irregular walls, but with scarcely any roof, almost the whole of the upper part being open to the sky.

A mass of broken rocks on the floor of this apartment showed that the roof had fallen in. The captain entered it and carefully examined it. A portion of the floor was level and unobstructed by rocks, and in the walls there was not the slightest sign of a doorway, except the one by which he had entered from the adjoining chamber.

"Hurrah!" cried Ralph. "Here is a suite of rooms. Isn't this grand? You and I can have that first one, Maka can sleep in the hall to keep out burglars, and Edna and Mrs. Cliff can have the middle room, and this open place here can be their garden, where they can take tea and sew. These rocks will make splendid tables and chairs."

The captain stood, breathing hard, a sense of relief coming over him like the warmth of fire. He had thought of what Ralph had said before the boy had spoken. Here was safety from wild beasts—here was immunity from the only danger he could imagine to those under his charge. It might be days yet before the mate returned,—he knew the probable difficulties of obtaining a vessel, even when a port should be reached,—but they would be safe here from the attacks of ferocious animals, principally to be feared in the night. They might well be thankful for such a good place as this in which to await the arrival of succor, if succor came before their water gave out. There were biscuits, salt meat, tea, and other things enough to supply their wants for perhaps a week longer, provided the three sailors did not return, but the supply of water, although they were very economical of it, must give out in a day or two. "But," thought the captain, "Rynders may be back before that, and, on the other hand, a family of jaguars might scent us out to-night."

"You are right, my boy," said he, speaking to Ralph. "Here is a suite of rooms, and we will occupy them just as you have said. They are dry and airy, and it will be far better for us to sleep here than out of doors."

As they returned, Ralph was full of talk about the grand find. But the captain made no answers to his remarks—his mind was busy contriving some means of barricading the narrow entrance at night.

When breakfast was over, and the entrance to the rocks had been made cleaner and easier by the efforts of Maka and Ralph, the ladies were conducted to the suite of rooms which Ralph had described in such glowing terms. Both were filled with curiosity to see these apartments, especially Miss Markham, who was fairly well read in the history of South America, and who had already imagined that the vast mass of rock by which they had camped might be in reality a temple of the ancient Peruvians, to which the stone face was a sacred sentinel. But when the three apartments had been thoroughly explored she was disappointed.

"There is not a sign or architectural adornment, or anything that seems to have the least religious significance, or significance of any sort," she said. "These are nothing but three stone rooms, with their roofs more or less broken in. They do not even suggest dungeons."

As for Mrs. Cliff, she did not hesitate to say that she should prefer to sleep in the open air.

"It would be dreadful," she said, "to awaken in the night and think of those great stone walls about me."

Even Ralph remarked that, on second thought, he believed he would rather sleep out of doors, for he liked to look up and see the stars before he went to sleep.

At first the captain was a little annoyed to find that this place of safety, the discovery of which had given him such satisfaction and relief, was looked upon with such disfavor by those who needed it so very much, but then the thought came to him, "Why should they care about a place of safety, when they have no idea of danger?" He did not now hesitate to settle the matter in the most straightforward and honest way. Having a place of refuge to offer, the time had come to speak of the danger. And so, standing in the larger apartment, and addressing his party, he told them of the fate he feared had overtaken the three sailors, and how anxious he had been lest the same fate should come upon some one or all of them.

Now vanished every spark of opposition to the captain's proffered lodgings.

"If we should be here but one night longer," cried Mrs. Cliff, echoing the captain's thought, "let us be safe."

In the course of the day the two rooms were made as comfortable as circumstances would allow with the blankets, shawls, and canvas which had been brought on shore, and that night they all slept in the rock chambers, the captain having made a barricade for the opening of the narrow passage with the four oars, which he brought up from the boat. Even should these be broken down by some wild beast, Captain Horn felt that, with his two guns at the end of the narrow passage, he might defend his party from the attacks of any of the savage animals of the country.

The captain slept soundly that night, for he had had but a nap of an hour or two on the previous morning, and, with Maka stretched in the passage outside the door of his room, he knew that he would have timely warning of danger, should any come. But Mrs. Cliff did not sleep well, spending a large part of the night imagining the descent of active carnivora down the lofty and perpendicular walls of the large adjoining apartment.

The next day was passed rather wearily by most of the party in looking out for signs of a vessel with the returning mate. Ralph had made a flag which he could wave from a high point near by, in case he should see a sail, for it would be a great misfortune should Mr. Rynders pass them without knowing it.

To the captain, however, came a new and terrible anxiety. He had looked into the water-keg, and saw that it held but a few quarts. It had not lasted as long as he had expected, for this was a thirsty climate.

The next night Mrs. Cliff slept, having been convinced that not even a cat could come down those walls. The captain woke very early, and when he went out he found, to his amazement, that the barricade had been removed, and he could not see Maka. He thought at first that perhaps the man had gone down to the sea-shore to get some water for washing purposes, but an hour passed, and Maka did not return. The whole party went down to the beach, for the captain insisted upon all keeping together. They shouted, they called, they did whatever they could to discover the lost African, but all without success.

They returned to camp, disheartened and depressed. This new loss had something terrible in it. What it meant no one could conjecture. There was no reason why Maka should run away, for there was no place to run to, and it was impossible that any wild beast should have removed the oars and carried off the man.

Twenty-Two

The Wreck of the *Citizen*

Lewis Holmes

The whale ship *Citizen*, of New Bedford, owned by J. Howland & Co., fitted for three or four years, and bound to the North Pacific on a whaling voyage, sailed from the port of New Bedford, October 29, 1851. She was commanded by Thomas Howes Norton, of Edgartown, Martha's Vineyard.

Her officers were the following, namely: first mate, Lewis H. Roey, of New Bedford; second mate, John P. Fisher, of Edgartown; third mate, Walter Smith, of New Bedford; fourth mate, William Collins, of New Bedford. Four boat steerers, namely: Abram Osborn, Jr., and John W. Norton, of Edgartown, John Blackadore and James W. Wentworth, of New Bedford.

The following were nearly all the names of her crew: Charles T. Heath, William E. Smith, Christopher Simmons, George W. Borth, Darius Aping, William Nye, Manuel Jose, Jose Joahim, Charles C. Dyer, Charles Noyes, Edmund Clifford, George Long, Charles Adams, Bernard Mitchell, Nicholas Powers, William H. May, Alpheus Townshend, Barney R. Kehoe, Joseph E. Mears, James Dougherty, and Peter M. Cox. The whole number on board when she sailed was thirty-three persons. In addition to the above, five seamen were shipped at the Verd Islands, which made thirty-eight, all told.

As is generally the case, the majority of these were strangers, and perhaps had never seen each other's countenances until they appeared on the deck of the ship, henceforth to be their new home for months, and it may be for years.

Besides, in this number there were representatives from different and distant sections of the country, and not unfrequently an assortment of nations, and even races.

Here were gathered for the first time many a wandering youth, attracted to the seaboard by the spirit of romantic adventure, to see the world of waters, and to share in the excitement of new scenes. His wayward history, in breaking away from the wholesome restraints and watchcare of home, may be found written, perhaps, in many sorrowful hearts which he has left behind. Years may pass away before either parents or relatives shall hear again from the absent one, and it may be never. Such instances are not uncommon.

How much interest there is centred in a whale ship, as she is about to leave port! It is felt not only by those who embark their property and lives in her, but there are other attractions towards the ship. They are found in the desolateness which is felt in many home circles, in bidding adieu to husbands, sons, and brothers. When the anchor is weighed, and the sails are spread to the faithful breeze, sadness reigns in many households and in many hearts. The thoughts are not only painfully busy concerning *present* separations, but they bound forward to the future, and anticipate what may be the experience of a few years to come. Changes! one hardly dares think of them! Amid the perils and dangers of the deep, how long will the ship's company remain unbroken? Will the *ship* ever return, and reenter her port again? Will those who have just released themselves from the embraces of friends, and wiped away the falling tear, and barred their hearts to the separation, will *they* ever return? Or, if they should, will they ever see again those whom they are now leaving? These inquiries and reflections find expression only in painful emotions, sadness, and sorrow. Time will make changes, and leave its ineffaceable footprints with every passing year.

The land was lost sight of in the evening of the day upon which we sailed, with a strong south-west wind. We were accompanied out of the bay by two other outward bound whale ships—the *Columbus*, of Fairhaven, Captain Crowell, and the *Hunter*, of New Bedford, Captain Holt.

After the usual passage, with variable winds, and no particular incident of marked importance, except the ordinary and certain amount of seasickness on board, which generally attends the uninitiated in their first interviews with "old Neptune," Cape Verd Islands were made on the 4th of December.

With seasickness, homesickness follows; and then it is that many of the inexperienced, having left good homes and quiet life, wish a thousand times that they had never "learned the trade." But all such wishes are now in vain. With a new life on shipboard and in the forecastle, romance passes away, and leaves in its place the stern outlines of a living reality. Seasickness, however, is only a temporary affair; in most cases, indeed, it soon subsides, and then spirits and hope revive with recruited and invigorated health.

We took our departure from the islands on the 6th, in company with the ship *Benjamin Tucker*, Captain Sands; strong breezes, north-east trades. The first whales were seen about lat. 30° S., lon. 31° 41′ W., distant about seven miles—light winds. We set signal for the *Benjamin Tucker*, four or five miles distant, to notify Captain Sands that whales were in sight—an agreement we made while sailing in company. Boats were lowered; the mate fastened to a whale, which brought the shoal to. The second mate was less successful; his boat was stoven by a whale, and his men were floating about upon scattered and broken pieces of the wreck. Other boats soon came up and rescued their companions. The ship now ran down to the boat which was fastened to the whale. The whale, however, was lost, in consequence of cutting the line in the act of lancing him. After a pursuit of an hour or more, the mate fastened to another whale, and finally secured it, though it proved to be of but little pecuniary value. At the same time the boats of the *Benjamin Tucker* captured a whale, but they could not boast of much superiority. It made them *three* barrels. Thus ended the first whaling scene on the voyage, and certainly not a very profitable day's work.

The *Citizen* was put on her course. We passed several ships—weather good. December 20, lat. 40° S., whales were raised again, but took no oil. Still in company with the *Benjamin Tucker*. On Christmas Eve, Captain Sands and his wife took tea on board of our ship, thus reviving remembrances of home and friends, though thousands of miles distant from our native port.

The next incident of more than ordinary interest was another whale scene, of sufficient excitement and peril to satisfy the most ardent and aspiring.

The *Benjamin Tucker* had luffed to, headed to the westward, with signal to the *Citizen* that whales were in sight. The ship *Columbus* was then in company. The three ships were in full pursuit of the monsters of the deep. The school was overtaken in course of an hour or two working to the leeward. At first, one of the boats was lowered from the *Citizen*, and then another, and another, until four boats were bounding over the waves, each seeking to be laid alongside of his victim, and join in the uncertain conflict. From the three ships there were twelve boats pressing forward with the utmost celerity to share in the encounter, and each emulous to bear off his prize. The fourth boat despatched from the *Citizen* fastened to a whale. He was shortly lanced, and spouted blood—a sure indication that he had received his death wound. In mortal agony, he plunged, and floundered, and mingled the warm current of his own life with the foaming waters around him. Conscious, apparently, of the authors of his sufferings, with rage and madness he at once attacked the boat, and with his ponderous jaws seized it, and in a moment bit it in two in the centre. Nor was there any time to be lost by the humble occupants of the boat. The rules of courtesy and ordinary politeness in entertaining a superior were for the time being laid entirely aside. Each seaman fled for his life— some from the stern, and others from the bow, while the cracking boards around and beneath them convinced them that the whale had every thing in his own way. Besides, the sensation was any thing but pleasant in expecting every moment to become fodder to the enraged leviathan of the deep. In quick succession those enormous jaws fell, accompanied with a deep, hollow moan or groan, which evinced intense pain, that sent a chill of terror to the stoutest hearts. They felt the feebleness of man when the monster arose in his fury and strength. A boat was soon sent to the rescue of their companions, who were swimming in every direction, to avoid contact with the enraged whale, which seemed bent on destroying every thing within his reach. He really asserted his original lordship in his own native element, and was determined to drive out all intruders. He therefore attacked the second boat, and would probably have ground it to atoms, had not a fortunate circumstance of two objects perhaps somewhat disconcerting him, and dividing his attention, turned him off from his purpose.

The captain of the *Citizen*, observing the affray from the beginning, was soon convinced that matters were taking rather a serious direction, and that not only the boats but the lives of his men were greatly imperilled. He therefore ordered the fifth boat to be instantly lowered, manned with "green hands," the command of which he himself assumed, and directed in pursuit of the whale. Five boats were now engaged in the contest, with the exception of the one stoven, and all the available crew and officers, including the captain, concentrated their efforts and energies in order to capture this "ugly customer." Just at the moment he was attacking or had already attacked the second boat, the captain's boat appeared on the ground, and from some cause best known to himself, the whale immediately left the former and assailed the latter. What the whale had already done, and what he appeared determined still to do, were by no means very flattering antecedents, and would very naturally impress the minds of "green hands," especially, that whaling, after all, was a reality, and not an imaginary affair or ordinary pastime.

On, therefore, the whale came to the captain's boat, ploughing the sea before him, jaws extended, with the fell purpose of destroying whatever he might chance to meet. As he approached near, the lance was thrust into his head and held in that position by the captain, and by this means he was kept at bay, while the boat was driven astern nearly half a mile. In this manner he was prevented from coming any nearer to the boat, the boat moving through the water as fast and as long as he pressed his head against the point of the lance. This was the only means of their defence. It was a most fortunate circumstance in a most trying situation. If the handle of the lance had broken, they would have been at the mercy of a desperate antagonist. The countenances of the boys were pallid with fear, and doubtless the very hair upon their heads stood erect.

It was a struggle for life. It was death presented to them under one of the most frightful forms. They were, however, as singularly and as suddenly relieved as they were unexpectedly attacked. The whale caught sight of the ship, as was supposed, which was running down towards the boats, and suddenly started for the new and larger object of attack. This was observed by the captain, who immediately made signal to keep the ship off the wind, which would give her more headway, and thus, if possible, escape a concussion which appeared at first sight inevitable.

The whale started on his new course towards the ship with the utmost velocity, with the intention of running into her. The consequences no one could predict; more than likely he would have either greatly disabled the ship, or even sunk her, had he struck her midships. To prevent such a catastrophe—the injury of the ship, and perhaps the ruin of the voyage—every thing now seemed to depend upon the direction of the ship and a favoring wind. Every eye was turned towards the ship; oars were resting over the gunwale of the boats, and each seaman instinctively fixed in his place, while anticipating a new encounter upon a larger scale, the results of which were fearfully problematical.

A good and merciful Providence, however, whose traces are easily discernible in the affairs of men both upon the ocean and upon the land, opportunely interfered. The ship was making considerable headway. The whale started on a bee line for the ship, but when he came up with her, in consequence of her increased speed before the wind, he fell short some ten or twelve feet from the stern. The crisis was passed. On he sped his way, dragging half of the boat still attached to the lines connected with the irons that were in his body. His death struggle was long and violent. In about half an hour he went into his "flurry, and turned up." Colors were set for the boats to return to the ship; the dead whale was brought alongside, cut in, boiled out, and seventy-five barrels of sperm oil were stowed away.

We copy the following whale incident from the *Vineyard Gazette* of October 14, 1853. The editor says,—

We are indebted to Captain Thomas A. Norton, of this town, one of the early commanders of the whale ship Hector, of New Bedford, for the following interesting particulars relative to an attack upon and

final capture of an ugly whale. Captain Norton was chief mate of the
Hector at the time.

"In October, 1832, when in lat. 12° S., lon. 80° W., the ship ninety
days from port, we raised a whale. The joyful cry was given of 'There
she blows!' and every thing on board at once assumed an aspect of
busy preparation for the capture. The boats were lowered, and chase
commenced. When we got within about three ships' lengths of him,
he turned and rushed furiously upon us. He struck us at the same
moment we fastened to him. He stove the boat badly; but with the
assistance of sails which were placed under her bottom, and constant
bailing, she was kept above water. The captain, John O. Morse, came
to our assistance. I told him he had better keep clear of the whale;
but he said he had a very long lance, and wanted to try it upon the
rascal. Captain Morse went up to the whale, when all at once he
turned upon the boat, which he took in his mouth, and held it 'right
up on end,' out of the water, and shook it all to pieces in a moment.

"The men were thrown in every direction, and Captain Morse
fell from a height of at least thirty feet into the water. Not being
satisfied with the total destruction of the boat, he set to work and
'chewed up' the boat kegs and lantern kegs, and whatever fragments of the boat he could find floating on the water. At this stage
of the 'fight,' I told Captain Morse that if he would give me the
choice of the ship's company, I would try him again. It was desperate work, to all appearance, and up to this time the vicious fellow
had had it all his own way. The captain was in favor of trying
him from the ship, but finally consented for us to attack him again
from a boat. With a picked crew, we again approached the whale,
now lying perfectly still, apparently ready for another attack, as
the event proved. Seeing our approach, he darted towards us with
his mouth wide open, his ponderous jaws coming together every
moment with tremendous energy. We gave the word to 'stern all,'
which was obeyed in good earnest. As we passed the ship, I heard
the captain exclaim, 'There goes another boat!' She did go, to be sure,

through the water with all speed, but fortunately not to destruction. The monster chased us in this way for half a mile or more, during most of which time his jaws were within six or eight inches of the head of the boat. Every time he brought them together, the concussion could be heard at the distance of at least a mile. I intended to jump overboard if he caught the boat. I told Mr. Mayhew, the third mate, who held the steering oar, that the whale would turn over soon to spout, and that then would be our time to kill him. After becoming exhausted, he turned over to spout, and at the same instant we stopped the boat, and buried our lances deep in 'his life.' One tremendous convulsion of his frame followed, and all was still. He never troubled us more. We towed him to the ship, tried him out, and took ninety barrels of sperm oil from him.

"When we were cutting him in, we found two irons in his body, marked with the name of the ship Barclay, and belonging to the mate's boat. We afterwards learned that three months before, when the same whale was in lat. 5° S., lon. 105° W., he was attacked by the mate of the ship Barclay, who had a desperate struggle with him, in which he lost his life."

Captain Norton, at the time of the adventure with this whale, had "seen some service," but he freely confesses that he never before nor since (though he has had his buttons bitten off his shirt by a whale) has come in contact with such an ugly customer as "the rogue whale," as he was termed in sailor parlance. He seemed to possess the spirit of a demon, and looked as savage as a hungry hyena. Our readers may imagine the effect such an encounter would have upon a crew of "green hands." During the frightful chase of the boat by the whale, their faces were of a livid whiteness, and their hair stood erect. On their arrival at the first port, they all took to the mountains, and few, if any of them, have ever been seen since.

The *Citizen* was put on her course again, with strong breezes and fair wind. About five days after, we spoke with the *Benjamin Tucker*, but Captain Sands had taken no oil. In lat. 47° S. another whale was raised;

three boats were lowered in pursuit, but before he could be reached by the irons, he turned flukes, and was seen no more. Lost sight of the *Benjamin Tucker*. We shaped our course for Statan Land. In lat. 48° S. we experienced a very heavy gale from the south-west, which continued with great severity for twenty-four hours. We spoke with the bark Oscar, Captain Dexter, bound round the cape.

Statan Land in sight, passed seventeen ships, all bound for the cape. The *Citizen* was eleven days in doubling the cape, and experienced very heavy weather. In lat. 54° S. we raised the first right whale, but, blowing hard, could not lower. Whales were in sight several days in succession, but we could not lower, on account of rugged weather. In lat. 47° S. a ship was discovered with her boats down in pursuit of whales; came up with her; lowered for right whales, and chased them for an hour or more, but took none. At this time we spoke with the ship *Columbus* again, with one of her boats fastened to a whale. She had one boat stoven.

Passed St. Felix Islands, on the coast of Chili, and sighted the Gallipagos. In crossing the equator, it was calm for twenty-seven days, and but little progress was made during that time. On the 20th of April, 1852, after a passage of more than five months from New Bedford, we entered the port of Hilo.

Hilo is a port on the Island of Hawaii, one of the cluster of islands in the North Pacific Ocean called Sandwich Islands. They were discovered by Captains Cook and King in 1778, who gave them their present name, in honor of the first lord of the admiralty. The group consists of ten islands, but all of them are not inhabited; they extend from lat. 18° 50' to 22° 20' N., and from lon. 154° 53' to 160° 15' W., lying about one third of the distance from the western coast of Mexico to the eastern coast of China. By the census of 1849, the population of seven of the islands is given as follows: Hawaii, 27,204; Oahu, 23,145; Maui, 18,671; Kauhai, 6,941; Molokai, 3,429; Nuhua, 723; Lanai, 523; amounting to 80,641.

Most of these islands are volcanic and mountainous. In several places the volcanoes are in activity. Some of the mountains are of great height, being estimated at fifteen thousand feet.

The climate is warm, but not unhealthy, the winter being marked only by the prevalence of heavy rains between December and March.

A meteorological table gives as the greatest heat during the year, 88° of Fahrenheit; as the least, 61°. Some of these islands are distinguished for the cultivation of the yam, which affords quite a valuable supply for ships.

The situation of the Sandwich Islands renders them important to vessels navigating the Northern Pacific, and especially so to whalemen. The ports of Hilo, Lahaina, Honolulu, and a few others, are the resort of a large number of whale ships, for the purpose of obtaining recruits. They may be considered as a central point, where ships meet both in the fall and spring, and from whence all matters of intelligence are transmitted to San Francisco, and from the latter place to the Atlantic States.

Formerly all ship news and letters were brought from the islands to the Atlantic States by homeward bound ships around the Horn, which required for their passages from three and a half to five months. But now, in consequence of mail communications across the isthmus to San Francisco, and from thence to the islands, letters and other public intelligence from the last-named place reach us in six weeks or two months from date.

At the port of Hilo the ship was recruited for the Arctic. We remained in port fifteen days, sailed for Honolulu, and left letters for owners and for home. We touched at another port before proceeding to the north, and there we took in an additional supply of provisions, and then directed our course towards the straits.

In lon. 180° W. we hauled to the north towards the coast of Kamtschatka. Passed Copper Island. We saw many ships on our passage thus far, but we took no whales until June.

About this time we captured two whales off shore, and found great quantities of ice. Spoke with Captain Crowell, of the ship *Columbus*, and Captain Crosby, of the ship *Cornelius Howland*.

We went into the ice with Captain Crosby, in search of whales, and soon found them; boats were lowered; pursuit commenced; several were struck; but our irons drew, and we therefore lost them.

A gale of wind coming on and increasing, we worked out of the ice as soon as possible. We were at that time, when the gale commenced, some fifteen or twenty miles in the floating and broken masses, of varied thickness and dimensions, greatly obstructing the course of the ship, and rendering her situation at times exceedingly dangerous. But by constant tacking and wearing in order to avoid concussion with the ice, or being jammed between opposite pieces, both ships were finally worked out of the ice in safety.

On the inside of Cape Thaddeus, we saw a large number of ships; spoke with several, but they reported that whales were scarce.

We now put the ship on her course for Behring Straits. We took one whale off the Bay of the Holy Cross, which made the fourth since we left port. We sailed along the coast towards the east; land frequently in sight; foggy; heard many guns from ships for their boats.

When off Plover Bay, ten miles from land, we picked up a dead whale, having no irons in him, nor anchored, and therefore a lawful prize. Many dead whales are found by ships in course of the season, and especially when ice is prevalent. They are struck by different boats, and if in the vicinity of ice, they will surely make for it, and go under it or among it; under these circumstances the lines must be cut. After some time, the badly wounded whales die, and are picked up as before stated.

We passed between St. Lawrence Island and the main land, or Indian Point. The huts of the natives were plainly seen from the ship's deck; still working our way towards the straits. At this time, we were in company with the ship *Montezuma*, Captain Tower, and the ship *Almira*, Captain Jenks. Whales were seen going towards the north, as it is usual for them to do so at this season of the year.

We anchored in St. Lawrence Bay; weather foggy. The natives came off to trade, and brought their accustomed articles for traffic, such as deer and walrus skins, furs, teeth, &c. They take in exchange needles, fancy articles, tobacco, &c.

After a few days, the fog having cleared away somewhat, we stood towards the north again; heard guns; saw whales; still in company with afore-mentioned ships; blowing heavy; all the ships in sight were under double-reefed topsails; beating.

Passed East Cape. Saw whales, but they were working quickly to the north; we followed them in their track with all the sail we could carry on the ship; they came to loose, floating ice, into which they went and shortly disappeared. A novel, and yet a common sight was now witnessed;

the ice was covered with a vast number of walruses, which, to appearance, extended many miles.

The weather being fine for the season, the last part of June, in company with the *Almira*, Captain Jenks, we concluded we would go into the ice again, and if good fortune would have it so, we might capture a few whales.

Accidents occur not unfrequently when least expected, and sad ones, too, arise sometimes from the slightest circumstance, or inattention. Contact with icebergs, or large masses of block ice, when a ship is under sail, is highly dangerous. A momentary relaxation of vigilance on the part of the mariner may bring the ship's bows on the submerged part of an iceberg, whose sharp, needle-like points, hard as rock, instantly pierce the planks and timbers of a ship, and perhaps open a fatal leak. Many lamentable shipwrecks have doubtless resulted from this cause. In the long, heavy swell, so common in the open sea, the peril of floating ice is greatly increased, as the huge angular masses are rolled and ground against each other with a force which nothing can resist.

The striking of the *Citizen* against a mass of ice, which nearly resulted in the loss of the ship and the destruction of the voyage, was simply inattention or misunderstanding the word of command.

The man at the wheel was ordered not to "luff" the ship any more, but "steady," as she was approaching a mass of ice; indeed, ice was all around us, which would have passed us on our larboard bow, and thus we should have escaped a concussion; but instead of doing this, he put the wheel down, which brought the ship into the wind, and the consequence was, a large hole was stoven in her larboard bow; the ship began to leak badly. Casks were immediately filled with water, and placed on the starboard side of the ship, and thus in a measure heeled the ship, which brought the leak to a considerable extent out of the water; otherwise, she must have sunk in a very little time. So far as we were able, we temporarily repaired the injury, and made all possible sail on the ship, in order to seek some place of safety, where the whole extent of the damage could be ascertained.

In the present disabled and crippled condition of the ship, we felt it was exceedingly perilous and unsafe to remain even a single day in the Arctic. We therefore left the whale ground, and though our progress was slow, yet we put upon the ship all the sail she would bear, since on account of the leak she was very much heeled, and we were obliged to sail her in that condition.

Nor was it safe for our ship to be left alone to beat her way back two hundred miles or more, unaccompanied by another vessel, lest by some unforeseen circumstance,—an event not altogether improbable,—the ship might founder at sea, and all on board perish.

Captain Jenks, of the ship *Almira*, therefore, kindly proffered his services, with whatever aid he could give, and accompanied our ship nearly to the point of her destination, to the Bay of St. Lawrence, which was about two hundred miles distant from the place where the accident occurred.

When off East Cape, we obtained some plank from the ship *Citizen*, Captain Bailey, of Nantucket. We passed the heads of the bay, and, with shortened sail, we worked our way up more than thirty miles beyond the direction of any chart, our boats being sent ahead, and sounding the depth of water. We finally reached a point, and came to anchor in a little basin, or inlet, about one hundred and sixty feet from the shore, in five fathoms of water, completely landlocked.

Here in good earnest we commenced breaking out the fore hold abreast of the leak, and took out casks, shooks, &c., and careened the ship still more, which exposed at once the full extent of the damage which the ship had sustained from the ice.

It was found that several planks and timbers were badly stoven. Repairs were made with the utmost expedition; and in seven days from the time the ship went into the bay, she was out again, and on her way towards the north, as strong, and perhaps stronger than she was before.

We passed through the straits, and came to anchor north of East Cape, in company with the ship *E. Frazer*, Captain Taber, and the bark *Martha*, Captain Crocker. After lying there three or four days, we got under weigh and stood towards north by west, with high winds, and foggy. We heard whales blowing in the night. The next day whales were seen going north; we followed, and finally passed the "school." We changed the course of the ship, beat back, found them again, and commenced taking oil.

About the first of August, the fog having cleared away, we saw a large number of ships "cutting in" and "boiling out," actively engaged in securing a good season's work. We took several whales at this time. All were busy, and at work as fast as possible, in capturing whales, cutting and boiling. The whole scene, in which were some forty or fifty ships taking whales and stowing away oil, was one of exciting and cheering interest.

Such times as these are the whalemen's harvests.

On the 15th of August, during a heavy blow, we lost run of the whales. We spoke with several ships about this time, among which were the *Benjamin Morgan*, Captain Capel, and the *General Scott*, Captain Alexander Fisher.

From this last date to the 22d of September, we spoke with a great number of ships; sometimes whales were plenty, and at other times scarce; and the weather equally changeable; sometimes heavy blows, rainy, and foggy; and then again mild and pleasant.

Among others we spoke with Captain Henry Jernegan, and Captain John Fisher, both of whom are now no more, having finished their earthly voyages, and gone to their "long home."

TWENTY-THREE

The Cruise of the *Wasp*

Henry Cabot Lodge and Theodore Roosevelt

A crash as when some swollen cloud
Cracks o'er the tangled trees!
With side to side, and spar to spar,
Whose smoking decks are these?
I know St. George's blood-red cross,
Thou mistress of the seas,
But what is she whose streaming bars
Roll out before the breeze?
Ah, well her iron ribs are knit,
Whose thunders strive to quell
The bellowing throats, the blazing lips,
That pealed the Armada's knell!
The mist was cleared,—a wreath of stars
Rose o'er the crimsoned swell,
And, wavering from its haughty peak,
The cross of England fell!
—Holmes.

IN THE WAR OF 1812 THE LITTLE AMERICAN NAVY, INCLUDING ONLY a dozen frigates and sloops of war, won a series of victories against the English, the hitherto undoubted masters of the sea, that attracted an attention altogether out of proportion to the force of the combatants or the actual damage done. For one hundred and fifty years the English ships of war had failed to find fit rivals in those of any other European power, although they had been matched against each in turn; and when the unknown navy of the new nation growing up across the Atlantic did what no European navy had ever been able to do, not only the English and Americans, but the people of Continental Europe as well, regarded the feat as important out of all proportion to the material aspects of the case. The Americans first proved that the English could be beaten at their own game on the sea. They did what the huge fleets of France, Spain, and Holland had failed to do, and the great modern writers on naval warfare in Continental Europe—men like Jurien de la Graviere—have paid the same attention to these contests of frigates and sloops that they give to whole fleet actions of other wars.

Among the famous ships of the Americans in this war were two named the *Wasp*. The first was an eighteen-gun ship-sloop, which at the very outset of the war captured a British brig-sloop of twenty guns, after an engagement in which the British fought with great gallantry, but were knocked to Pieces, while the Americans escaped comparatively unscathed. Immediately afterward a British seventy-four captured the victor. In memory of her the Americans gave the same name to one of the new sloops they were building. These sloops were stoutly made, speedy vessels which in strength and swiftness compared favorably with any ships of their class in any other navy of the day, for the American shipwrights were already as famous as the American gunners and seamen. The new *Wasp*, like her sister ships, carried twenty-two guns and a crew of one hundred and seventy men, and was ship-rigged. Twenty of her guns were 32-pound carronades, while for bow-chasers she had two "long Toms." It was in the year 1814 that the *Wasp* sailed from the United States to prey on the navy and commerce of Great Britain. Her commander was a gallant South Carolinian named Captain Johnson Blakeley. Her crew were nearly all native Americans, and were an exceptionally fine set of men. Instead of staying near the American coasts or of sailing the high seas, the *Wasp* at once headed boldly for the English Channel, to carry the war to the very doors of the enemy.

At that time the English fleets had destroyed the navies of every other power of Europe, and had obtained such complete supremacy over the French that the French fleets were kept in port. Off these ports lay the great squadrons of the English ships of the line, never, in gale or in calm, relaxing their watch upon the rival war-ships of the French emperor. So close was the blockade of the French ports, and so hopeless were the French of making headway in battle with their antagonists, that not only the great French three-deckers and two-deckers, but their frigates and sloops as well, lay harmless in their harbors, and the English ships patroled the seas unchecked in every direction. A few French privateers still slipped out now and then, and the far bolder and more formidable American privateersmen drove hither and thither across the ocean in their swift schooners and brigantines, and harried the English commerce without mercy.

The *Wasp* proceeded at once to cruise in the English Channel and off the coasts of England, France, and Spain. Here the water was traversed continually by English fleets and squadrons and single ships of war, which were sometimes convoying detachments of troops for Wellington's Peninsular army, sometimes guarding fleets of merchant vessels bound homeward, and sometimes merely cruising for foes. It was this spot, right in the teeth of the British naval power, that the *Wasp* chose for her cruising ground. Hither and thither she sailed through the narrow seas, capturing and destroying the merchantmen, and by the seamanship of her crew and the skill and vigilance of her commander, escaping the pursuit of frigate and ship of the line. Before she had been long on the ground, one June morning, while in chase of a couple of merchant ships, she spied a sloop of war, the British brig *Reindeer*, of eighteen guns and a hundred and twenty men. The *Reindeer* was a weaker ship than the *Wasp*, her guns were lighter, and her men fewer; but her commander, Captain Manners, was one of the most gallant men in the splendid British navy, and he promptly took up the gage of battle which the *Wasp* threw down.

The day was calm and nearly still; only a light wind stirred across the sea. At one o'clock the *Wasp*'s drum beat to quarters, and the sailors and marines gathered at their appointed posts. The drum of the *Reindeer* responded to the challenge, and with her sails reduced to fighting trim, her guns run out, and every man ready, she came down upon the Yankee ship. On her forecastle she had rigged a light carronade, and coming up from behind, she five times discharged this pointblank into the American sloop; then in the light air the latter luffed round, firing her guns as they bore, and the two ships engaged yard-arm to yard-arm.

The guns leaped and thundered as the grimy gunners hurled them out to fire and back again to load, working like demons. For a few minutes the cannonade was tremendous, and the men in the tops could hardly see the decks for the wreck of flying splinters. Then the vessels ground together, and through the open ports the rival gunners hewed, hacked, and thrust at one another, while the black smoke curled up from between the hulls. The English were suffering terribly. Captain Manners himself was wounded, and realizing that he was doomed to defeat unless by some desperate effort he could avert it, he gave the signal to board. At the call the boarders gathered, naked to the waist, black with powder and spattered with blood, cutlass and pistol in hand. But the Americans were ready. Their marines were drawn up on deck, the pikemen stood behind the bulwarks, and the officers watched, cool and alert, every movement of the foe. Then the British sea-dogs tumbled aboard, only to perish by shot or steel. The combatants slashed and stabbed with savage fury, and the assailants were driven back. Manners sprang to their head to lead them again himself, when a ball fired by one of the sailors in the American tops crashed through his skull, and he fell, sword in hand, with his face to the foe, dying as honorable a death as ever a brave man died in fighting against odds for the flag of his country. As he fell the American officers passed the word to board. With wild cheers the fighting sailormen sprang forward, sweeping the wreck of the British force before them, and in a minute the *Reindeer* was in their possession. All of her officers, and nearly two thirds of the crew, were killed or wounded; but they had proved themselves as skilful as they were brave, and twenty-six of the Americans had been killed or wounded.

The *Wasp* set fire to her prize, and after retiring to a French port to refit, came out again to cruise. For some time she met no antagonist of her own size with which to wage war, and she had to exercise the sharpest vigilance to escape capture. Late one September afternoon, when she could see ships of war all around her, she selected one which was isolated from the others, and decided to run alongside her and try to sink her after nightfall.

Accordingly she set her sails in pursuit, and drew steadily toward her antagonist, a big eighteen-gun brig, the *Avon*, a ship more powerful than the *Reindeer*. The Avon kept signaling to two other British war vessels which were in sight—one an eighteen-gun brigand the other a twenty-gun ship; they were so close that the *Wasp* was afraid they would interfere before the combat could be ended. Nevertheless, Blakeley persevered, and made his attack with equal skill and daring.

It was after dark when he ran alongside his opponent, and they began forthwith to exchange furious broadsides. As the ships plunged and wallowed in the seas, the Americans could see the clusters of topmen in the rigging of their opponent, but they knew nothing of the vessel's name or of her force, save only so far as they felt it. The firing was fast and furious, but the British shot with bad aim, while the skilled American gunners hulled their opponent at almost every discharge. In a very few minutes the *Avon* was in a sinking condition, and she struck her flag and cried for quarter, having lost forty or fifty men, while but three of the Americans had fallen.

Before the *Wasp* could take possession of her opponent, however, the two war vessels to which the Avon had been signaling came up. One of them fired at the *Wasp*, and as the latter could not fight two new foes, she ran off easily before the wind. Neither of her new antagonists followed her, devoting themselves to picking up the crew of the sinking *Avon*.

It would be hard to find a braver feat more skilfully performed than this; for Captain Blakeley, with hostile foes all round him, had closed with and sunk one antagonist not greatly his inferior in force, suffering hardly any loss himself, while two of her friends were coming to her help.

Both before and after this the *Wasp* cruised hither and thither making prizes. Once she came across a convoy of ships bearing arms and munitions to Wellington's army, under the care of a great two-decker. Hovering about, the swift sloop evaded the two-decker's movements, and actually cut out and captured one of the transports she was guarding, making her escape unharmed. Then she sailed for the high seas. She made several other prizes, and on October 9 spoke a Swedish brig.

This was the last that was ever heard of the gallant *Wasp*. She never again appeared, and no trace of any of those aboard her was ever found. Whether she was wrecked on some desert coast, whether she foundered in some furious gale, or what befell her none ever knew. All that is certain is that she perished, and that all on board her met death in some one of the myriad forms in which it must always be faced by those who go down to the sea in ships; and when she sank there sank one of the most gallant ships of the American navy, with as brave a captain and crew as ever sailed from any port of the New World.

Twenty-Four

Dirty Weather

Joseph Conrad

Observing the steady fall of the barometer, Captain MacWhirr thought, "There's some dirty weather knocking about." This is precisely what he thought. He had had an experience of moderately dirty weather—the term dirty as applied to the weather implying only moderate discomfort to the seaman. Had he been informed by an indisputable authority that the end of the world was to be finally accomplished by a catastrophic disturbance of the atmosphere, he would have assimilated the information under the simple idea of dirty weather, and no other, because he had no experience of cataclysms, and belief does not necessarily imply comprehension. The wisdom of his county had pronounced by means of an Act of Parliament that before he could be considered as fit to take charge of a ship he should be able to answer certain simple questions on the subject of circular storms such as hurricanes, cyclones, typhoons; and apparently he had answered them, since he was now in command of the *Nan-Shan* in the China seas during the season of typhoons. But if he had answered he remembered nothing of it. He was, however, conscious of being made uncomfortable by the clammy heat. He came out on the bridge, and found no relief to this oppression. The air seemed thick. He gasped like a fish, and began to believe himself greatly out of sorts.

The *Nan-Shan* was ploughing a vanishing furrow upon the circle of the sea that had the surface and the shimmer of an undulating piece of gray silk. The sun, pale and without rays, poured down leaden heat in a strangely indecisive light, and the Chinamen were lying prostrate about the decks. Their bloodless, pinched, yellow faces were like the faces of bilious invalids. Captain MacWhirr noticed two of them especially, stretched out on their backs below the bridge. As soon as they had closed their eyes they seemed dead. Three others, however, were quarrelling barbarously away forward; and one big fellow, half naked, with herculean shoulders, was hanging limply over a winch; another, sitting on the deck, his knees up and his head drooping sideways in a girlish attitude, was plaiting his pigtail with infinite languor depicted in his whole person and in the very movement of his fingers. The smoke struggled with difficulty out of the funnel, and instead of streaming away spread itself out like an infernal sort of cloud, smelling of sulphur and raining soot all over the decks.

"What the devil are you doing there, Mr. Jukes?" asked Captain MacWhirr. This unusual form of address, though mumbled rather than spoken, caused the body of Mr. Jukes to start as though it had been prodded under the fifth rib. He had had a low bench brought on the bridge, and sitting on it, with a length of rope curled about his feet and a piece of canvas stretched over his knees, was pushing a sail needle vigorously. He looked up, and his surprise gave to his eyes an expression of innocence and candour.

"I am only roping some of that new set of bags we made last trip for whipping up coals," he remonstrated, gently. "We shall want them for the next coaling, sir."

"What became of the others?"

"Why, worn out of course, sir."

Captain MacWhirr, after glaring down irresolutely at his chief mate, disclosed the gloomy and cynical conviction that more than half of them had been lost overboard, "if only the truth was known," and retired to the other end of the bridge. Jukes, exasperated by this unprovoked attack, broke the needle at the second stitch, and dropping his work got up and cursed the heat in a violent undertone.

The propeller thumped, the three Chinamen forward had given up squabbling very suddenly, and the one who had been plaiting his tail clasped his legs and stared dejectedly over his knees. The lurid sunshine cast faint and sickly shadows. The swell ran higher and swifter every moment, and the ship lurched heavily in the smooth, deep hollows of the sea.

"I wonder where that beastly swell comes from," said Jukes aloud, recovering himself after a stagger.

"Northeast," grunted the literal MacWhirr, from his side of the bridge. "There's some dirty weather knocking about. Go and look at the glass."

When Jukes came out of the chart-room, the cast of his countenance had changed to thoughtfulness and concern. He caught hold of the bridge-rail and stared ahead.

The temperature in the engine room had gone up to a hundred and seventeen degrees. Irritated voices were ascending through the skylight and through the fiddle of the stokehold in a harsh and resonant uproar, mingled with angry clangs and scrapes of metal, as if men with limbs of iron and throats of bronze had been quarrelling down there. The second engineer was falling foul of the stokers for letting the steam go down. He was a man with arms like a blacksmith, and generally feared; but that afternoon the stokers were answering him back recklessly, and slammed the furnace doors with the fury of despair. Then the noise ceased suddenly, and the second engineer appeared, emerging out of the stokehold streaked with grime and soaking wet like a chimney sweep coming out of a well. As soon as his head was clear of the fiddle he began to scold Jukes for not trimming properly the stokehold ventilators; and in answer Jukes made with his hands deprecatory soothing signs meaning:

"No wind—can't be helped—you can see for yourself."

But the other wouldn't hear reason. His teeth flashed angrily in his dirty face. He didn't mind, he said, the trouble of punching their blanked heads down there, blank his soul, but did the condemned sailors think you could keep steam up in the Godforsaken boilers simply by knocking the blanked stokers about? No, by George! You had to get some draught, too—may he be everlastingly blanked for a swabheaded deckhand if you didn't! And the chief, too, rampaging before the steam gauge and carrying on like a lunatic up and down the engine room ever since noon. What did Jukes think he was stuck up there for, if he couldn't get one of his decayed, good-for-nothing deckcripples to turn the ventilators to the wind?

The relations of the "engine room" and the "deck" of the *Nan-Shan* were, as is known, of a brotherly nature; therefore Jukes leaned over and begged the other in a restrained tone not to make a disgusting ass of himself; the skipper was on the other side of the bridge. But the second declared mutinously that he didn't care a rap who was on the other side of the bridge, and Jukes, passing in a flash from lofty disapproval into a state of exaltation, invited him in unflattering terms to come up and twist the beastly things to please himself, and catch such wind as a donkey of his sort could find. The second rushed up to the fray. He flung himself at the port ventilator as though he meant to tear it out bodily and toss it overboard. All he did was to move the cowl round a few inches, with an enormous expenditure of force, and seemed spent in the effort. He leaned against the back of the wheelhouse, and Jukes walked up to him.

"Oh, Heavens!" ejaculated the engineer in a feeble voice. He lifted his eyes to the sky, and then let his glassy stare descend to meet the horizon that, tilting up to an angle of forty degrees, seemed to hang on a slant for a while and settled down slowly.

"Heavens! Phew! What's up, anyhow?"

Jukes, straddling his long legs like a pair of compasses, put on an air of superiority. "We're going to catch it this time," he said. "The barometer is tumbling down like anything, Harry. And you trying to kick up that silly row. . . ."

The word "barometer" seemed to revive the second engineer's mad animosity. Collecting afresh all his energies, he directed Jukes in a low and brutal tone to shove the unmentionable instrument down his gory throat. Who cared for his crimson barometer? It was the steam—the steam—that was going down; and what between the firemen going faint and the chief going silly, it was worse than a dog's life for him; he didn't care a tinker's curse how soon the whole show was blown out of the water. He seemed on the point of having a cry, but after regaining his breath he muttered darkly, "I'll faint them," and dashed off. He stopped upon the fiddle long enough to shake his fist at the unnatural daylight, and dropped into the dark hole with a whoop.

When Jukes turned, his eyes fell upon the rounded back and the big red ears of Captain MacWhirr, who had come across. He did not look at his chief officer, but said at once,

"That's a very violent man, that second engineer."

"Jolly good second, anyhow," grunted Jukes. "They can't keep up steam," he added, rapidly, and made a grab at the rail against the coming lurch. Captain MacWhirr, unprepared, took a run and brought himself up with a jerk by an awning stanchion.

"A profane man," he said, obstinately. "If this goes on, I'll have to get rid of him the first chance."

"It's the heat," said Jukes. "The weather's awful. It would make a saint swear. Even up here I feel exactly as if I had my head tied up in a woollen blanket."

Captain MacWhirr looked up.

"D'ye mean to say, Mr. Jukes, you ever had your head tied up in a blanket? What was that for?"

"It's a manner of speaking, sir," said Jukes, stolidly.

"Some of you fellows do go on! What's that about saints swearing? I wish you wouldn't talk so wild. What sort of saint would that be that would swear? No more saint than yourself, I expect. And what's a blanket got to do with it—or the weather either. . . . The heat does not make me swear—does it? It's filthy bad temper. That's what it is. And what's the good of your talking like this?"

Thus Captain MacWhirr expostulated against the use of images in speech, and at the end electrified Jukes by a contemptuous snort, followed by words of passion and resentment: "Damme! I'll fire him out of the ship if he don't look out."

And Jukes, incorrigible, thought: "Goodness me! Somebody's put a new inside to my old man. Here's temper, if you like. Of course it's the weather; what else? It would make an angel quarrelsome—let alone a saint."

All the Chinamen on deck appeared at their last gasp. At its setting the sun had a diminished diameter and an expiring brown, rayless glow, as if millions of centuries elapsing since the morning had brought it near its end. A dense bank of cloud became visible to the northward; it had a sinister dark olive tint, and lay low and motionless upon the sea, resembling a solid obstacle in the path of the ship. She went floundering towards it like an exhausted creature driven to its death. The coppery twilight retired slowly, and the darkness brought out overhead a swarm of unsteady, big stars, that, as if blown upon, flickered exceedingly and seemed to hang very near the earth. At eight o'clock Jukes went into the chartroom to write up the ship's log.

He copied neatly out of the rough-book the number of miles, the course of the ship, and in the column for "wind" scrawled the word "calm" from top to bottom of the eight hours since noon. He was exasperated by the continuous, monotonous rolling of the ship. The heavy inkstand would slide away in a manner that suggested perverse intelligence in dodging the pen. Having written in the large space under the head of "Remarks," "Heat very oppressive," he stuck the end of the penholder in his teeth, pipe fashion, and mopped his face carefully.

"Ship rolling heavily in a high cross swell," he began again, and commented to himself, "Heavily is no word for it." Then he wrote: "Sunset threatening, with a low bank of clouds to N. and E. Sky clear overhead."

Sprawling over the table with arrested pen, he glanced out of the door, and in that frame of his vision he saw all the stars flying upwards between the teakwood jambs on a black sky. The whole lot took flight together and disappeared, leaving only a blackness flecked with white flashes, for the sea was as black as the sky and speckled with foam afar. The stars that had flown to the roll came back on the return swing of the ship, rushing downwards in their glittering multitude, not of fiery points, but enlarged to tiny discs brilliant with a clear wet sheen.

Jukes watched the flying big stars for a moment, and then wrote: "8 p.m. Swell increasing. Ship labouring and taking water on her decks. Battened down the hatches for the night. Barometer still falling." He paused, and thought to himself, "Perhaps nothing whatever'll come of it." And then he closed resolutely his entries: "Every appearance of a typhoon coming on."

On going out he had to stand aside, and Captain MacWhirr strode over the doorstep without saying a word or making a sign.

"Shut the door, Mr. Jukes, will you?" he cried from within. Jukes turned back to do so, muttering ironically: "Afraid to catch cold, I suppose." It was his watch below, but he yearned for communion with his kind; and he remarked cheerily to the second mate: "Doesn't look so bad, after all—does it?"

The second mate was marching to and fro on the bridge, tripping down with small steps one moment, and the next climbing with difficulty the shifting slope of the deck. At the sound of Jukes' voice he stood still, facing forward, but made no reply.

"Hallo! That's a heavy one," said Jukes, swaying to meet the long roll till his lowered hand touched the planks. This time the second mate made in his throat a noise of an unfriendly nature.

He was an oldish, shabby little fellow, with bad teeth and no hair on his face. He had been shipped in a hurry in Shanghai, that trip when the second officer brought from home had delayed the ship three hours in port by contriving (in some manner Captain MacWhirr could never understand) to fall overboard into an empty coallighter lying alongside, and had to be sent ashore to the hospital with concussion of the brain and a broken limb or two.

Jukes was not discouraged by the unsympathetic sound. "The Chinamen must be having a lovely time of it down there," he said. "It's lucky for them the old girl has the easiest roll of any ship I've ever been in. There now! This one wasn't so bad."

"You wait," snarled the second mate.

With his sharp nose, red at the tip, and his thin pinched lips, he always looked as though he were raging inwardly; and he was concise in his speech to the point of rudeness. All his time off duty he spent in his cabin with the door shut, keeping so still in there that he was supposed to fall asleep as soon as he had disappeared; but the man who came in to wake him for his watch on deck would invariably find him with his eyes wide open, flat on his back in the bunk, and glaring irritably from a soiled pillow. He never wrote any letters, did not seem to hope for news from anywhere; and though he had been heard once to mention West Hartlepool, it was with extreme bitterness, and only in connection with the extortionate charges of a boarding-house. He was one of those men who are picked up at need in the ports of the world. They are competent enough, appear hopelessly hard up, show no evidence of any sort of vice, and carry about them all the signs of manifest failure. They come aboard on an emergency, care for no ship afloat, live in their own atmosphere of casual connection amongst their shipmates who know nothing of them, and make up their minds to leave at inconvenient times. They clear out with no words of leavetaking in some God-forsaken port other men would fear to be stranded in, and go ashore in company of a shabby sea-chest, corded like a treasure-box, and with an air of shaking the ship's dust off their feet.

"You wait," he repeated, balanced in great swings with his back to Jukes, motionless and implacable.

"Do you mean to say we are going to catch it hot?" asked Jukes with boyish interest.

"Say? . . . I say nothing. You don't catch me," snapped the little second mate, with a mixture of pride, scorn, and cunning, as if Jukes' question had been a trap cleverly detected. "Oh, no! None of you here shall make a fool of me if I know it," he mumbled to himself.

Jukes reflected rapidly that this second mate was a mean little beast, and in his heart he wished poor Jack Allen had never smashed himself up in the coal-lighter. The far-off blackness ahead of the ship was like another night seen through the starry night of the earth—the starless night of the immensities beyond the created universe, revealed in its appalling stillness through a low fissure in the glittering sphere of which the earth is the kernel.

"Whatever there might be about," said Jukes, "we are steaming straight into it."

"You've said it," caught up the second mate, always with his back to Jukes. "You've said it, mind—not I."

"Oh, go to Jericho!" said Jukes, frankly; and the other emitted a triumphant little chuckle.

"You've said it," he repeated.

"And what of that?"

"I've known some real good men get into trouble with their skippers for saying a dam' sight less," answered the second mate feverishly. "Oh, no! You don't catch me."

"You seem deucedly anxious not to give yourself away," said Jukes, completely soured by such absurdity. "I wouldn't be afraid to say what I think."

"Aye, to me! That's no great trick. I am nobody, and well I know it."

The ship, after a pause of comparative steadiness, started upon a series of rolls, one worse than the other, and for a time Jukes, preserving his equilibrium, was too busy to open his mouth. As soon as the violent swinging had quieted down somewhat, he said: "This is a bit too much of a good thing. Whether anything is coming or not I think she ought to be put head on to that swell. The old man is just gone in to lie down. Hang me if I don't speak to him."

But when he opened the door of the chartroom he saw his captain reading a book. Captain MacWhirr was not lying down: he was standing up with one hand grasping the edge of the bookshelf and the other holding open before his face a thick volume. The lamp wriggled in the gimbals, the loosened books toppled from side to side on the shelf, the long barometer swung in jerky circles, the table altered its slant every moment. In the midst of all this stir and movement Captain MacWhirr, holding on, showed his eyes above the upper edge, and asked, "What's the matter?"

"Swell getting worse, sir."

"Noticed that in here," muttered Captain MacWhirr. "Anything wrong?"

Jukes, inwardly disconcerted by the seriousness of the eyes looking at him over the top of the book, produced an embarrassed grin.

"Rolling like old boots," he said, sheepishly.

"Aye! Very heavy—very heavy. What do you want?"

At this Jukes lost his footing and began to flounder. "I was thinking of our passengers," he said, in the manner of a man clutching at a straw.

"Passengers?" wondered the Captain, gravely. "What passengers?"

"Why, the Chinamen, sir," explained Jukes, very sick of this conversation.

"The Chinamen! Why don't you speak plainly? Couldn't tell what you meant. Never heard a lot of Chinamen spoken of as passengers before. Passengers, indeed! What's come to you?"

Captain MacWhirr, closing the book on his forefinger, lowered his arm and looked completely mystified. "Why are you thinking of the Chinamen, Mr. Jukes?" he inquired.

Jukes took a plunge, like a man driven to it. "She's rolling her decks full of water, sir. Thought you might put her head on perhaps—for a while. Till this goes down a bit—very soon, I dare say. Head to the eastward. I never knew a ship roll like this."

He held on in the doorway, and Captain MacWhirr, feeling his grip on the shelf inadequate, made up his mind to let go in a hurry, and fell heavily on the couch.

"Head to the eastward?" he said, struggling to sit up. "That's more than four points off her course."

"Yes, sir. Fifty degrees.... Would just bring her head far enough round to meet this...."

Captain MacWhirr was now sitting up. He had not dropped the book, and he had not lost his place.

"To the eastward?" he repeated, with dawning astonishment. "To the ... Where do you think we are bound to? You want me to haul a fullpowered steamship four points off her course to make the Chinamen comfortable! Now, I've heard more than enough of mad things done in the world—but this.... If I didn't know you, Jukes, I would think you were in liquor. Steer four points off.... And what afterwards? Steer four points over the other way, I suppose, to make the course good. What put it into your head that I would start to tack a steamer as if she were a sailingship?"

"Jolly good thing she isn't," threw in Jukes, with bitter readiness. "She would have rolled every blessed stick out of her this afternoon."

"Aye! And you just would have had to stand and see them go," said Captain MacWhirr, showing a certain animation. "It's a dead calm, isn't it?"

"It is, sir. But there's something out of the common coming, for sure."

"Maybe. I suppose you have a notion I should be getting out of the way of that dirt," said Captain MacWhirr, speaking with the utmost simplicity of manner and tone, and fixing the oilcloth on the floor with a heavy stare. Thus he noticed neither Jukes' discomfiture nor the mixture of vexation and astonished respect on his face.

"Now, here's this book," he continued with deliberation, slapping his thigh with the closed volume. "I've been reading the chapter on the storms there."

This was true. He had been reading the chapter on the storms. When he had entered the chartroom, it was with no intention of taking the book down. Some influence in the air—the same influence, probably, that caused the steward to bring without orders the Captain's sea-boots and oilskin coat up to the chart-room—had as it were guided his hand to the shelf; and without taking the time to sit down he had waded with a conscious effort into the terminology of the subject. He lost himself amongst advancing semi-circles, left-and right-hand quadrants, the curves of the tracks, the probable bearing of the centre, the shifts of wind and the readings of barometer. He tried to bring all these things into a definite relation to himself, and ended by becoming contemptuously angry with such a lot of words, and with so much advice, all headwork and supposition, without a glimmer of certitude.

"It's the damnedest thing, Jukes," he said. "If a fellow was to believe all that's in there, he would be running most of his time all over the sea trying to get behind the weather."

Again he slapped his leg with the book; and Jukes opened his mouth, but said nothing.

"Running to get behind the weather! Do you understand that, Mr. Jukes? It's the maddest thing!" ejaculated Captain MacWhirr, with pauses, gazing at the floor profoundly. "You would think an old woman had been writing this. It passes me. If that thing means anything useful, then it means that I should at once alter the course away, away to the devil somewhere, and come booming down on Fuchau from the northward at the tail of this dirty weather that's supposed to be knocking about in our way. From the north! Do you understand, Mr. Jukes? Three hundred extra miles to the distance, and a pretty coal bill to show. I couldn't bring myself to do that if every word in there was gospel truth, Mr. Jukes. Don't you expect me. . . ."

And Jukes, silent, marvelled at this display of feeling and loquacity.

"But the truth is that you don't know if the fellow is right, anyhow. How can you tell what a gale is made of till you get it?

"He isn't aboard here, is he? Very well. Here he says that the centre of them things bears eight points off the wind; but we haven't got any wind, for all the barometer falling. Where's his centre now?"

"We will get the wind presently," mumbled Jukes.

"Let it come, then," said Captain MacWhirr, with dignified indignation. "It's only to let you see, Mr. Jukes, that you don't find everything in books. All these rules for dodging breezes and circumventing the winds of heaven, Mr. Jukes, seem to me the maddest thing, when you come to look at it sensibly."

He raised his eyes, saw Jukes gazing at him dubiously, and tried to illustrate his meaning.

"About as queer as your extraordinary notion of dodging the ship head to sea, for I don't know how long, to make the Chinamen comfortable; whereas all we've got to do is to take them to Fuchau, being timed to get there before noon on Friday. If the weather delays me—very well. There's your logbook to talk straight about the weather. But suppose I went swinging off my course and came in two days late, and they asked me: 'Where have you been all that time, Captain?' What could I say to that?

'Went around to dodge the bad weather,' I would say. 'It must've been dam' bad,' they would say. 'Don't know,' I would have to say; 'I've dodged clear of it.' See that, Jukes? I have been thinking it all out this afternoon."

He looked up again in his unseeing, unimaginative way. No one had ever heard him say so much at one time. Jukes, with his arms open in the doorway, was like a man invited to behold a miracle. Unbounded wonder was the intellectual meaning of his eye, while incredulity was seated in his whole countenance.

"A gale is a gale, Mr. Jukes," resumed the Captain, "and a fullpowered steamship has got to face it. There's just so much dirty weather knocking about the world, and the proper thing is to go through it with none of what old Captain Wilson of the Melita calls 'storm strategy.' The other day ashore I heard him hold forth about it to a lot of shipmasters who came in and sat at a table next to mine. It seemed to me the greatest nonsense.

"He was telling them how he outmanoeuvred, I think he said, a terrific gale, so that it never came nearer than fifty miles to him. A neat piece of headwork he called it. How he knew there was a terrific gale fifty miles off beats me altogether. It was like listening to a crazy man. I would have thought Captain Wilson was old enough to know better."

Captain MacWhirr ceased for a moment, then said, "It's your watch below, Mr. Jukes?"

Jukes came to himself with a start. "Yes, sir."

"Leave orders to call me at the slightest change," said the Captain. He reached up to put the book away, and tucked his legs upon the couch. "Shut the door so that it don't fly open, will you? I can't stand a door banging. They've put a lot of rubbishy locks into this ship, I must say."

Captain MacWhirr closed his eyes.

He did so to rest himself. He was tired, and he experienced that state of mental vacuity which comes at the end of an exhaustive discussion that has liberated some belief matured in the course of meditative years. He had indeed been making his confession of faith, had he only known it; and its effect was to make Jukes, on the other side of the door, stand scratching his head for a good while.

Captain MacWhirr opened his eyes.

He thought he must have been asleep. What was that loud noise? Wind? Why had he not been called? The lamp wriggled in its gimbals, the barometer swung in circles, the table altered its slant every moment; a pair of limp seaboots with collapsed tops went sliding past the couch. He put out his hand instantly, and captured one.

Jukes' face appeared in a crack of the door: only his face, very red, with staring eyes. The flame of the lamp leaped, a piece of paper flew up, a rush of air enveloped Captain MacWhirr.

Beginning to draw on the boot, he directed an expectant gaze at Jukes' swollen, excited features.

"Came on like this," shouted Jukes, "five minutes ago . . . all of a sudden."

The head disappeared with a bang, and a heavy splash and patter of drops swept past the closed door as if a pailful of melted lead had been flung against the house. A whistling could be heard now upon the deep vibrating noise outside. The stuffy chartroom seemed as full of draughts as a shed. Captain MacWhirr collared the other seaboot on its violent passage along the floor. He was not flustered, but he could not find at once the opening for inserting his foot. The shoes he had flung off were scurrying from end to end of the cabin, gambolling playfully over each other like puppies. As soon as he stood up he kicked at them viciously, but without effect.

He threw himself into the attitude of a lunging fencer, to reach after his oilskin coat; and afterwards he staggered all over the confined space while he jerked himself into it. Very grave, straddling his legs far apart, and stretching his neck, he started to tie deliberately the strings of his sou'wester under his chin, with thick fingers that trembled slightly. He went through all the movements of a woman putting on her bonnet before a glass, with a strained, listening attention, as though he had expected every moment to hear the shout of his name in the confused clamour that had suddenly beset his ship. Its increase filled his ears while he was getting ready to go out and confront whatever it might mean. It was tumultuous and very loud—made up of the rush of the wind, the crashes of the sea, with that prolonged deep vibration of the air, like the roll of an immense and remote drum beating the charge of the gale.

He stood for a moment in the light of the lamp, thick, clumsy, shape-less in his panoply of combat, vigilant and red-faced.

"There's a lot of weight in this," he muttered.

As soon as he attempted to open the door the wind caught it. Cling-ing to the handle, he was dragged out over the doorstep, and at once found himself engaged with the wind in a sort of personal scuffle whose object was the shutting of that door. At the last moment a tongue of air scurried in and licked out the flame of the lamp.

Ahead of the ship he perceived a great darkness lying upon a multi-tude of white flashes; on the starboard beam a few amazing stars drooped, dim and fitful, above an immense waste of broken seas, as if seen through a mad drift of smoke.

On the bridge a knot of men, indistinct and toiling, were making great efforts in the light of the wheelhouse windows that shone mistily on their heads and backs. Suddenly darkness closed upon one pane, then on another. The voices of the lost group reached him after the manner of men's voices in a gale, in shreds and fragments of forlorn shouting snatched past the ear. All at once Jukes appeared at his side, yelling, with his head down.

"Watch—put in—wheelhouse shutters—glass—afraid—blow in."

Jukes heard his commander upbraiding.

"This—come—anything—warning—call me."

He tried to explain, with the uproar pressing on his lips.

"Light—air—remained—bridge—sudden—northeast—could turn—thought—you—sure—hear."

They had gained the shelter of the weathercloth, and could converse with raised voices, as people quarrel.

"I got the hands along to cover up all the ventilators. Good job I had remained on deck. I didn't think you would be asleep, and so . . . What did you say, sir? What?"

"Nothing," cried Captain MacWhirr. "I said—all right."

"By all the powers! We've got it this time," observed Jukes in a howl.

"You haven't altered her course?" inquired Captain MacWhirr, straining his voice.

"No, sir. Certainly not. Wind came out right ahead. And here comes the head sea."

A plunge of the ship ended in a shock as if she had landed her forefoot upon something solid. After a moment of stillness a lofty flight of sprays drove hard with the wind upon their faces.

"Keep her at it as long as we can," shouted Captain MacWhirr.

Before Jukes had squeezed the salt water out of his eyes all the stars had disappeared.

Twenty-Five

An Introduction to Informality

Erskine Childers

FROM FLUSHING EASTWARD TO HAMBURG, THEN NORTHWARD TO Flensburg, I cut short the next day's sultry story. Past dyke and windmill and still canals, on to blazing stubbles and roaring towns; at the last, after dusk, through a quiet level region where the train pottered from one lazy little station to another, and at ten o'clock I found myself, stiff and stuffy, on the platform at Flensburg, exchanging greetings with Davies.

"It's awfully good of you to come."

"Not at all; it's very good of you to ask me."

We were both of us ill at ease. Even in the dim gaslight he clashed on my notions of a yachtsman—no cool white ducks or neat blue serge; and where was the snowy crowned yachting cap, that precious charm that so easily converts a landsman into a dashing mariner? Conscious that this impressive uniform, in high perfection, was lying ready in my portmanteau, I felt oddly guilty. He wore an old Norfolk jacket, muddy brown shoes, grey flannel trousers (or had they been white?), and an ordinary tweed cap. The hand he gave me was horny, and appeared to be stained with paint; the other one, which carried a parcel, had a bandage on it which would have borne renewal. There was an instant of mutual inspection. I thought he gave me a shy, hurried scrutiny as though to test past conjectures, with something of anxiety in it, and perhaps (save the mark!) a tinge of admiration. The face was familiar, and yet not familiar; the pleasant blue eyes, open, clean-cut features, unintellectual forehead were the same; so were the brisk and impulsive movements; there was some change; but the moment of awkward hesitation was over and the light was bad; and, while strolling down the platform for my luggage, we chatted with constraint about trivial things.

"By the way," he suddenly said, laughing, "I'm afraid I'm not fit to be seen; but it's so late it doesn't matter. I've been painting hard all day, and just got it finished. I only hope we shall have some wind tomorrow—it's been hopelessly calm lately. I say, you've brought a good deal of stuff," he concluded, as my belongings began to collect.

Here was a reward for my submissive exertions in the far east!

"You gave me a good many commissions!"

"Oh, I didn't mean those things," he said, absently. "Thanks for bringing them, by the way. That's the stove, I suppose; cartridges, this one, by the weight. You got the rigging screws all right, I hope? They're not really necessary, of course" (I nodded vacantly, and felt a little hurt); "but they're simpler than lanyards, and you can't get them here. It's that portmanteau," he said, slowly, measuring it with a doubtful eye. "Never mind! we'll try. You couldn't do with the Gladstone only, I suppose? You see, the dinghy—h'm, and there's the hatchway, too"—he was lost in thought.

"Anyhow, we'll try. I'm afraid there are no cabs; but it's quite near, and the porter'll help."

Sickening forebodings crept over me, while Davies shouldered my Gladstone and clutched at the parcels.

"Aren't your men here?" I asked, faintly.

"Men?" He looked confused. "Oh, perhaps I ought to have told you, I never have any paid hands; it's quite a small boat, you know—I hope you didn't expect luxury. I've managed her single-handed for some time. A man would be no use, and a horrible nuisance." He revealed these appalling truths with a cheerful assurance, which did nothing to hide a naive apprehension of their effect on me. There was a check in our mobilization.

"It's rather late to go on board, isn't it?" I said, in a wooden voice. Someone was turning out the gaslights, and the porter yawned ostentatiously. "I think I'd rather sleep at an hotel to-night." A strained pause.

"Oh, of course you can do that, if you like," said Davies, in transparent distress of mind. "But it seems hardly worth while to cart this stuff all the way to an hotel (I believe they're all on the other side of the harbour), and back again to the boat tomorrow. She's quite comfortable, and you're sure to sleep well, as you're tired."

"We can leave the things here," I argued feebly, "and walk over with my bag."

"Oh, I shall have to go aboard anyhow," he rejoined; "I never sleep on shore."

He seemed to be clinging timidly, but desperately, to some diplomatic end. A stony despair was invading me and paralysing resistance. Better face the worst and be done with it.

"Come on," I said, grimly.

Heavily loaded, we stumbled over railway lines and rubble heaps, and came on the harbour. Davies led the way to a stairway, whose weedy steps disappeared below in gloom.

"If you'll get into the dinghy," he said, all briskness now, "I'll pass the things down."

I descended gingerly, holding as a guide a sodden painter which ended in a small boat, and conscious that I was collecting slime on cuffs and trousers.

"Hold up!" shouted Davies, cheerfully, as I sat down suddenly near the bottom, with one foot in the water.

I climbed wretchedly into the dinghy and awaited events.

"Now float her up close under the quay wall, and make fast to the ring down there," came down from above, followed by the slack of the sodden painter, which knocked my cap off as it fell. "All fast? Any knot'll do," I heard, as I grappled with this loathsome task, and then a big, dark object loomed overhead and was lowered into the dinghy. It was my portmanteau, and, placed athwart, exactly filled all the space amidships. "Does it fit?" was the anxious inquiry from aloft.

"Beautifully."

"Capital!"

Scratching at the greasy wall to keep the dinghy close to it, I received in succession our stores, and stowed the cargo as best I could, while the dinghy sank lower and lower in the water, and its precarious superstructure grew higher.

"Catch!" was the final direction from above, and a damp soft parcel hit me in the chest. "Be careful of that, it's meat. Now back to the stairs!"

I painfully acquiesced, and Davies appeared.

"It's a bit of a load, and she's rather deep; but I think we shall manage," he reflected. "You sit right aft, and I'll row."

I was too far gone for curiosity as to how this monstrous pyramid was to be rowed, or even for surmises as to its foundering by the way. I crawled to my appointed seat, and Davies extricated the buried sculls by a series of tugs, which shook the whole structure, and made us roll alarmingly. How he stowed himself into rowing posture I have not the least idea, but eventually we were moving sluggishly out into the open water, his head just visible in the bows. We had started from what appeared to be the head of a narrow loch, and were leaving behind us the lights of a big town. A long frontage of lamplit quays was on our left, with here and there the vague hull of a steamer alongside. We passed the last of the lights and came out into a broader stretch of water, when a light breeze was blowing and dark hills could be seen on either shore.

"I'm lying a little way down the fiord, you see," said Davies. "I hate to be too near a town, and I found a carpenter handy here—There she is! I wonder how you'll like her!"

I roused myself. We were entering a little cove encircled by trees, and approaching a light which flickered in the rigging of a small vessel, whose outline gradually defined itself.

"Keep her off," said Davies, as we drew alongside.

In a moment he had jumped on deck, tied the painter, and was round at my end.

"You hand them up," he ordered, "and I'll take them."

It was a laborious task, with the one relief that it was not far to hand them—a doubtful compensation, for other reasons distantly shaping themselves. When the stack was transferred to the deck I followed it, tripping over the flabby meat parcel, which was already showing ghastly signs of disintegration under the dew. Hazily there floated through my mind my last embarkation on a yacht; my faultless attire, the trim gig and obsequious sailors, the accommodation ladder flashing with varnish and brass in the August sun; the orderly, snowy decks and basket chairs under the awning aft. What a contrast with this sordid midnight scramble, over damp meat and littered packing cases! The bitterest touch of all was a growing sense of inferiority and ignorance which I had never before been allowed to feel in my experience of yachts.

Davies awoke from another reverie over my portmanteau to say, cheerily: "I'll just show you round down below first, and then we'll stow things away and get to bed."

He dived down a companion ladder, and I followed cautiously. A complex odour of paraffin, past cookery, tobacco, and tar saluted my nostrils.

"Mind your head," said Davies, striking a match and lighting a candle, while I groped into the cabin. "You'd better sit down; it's easier to look round."

There might well have been sarcasm in this piece of advice, for I must have cut a ridiculous figure, peering awkwardly and suspiciously round, with shoulders and head bent to avoid the ceiling, which seemed in the halflight to be even nearer the floor than it was.

"You see," were Davies's reassuring words, "there's plenty of room to sit upright" (which was strictly true; but I am not very tall, and he is short). "Some people make a point of headroom, but I never mind much about it. That's the centreboard case," he explained, as, in stretching my legs out, my knee came into contact with a sharp edge.

I had not seen this devilish obstruction, as it was hidden beneath the table, which indeed rested on it at one end. It appeared to be a long, low triangle, running lengthways with the boat and dividing the naturally limited space into two.

"You see, she's a flatbottomed boat, drawing very little water without the plate; that's why there's so little headroom. For deep water you lower the plate; so, in one way or another, you can go practically anywhere."

I was not nautical enough to draw any very definite conclusions from this, but what I did draw were not promising. The latter sentences were spoken from the forecastle, whither Davies had crept through a low sliding door, like that of a rabbithutch, and was already busy with a kettle over a stove which I made out to be a battered and disreputable twin brother of the No. 3 Rippingille.

"It'll be boiling soon," he remarked, "and we'll have some grog."

My eyes were used to the light now, and I took in the rest of my surroundings, which may be very simply described. Two long cushion-covered seats flanked the cabin, bounded at the after end by cupboards, one of which was cut low to form a sort of miniature sideboard, with glasses hung in a rack above it. The deck overhead was very low at each side but rose shoulder high for a space in the middle, where a "coachhouse roof" with a skylight gave additional cabin space. Just outside the door was a fold-up washingstand. On either wall were long net-racks holding a medley of flags, charts, caps, cigarboxes, banks of yam, and such like. Across the forward bulkhead was a bookshelf crammed to overflowing with volumes of all sizes, many upside down and some coverless. Below this were a pipe-rack, an aneroid, and a clock with a hearty tick. All the woodwork was painted white, and to a less jaundiced eye than mine the interior might have had an enticing look of snugness. Some Kodak prints were nailed roughly on the after bulkhead, and just over the doorway was the photograph of a young girl.

"That's my sister," said Davies, who had emerged and saw me looking at it. "Now, let's get the stuff down." He ran up the ladder, and soon my portmanteau blackened the hatchway, and a great straining and squeezing began. "I was afraid it was too big," came down; "I'm sorry, but you'll have to unpack on deck—we may be able to squash it down when it's empty." Then the wearisome tail of packages began to form a fresh stack in the cramped space at my feet, and my back ached with stooping and moiling in unfamiliar places. Davies came down, and with unconcealed pride introduced me to the sleeping cabin (he called the other one "the saloon"). Another candle was lit and showed two short and narrow berths with blankets, but no sign of sheets; beneath these were drawers, one set of which Davies made me master of, evidently thinking them a princely allowance of space for my wardrobe.

"You can chuck your things down the skylight on to your berth as you unpack them," he remarked. "By the way, I doubt if there's room for all you've got. I suppose you couldn't manage—"

"No, I couldn't," I said shortly.

The absurdity of argument struck me; two men, doubled up like monkeys, cannot argue.

"If you'll go out I shall be able to get out too," I added. He seemed miserable at this ghost of an altercation, but I pushed past, mounted the ladder, and in the expiring moonlight unstrapped that accursed portmanteau and, brimming over with irritation, groped among its contents, sorting some into the skylight with the same feeling that nothing mattered much now, and it was best to be done with it; repacking the rest with guilty stealth ere Davies should discover their character, and strapping up the whole again. Then I sat down upon my white elephant and shivered, for the chill of autumn was in the air. It suddenly struck me that if it had been raining things might have been worse still. The notion made me look round. The little cove was still as glass; stars above and stars below; a few white cottages glimmering at one point on the shore; in the west the lights of Flensburg; to the east the fiord broadening into unknown gloom. From Davies toiling below there were muffled sounds of wrenching, pushing, and hammering, punctuated occasionally by a heavy splash as something shot up from the hatchway and fell into the water.

How it came about I do not know. Whether it was something pathetic in the look I had last seen on his face—a look which I associated for no reason whatever with his bandaged hand; whether it was one of those instants of clear vision in which our separate selves are seen divided, the baser from the better, and I saw my silly egotism in contrast with a simple generous nature; whether it was an impalpable air of mystery which pervaded the whole enterprise and refused to be dissipated by its most mortifying and vulgarizing incidents—a mystery dimly connected with my companion's obvious consciousness of having misled me into joining him; whether it was only the stars and the cool air rousing atrophied instincts of youth and spirits; probably, indeed, it was all these influences, cemented into strength by a ruthless sense of humour which whispered that I was in danger of making a mere commonplace fool of myself in spite of all my laboured calculations; but whatever it was, in a flash my mood changed. The crown of martyrdom disappeared, the wounded vanity healed; that precious fund of fictitious resignation drained away, but left no void. There was left a fashionable and dishevelled young man sitting in the dew and in the dark on a ridiculous portmanteau which dwarfed the yacht that was to carry it; a youth acutely sensible of ignorance in a strange and strenuous atmosphere; still feeling sore and victimized; but withal sanely ashamed and sanely resolved to enjoy himself. I anticipate; for though the change was radical its full growth was slow. But in any case it was here and now that it took its birth.

"Grog's ready!" came from below. Bunching myself for the descent I found to my astonishment that all trace of litter had miraculously vanished, and a cosy neatness reigned. Glasses and lemons were on the table, and a fragrant smell of punch had deadened previous odours. I showed little emotion at these amenities, but enough to give intense relief to Davies, who delightedly showed me his devices for storage, praising the "roominess" of his floating den. "There's your stove, you see," he ended; "I've chucked the old one overboard." It was a weakness of his, I should say here, to rejoice in throwing things overboard on the flimsiest pretexts. I afterwards suspected that the new stove had not been "really necessary" any more than the rigging screws, but was an excuse for gratifying this curious taste.

We smoked and chatted for a little, and then came the problem of going to bed. After much bumping of knuckles and head, and many giddy writhings, I mastered it, and lay between the rough blankets. Davies, moving swiftly and deftly, was soon in his.

"It's quite comfortable, isn't it?" he said, as he blew out the light from where he lay, with an accuracy which must have been the fruit of long practice.

I felt prickly all over, and there was a damp patch on the pillow, which was soon explained by a heavy drop of moisture falling on my forehead.

"I suppose the deck's not leaking?" I said, as mildly as I could.

"I'm awfully sorry," said Davies, earnestly, tumbling out of his bunk. "It must be the heavy dew. I did a lot of caulking yesterday, but I suppose I missed that place. I'll run up and square it with an oilskin."

"What's wrong with your hand?" I asked, sleepily, on his return, for gratitude reminded me of that bandage.

"Nothing much; I strained it the other day," was the reply; and then the seemingly inconsequent remark: "I'm glad you brought that prismatic compass. It's not really necessary, of course; but" (muffled by blankets) "it may come in useful."

I dozed but fitfully, with a fretful sense of sore elbows and neck and many a draughty hiatus among the blankets. It was broad daylight before I had reached the stage of torpor in which such slumber merges. That was finally broken by the descent through the skylight of a torrent of water. I started up, bumped my head hard against the decks, and blinked leaden-eyed upwards.

"Sorry! I'm scrubbing decks. Come up and bathe. Slept well?" I heard a voice saying from aloft.

"Fairly well," I growled, stepping out into a pool of water on the oilcloth. Thence I stumbled up the ladder, dived overboard, and buried bad dreams, stiffness, frowsiness, and tormented nerves in the loveliest fiord of the lovely Baltic. A short and furious swim and I was back again, searching for a means of ascent up the smooth black side, which, low as it was, was slippery and unsympathetic. Davies, in a loose canvas shirt, with the sleeves tucked up, and flannels rolled up to the knee, hung over me with a rope's end, and chatted unconcernedly about the easiness of the job when you know how, adjuring me to mind the paint, and talking about an accommodation ladder he had once had, but had thrown overboard because it was so horribly in the way. When I arrived, my knees and elbows were picked out in black paint, to his consternation. Nevertheless, as I plied the towel, I knew that I had left in those limpid depths yet another crust of discontent and self-conceit.

As I dressed into flannels and blazer, I looked round the deck, and with an unskilled and doubtful eye took in all that the darkness had hitherto hidden. She seemed very small (in point of fact she was seven tons), something over thirty feet in length and nine in beam, a size very suitable to weekends in the Solent, for such as liked that sort of thing; but that she should have come from Dover to the Baltic suggested a world of physical endeavour of which I had never dreamed.

I passed to the aesthetic side. Smartness and beauty were essential to yachts, in my mind, but with the best resolves to be pleased I found little encouragement here. The hull seemed too low, and the mainmast too high; the cabin roof looked clumsy, and the skylights saddened the eye with dull iron and plebeian graining. What brass there was, on the tiller-head and elsewhere, was tarnished with sickly green. The decks had none of that creamy purity which Cowes expects, but were rough and grey, and showed tarry exhalations round the seams and rusty stains near the bows. The ropes and rigging were in mourning when contrasted with the delicate buff manilla so satisfying to the artistic eye as seen against the blue of a June sky at Southsea. Nor was the whole effect bettered by many signs of recent refitting. An impression of paint, varnish, and carpentry was in the air; a gaudy new burgee fluttered aloft; there seemed to be a new rope or two, especially round the diminutive mizzenmast, which itself looked altogether new. But all this only emphasized the general plainness, reminding one of a respectable woman of the working classes trying to dress above her station, and soon likely to give it up.

That the ensemble was businesslike and solid even my untrained eye could see. Many of the deck fittings seemed disproportionately substantial. The anchor-chain looked contemptuous of its charge; the binnacle with its compass was of a size and prominence almost comically impressive, and was, moreover, the only piece of brass which was burnished and showed traces of reverent care. Two huge coils of stout and dingy warp lay just abaft the mainmast, and summed up the weather-beaten aspect of the little ship. I should add here that in the distant past she had been a lifeboat, and had been clumsily converted into a yacht by the addition of a counter, deck, and the necessary spars. She was built, as all lifeboats are, diagonally, of two skins of teak, and thus had immense strength, though, in the matter of looks, all a hybrid's failings.

Hunger and "Tea's made!" from below brought me down to the cabin, where I found breakfast laid out on the table over the centreboard case, with Davies earnestly presiding, rather flushed as to the face, and sooty as to the fingers. There was a slight shortage of plate and crockery, but I praised the bacon and could do so truthfully, for its crisp and steaming shavings would have put to shame the efforts of my London cook. Indeed, I should have enjoyed the meal heartily were it not for the lowness of the sofa and table, causing a curvature of the body which made swallowing a more lengthy process than usual, and induced a periodical yearning to get up and stretch—a relief which spelt disaster to the skull. I noticed, too, that Davies spoke with a zest, sinister to me, of the delights of white bread and fresh milk, which he seemed to consider unusual luxuries, though suitable to an inaugural banquet in honour of a fastidious stranger. "One can't be always going on shore," he said, when I showed a discreet interest in these things. "I lived for ten days on a big rye loaf over in the Frisian Islands."

"And it died hard, I suppose?"

"Very hard, but" (gravely) "quite good. After that I taught myself to make rolls; had no baking powder at first, so used Eno's fruit salt, but they wouldn't rise much with that. As for milk, condensed is—I hope you don't mind it?"

I changed the subject, and asked about his plans.

"Let's get under way at once," he said, "and sail down the fiord." I tried for something more specific, but he was gone, and his voice drowned in the fo'c'sle by the clatter and swish of washing up. Thenceforward events moved with bewildering rapidity. Humbly desirous of being useful I joined him on deck, only to find that he scarcely noticed me, save as a new and unexpected obstacle in his round of activity. He was everywhere at once—heaving in chain, hooking on halyards, hauling ropes; while my part became that of the clown who does things after they are already done, for my knowledge of a yacht was of that floating and inaccurate kind which is useless in practice. Soon the anchor was up (a great rusty monster it was!), the sails set, and Davies was darting swiftly to and fro between the tiller and jib-sheets, while the *Dulcibella* bowed a lingering farewell to the shore and headed for the open fiord. Erratic puffs from the high land behind made her progress timorous at first, but soon the fairway was reached and a true breeze from Flensburg and the west took her in its friendly grip. Steadily she rustled down the calm blue highway whose soft beauty was the introduction to a passage in my life, short, but pregnant with moulding force, through stress and strain, for me and others.

Davies was gradually resuming his natural self, with abstracted intervals, in which he lashed the helm to finger a distant rope, with such speed that the movements seemed simultaneous. Once he vanished, only to reappear in an instant with a chart, which he studied, while steering, with a success that its reluctant folds seemed to render impossible. Waiting respectfully for his revival I had full time to look about. The fiord here was about a mile broad. From the shore we had left the hills rose steeply, but with no rugged grandeur; the outlines were soft; there were green spaces and rich woods on the lower slopes; a little white town was opening up in one place, and scattered farms dotted the prospect. The other shore, which I could just see, framed between the gunwale and the mainsail, as I sat leaning against the hatchway, and sadly missing a deck chair, was lower and lonelier, though prosperous and pleasing to the eye.

Spacious pastures led up by slow degrees to ordered clusters of wood, which hinted at the presence of some great manor house. Behind us, Flensburg was settling into haze. Ahead, the scene was shut in by the contours of hills, some clear, some dreamy and distant. Lastly, a single glimpse of water shining between the folds of hill far away hinted at spaces of distant sea of which this was but a secluded inlet. Everywhere was that peculiar charm engendered by the association of quiet pastoral country and a homely human atmosphere with a branch of the great ocean that bathes all the shores of our globe.

There was another charm in the scene, due to the way in which I was viewing it—not as a pampered passenger on a "fine steam yacht," or even on "a powerful modern schooner," as the yacht agents advertise, but from the deck of a scrubby little craft of doubtful build and distressing plainness, which yet had smelt her persistent way to this distant fiord through I knew not what of difficulty and danger, with no apparent motive in her single occupant, who talked as vaguely and unconcernedly about his adventurous cruise as though it were all a protracted afternoon on Southampton Water.

I glanced round at Davies. He had dropped the chart and was sitting, or rather half lying, on the deck with one bronzed arm over the tiller, gazing fixedly ahead, with just an occasional glance around and aloft. He still seemed absorbed in himself, and for a moment or two I studied his face with an attention I had never, since I had known him, given it. I had always thought it commonplace, as I had thought him commonplace, so far as I had thought at all about either.

It had always rather irritated me by an excess of candour and boyishness. These qualities it had kept, but the scales were falling from my eyes, and I saw others. I saw strength to obstinacy and courage to recklessness, in the firm lines of the chin; an older and deeper look in the eyes. Those odd transitions from bright mobility to detached earnestness, which had partly amused and chiefly annoyed me hitherto, seemed now to be lost in a sensitive reserve, not cold or egotistic, but strangely winning from its paradoxical frankness.

Sincerity was stamped on every lineament. A deep misgiving stirred me that, clever as I thought myself, nicely perceptive of the right and congenial men to know, I had made some big mistakes—how many, I wondered? A relief, scarcely less deep because it was unconfessed, stole in on me with the suspicion that, little as I deserved it, the patient fates were offering me a golden chance of repairing at least one. And yet, I mused, the patient fates have crooked methods, besides a certain mischievous humour, for it was Davies who had asked me out—though now he scarcely seemed to need me—almost tricked me into coming out, for he might have known I was not suited to such a life; yet trickery and Davies sounded an odd conjuncture.

Probably it was the growing discomfort of my attitude which produced this backsliding. My night's rest and the "ascent from the bath" had, in fact, done little to prepare me for contact with sharp edges and hard surfaces. But Davies had suddenly come to himself, and with an "I say, are you comfortable? Have something to sit on?" jerked the helm a little to windward, felt it like a pulse for a moment, with a rapid look to windward, and dived below, whence he returned with a couple of cushions, which he threw to me. I felt perversely resentful of these luxuries, and asked:

"Can't I be of any use?"

"Oh, don't you bother," he answered. "I expect you're tired. Aren't we having a splendid sail? That must be Ekken on the port bow," peering under the sail, "where the trees run in. I say, do you mind looking at the chart?" He tossed it over to me. I spread it out painfully, for it curled up like a watchspring at the least slackening of pressure. I was not familiar with charts, and this sudden trust reposed in me, after a good deal of neglect, made me nervous.

"You see Flensburg, don't you?" he said. "That's where we are," dabbing with a long reach at an indefinite space on the crowded sheet. "Now which side of that buoy off the point do we pass?"

I had scarcely taken in which was land and which was water, much less the significance of the buoy, when he resumed:

"Never mind; I'm pretty sure it's all deep water about here. I expect that marks the fairway for steamers."

In a minute or two we were passing the buoy in question, on the wrong side I am pretty certain, for weeds and sand came suddenly into view below us with uncomfortable distinctness. But all Davies said was: "There's never any sea here, and the plate's not down," a dark utterance which I pondered doubtfully. "The best of these Schleswigwaters," he went on, "is that a boat of this size can go almost anywhere. There's no navigation required. Why—" At this moment a faint scraping was felt, rather than heard, beneath us.

"Aren't we aground?" I asked, with great calmness.

"Oh, she'll blow over," he replied, wincing a little.

She "blew over," but the episode caused a little naive vexation in Davies. I relate it as a good instance of one of his minor peculiarities. He was utterly without that didactic pedantry which yachting has a fatal tendency to engender in men who profess it. He had tossed me the chart without a thought that I was an ignoramus, to whom it would be Greek, and who would provide him with an admirable subject to drill and lecture, just as his neglect of me throughout the morning had been merely habitual and unconscious independence. In the second place, master of his métier as I knew him afterwards to be, resourceful, skilful, and alert, he was liable to lapse into a certain amateurish vagueness, half irritating and half amusing. I think truly that both these peculiarities came from the same source, a hatred of any sort of affectation. To the same source I traced the fact that he and his yacht observed none of the superficial etiquette of yachts and yachtsmen, that she never, for instance, flew a national ensign, and he never wore a "yachting suit."

We rounded a low green point which I had scarcely noticed before.

"We must jibe," said Davies. "Just take the helm, will you?" and, without waiting for my cooperation, he began hauling in the mainsheet with great vigour. I had rude notions of steering, but jibing is a delicate operation. No yachtsman will be surprised to hear that the boom saw its opportunity and swung over with a mighty crash, with the mainsheet entangled round me and the tiller.

"Jibed all standing," was his sorrowful comment. "You're not used to her yet. She's very quick on the helm."

"Where am I to steer for?" I asked, wildly.

"Oh, don't trouble, I'll take her now," he replied.

I felt it was time to make my position clear. "I'm an utter duffer at sailing," I began. "You'll have a lot to teach me, or one of these days I shall be wrecking you. You see, there's always been a crew—"

"Crew!"—with sovereign contempt—"why, the whole fun of the thing is to do everything oneself."

"Well, I've felt in the way the whole morning."

"I'm awfully sorry!" His dismay and repentance were comical. "Why, it's just the other way; you may be all the use in the world." He became absent.

We were following the inward trend of a small bay towards a cleft in the low shore.

"That's Ekken Sound," said Davies; "let's look into it," and a minute or two later we were drifting through a dainty little strait, with a peep of open water at the end of it. Cottages bordered either side, some overhanging the very water, some connecting with it by a rickety wooden staircase or a miniature landing-stage. Creepers and roses rioted over the walls and tiny porches. For a space on one side, a rude quay, with small smacks floating off it, spoke of some minute commercial interests; a very small tea-garden, with neglected-looking bowers and leaf-strewn tables, hinted at some equally minute tripping interest. A pervading hue of mingled bronze and rose came partly from the weather-mellowed woodwork of the cottages and stages, and partly from the creepers and the trees behind, where autumn's subtle fingers were already at work. Down this exquisite sealane we glided till it ended in a broad mere, where our sails, which had been shivering and complaining, filled into contented silence.

"Ready about!" said Davies, callously. "We must get out of this again." And round we swung.

"Why not anchor and stop here?" I protested; for a view of tantalizing loveliness was unfolding itself.

"Oh, we've seen all there is to be seen, and we must take this breeze while we've got it." It was always torture to Davies to feel a good breeze running to waste while he was inactive at anchor or on shore.

The "shore" to him was an inferior element, merely serving as a useful annexe to the water—a source of necessary supplies.

"Let's have lunch," he pursued, as we resumed our way down the fiord. A vision of iced drinks, tempting salads, white napery, and an attentive steward mocked me with past recollections.

"You'll find a tongue," said the voice of doom, "in the starboard sofa-locker; beer under the floor in the bilge. I'll see her round that buoy, if you wouldn't mind beginning." I obeyed with a bad grace, but the close air and cramped posture must have benumbed my faculties, for I opened the portside locker, reached down, and grasped a sticky body, which turned out to be a pot of varnish.

Recoiling wretchedly, I tried the opposite one, combating the embarrassing heel of the boat and the obstructive edges of the centreboard case. A medley of damp tins of varied sizes showed in the gloom, exuding a mouldy odour. Faded legends on dissolving paper, like the remnants of old posters on a disused hoarding, spoke of soups, curries, beefs, potted meats, and other hidden delicacies. I picked out a tongue, reimprisoned the odour, and explored for beer.

It was true, I supposed, that bilge didn't hurt it, as I tugged at the plank on my hands and knees, but I should have myself preferred a more accessible and less humid wine-cellar than the cavities among slimy ballast from which I dug the bottles. I regarded my hard-won and ill-favoured pledges of a meal with giddiness and discouragement.

"How are you getting on?" shouted Davies; "the tin opener's hanging up on the bulkhead; the plates and knives are in the cupboard."

I doggedly pursued my functions. The plates and knives met me half-way, for, being on the weather side, and thus having a downward slant, its contents, when I slipped the latch, slid affectionately into my bosom, and overflowed with a clatter and jingle on to the floor.

"That often happens," I heard from above. "Never mind! There are no breakables. I'm coming down to help." And down he came, leaving the *Dulcibella* to her own devices.

"I think I'll go on deck," I said. "Why in the world couldn't you lunch comfortably at Ekken and save this infernal pandemonium of a picnic? Where's the yacht going to meanwhile? And how are we to lunch on that slanting table? I'm covered with varnish and mud, and ankle-deep in crockery. There goes the beer!"

"You shouldn't have stood it on the table with this list on," said Davies, with intense composure, "but it won't do any harm; it'll drain into the bilge" (ashes to ashes, dust to dust, I thought). "You go on deck now, and I'll finish getting ready." I regretted my explosion, though wrung from me under great provocation.

"Keep her straight on as she's going," said Davies, as I clambered up out of the chaos, brushing the dust off my trousers and varnishing the ladder with my hands. I unlashed the helm and kept her as she was going.

We had rounded a sharp bend in the fiord, and were sailing up a broad and straight reach which every moment disclosed new beauties, sights fair enough to be balm to the angriest spirit. A red-roofed hamlet was on our left, on the right an ivied ruin, close to the water, where some contemplative cattle stood kneedeep. The view ahead was a white strand which fringed both shores, and to it fell wooded slopes, interrupted here and there by low sandstone cliffs of warm red colouring, and now and again by a dingle with cracks of greensward.

I forgot petty squalors and enjoyed things—the coy tremble of the tiller and the backwash of air from the dingy mainsail, and, with a somewhat chastened rapture, the lunch which Davies brought up to me and solicitously watched me eat.

Later, as the wind sank to lazy airs, he became busy with a larger topsail and jib; but I was content to doze away the afternoon, drenching brain and body in the sweet and novel foreign atmosphere, and dreamily watching the fringe of glen cliff and cool white sand as they passed ever more slowly by.

Twenty-Six

The Siege of the Round-House

Robert Louis Stevenson

But now our time of truce was come to an end. Those on deck had waited for my coming till they grew impatient; and scarce had Alan spoken, when the captain showed face in the open door.

"Stand!" cried Alan, and pointed his sword at him. The captain stood, indeed; but he neither winced nor drew back a foot.

"A naked sword?" says he. "This is a strange return for hospitality."

"Do ye see me?" said Alan. "I am come of kings; I bear a king's name. My badge is the oak. Do ye see my sword? It has slashed the heads off mair Whigamores than you have toes upon your feet. Call up your vermin to your back, sir, and fall on! The sooner the clash begins, the sooner ye'll taste this steel throughout your vitals."

The captain said nothing to Alan, but he looked over at me with an ugly look. "David," said he, "I'll mind this"; and the sound of his voice went through me with a jar.

Next moment he was gone.

"And now," said Alan, "let your hand keep your head, for the grip is coming."

Alan drew a dirk, which he held in his left hand in case they should run in under his sword. I, on my part, clambered up into the berth with an armful of pistols and something of a heavy heart, and set open the window where I was to watch. It was a small part of the deck that I could overlook, but enough for our purpose. The sea had gone down, and the wind was steady and kept the sails quiet; so that there was a great stillness in the ship, in which I made sure I heard the sound of muttering voices. A little after, and there came a clash of steel upon the deck, by which I knew they were dealing out the cutlasses and one had been let fall; and after that, silence again.

I do not know if I was what you call afraid; but my heart beat like a bird's, both quick and little; and there was a dimness came before my eyes which I continually rubbed away, and which continually returned. As for hope, I had none; but only a darkness of despair and a sort of anger against all the world that made me long to sell my life as dear as I was able. I tried to pray, I remember, but that same hurry of my mind, like a man running, would not suffer me to think upon the words; and my chief wish was to have the thing begin and be done with it.

It came all of a sudden when it did, with a rush of feet and a roar, and then a shout from Alan, and a sound of blows and some one crying out as if hurt. I looked back over my shoulder, and saw Mr. Shuan in the doorway, crossing blades with Alan.

"That's him that killed the boy!" I cried.

"Look to your window!" said Alan; and as I turned back to my place, I saw him pass his sword through the mate's body.

It was none too soon for me to look to my own part; for my head was scarce back at the window, before five men, carrying a spare yard for a battering-ram, ran past me and took post to drive the door in. I had never fired with a pistol in my life, and not often with a gun; far less against a fellow creature. But it was now or never; and just as they swang the yard, I cried out: "Take that!" and shot into their midst.

I must have hit one of them, for he sang out and gave back a step, and the rest stopped as if a little disconcerted. Before they had time to recover, I sent another ball over their heads; and at my third shot (which went as wide as the second) the whole party threw down the yard and ran for it.

Then I looked round again into the deckhouse. The whole place was full of the smoke of my own firing, just as my ears seemed to be burst with the noise of the shots. But there was Alan, standing as before; only now his sword was running blood to the hilt, and himself so swelled with triumph and fallen into so fine an attitude, that he looked to be invincible. Right before him on the floor was Mr. Shuan, on his hands and knees; the blood was pouring from his mouth, and he was sinking slowly lower, with a terrible, white face; and just as I looked, some of those from behind caught hold of him by the heels and dragged him bodily out of the round-house. I believe he died as they were doing it.

"There's one of your Whigs for ye!" cried Alan; and then turning to me, he asked if I had done much execution.

I told him I had winged one, and thought it was the captain.

"And I've settled two," says he. "No, there's not enough blood let; they'll be back again. To your watch, David. This was but a dram before meat."

I settled back to my place, re-charging the three pistols I had fired, and keeping watch with both eye and ear.

Our enemies were disputing not far off upon the deck, and that so loudly that I could hear a word or two above the washing of the seas.

"It was Shuan bauchled it," I heard one say.

And another answered him with a "Wheesht, man! He's paid the piper."

After that the voices fell again into the same muttering as before. Only now, one person spoke most of the time, as though laying down a plan, and first one and then another answered him briefly, like men taking orders. By this, I made sure they were coming on again, and told Alan.

"It's what we have to pray for," said he. "Unless we can give them a good distaste of us, and be done with it, there'll be nae sleep for either you or me. But this time, mind, they'll be in earnest."

By this, my pistols were ready, and there was nothing to do but listen and wait. While the brush lasted, I had not the time to think if I was frighted; but now, when all was still again, my mind ran upon nothing else. The thought of the sharp swords and the cold steel was strong in me; and presently, when I began to hear stealthy steps and a brushing of men's clothes against the round-house wall, and knew they were taking their places in the dark, I could have found it in my mind to cry out aloud.

All this was upon Alan's side; and I had begun to think my share of the fight was at an end, when I heard some one drop softly on the roof above me. Then there came a single call on the seapipe, and that was the signal. A knot of them made one rush of it, cutlass in hand, against the door; and at the same moment, the glass of the skylight was dashed in a thousand pieces, and a man leaped through and landed on the floor. Before he got his feet, I had clapped a pistol to his back, and might have shot him, too; only at the touch of him (and him alive) my whole flesh misgave me, and I could no more pull the trigger than I could have flown.

He had dropped his cutlass as he jumped, and when he felt the pistol, whipped straight round and laid hold of me, roaring out an oath; and at that either my courage came again, or I grew so much afraid as came to the same thing; for I gave a shriek and shot him in the midst of the body. He gave the most horrible, ugly groan and fell to the floor. The foot of a second fellow, whose legs were dangling through the skylight, struck me at the same time upon the head; and at that I snatched another pistol and shot this one through the thigh, so that he slipped through and tumbled in a lump on his companion's body. There was no talk of missing, any more than there was time to aim; I clapped the muzzle to the very place and fired. I might have stood and stared at them for long, but I heard Alan shout as if for help, and that brought me to my senses. He had kept the door so long; but one of the seamen, while he was engaged with others, had run in under his guard and caught him about the body. Alan was dirking him with his left hand, but the fellow clung like a leech. Another had broken in and had his cutlass raised. The door was thronged with their faces. I thought we were lost, and catching up my cutlass, fell on them in flank.

But I had not time to be of help. The wrestler dropped at last; and Alan, leaping back to get his distance, ran upon the others like a bull, roaring as he went. They broke before him like water, turning, and running, and falling one against another in their haste. The sword in his hands flashed like quicksilver into the huddle of our fleeing enemies; and at every flash there came the scream of a man hurt. I was still thinking we were lost, when lo! they were all gone, and Alan was driving them along the deck as a sheepdog chases sheep.

Yet he was no sooner out than he was back again, being as cautious as he was brave; and meanwhile the seamen continued running and crying out as if he was still behind them; and we heard them tumble one upon another into the forecastle, and clapto the hatch upon the top.

The round-house was like a shambles; three were dead inside, another lay in his death agony across the threshold; and there were Alan and I victorious and unhurt.

He came up to me with open arms. "Come to my arms!" he cried, and embraced and kissed me hard upon both cheeks. "David," said he, "I love you like a brother. And O, man," he cried in a kind of ecstasy, "am I no a bonny fighter?"

Thereupon he turned to the four enemies, passed his sword clean through each of them, and tumbled them out of doors one after the other. As he did so, he kept humming and singing and whistling to himself, like a man trying to recall an air; only what he was trying was to make one. All the while, the flush was in his face, and his eyes were as bright as a five-year-old child's with a new toy. And presently he sat down upon the table, sword in hand; the air that he was making all the time began to run a little clearer, and then clearer still; and then out he burst with a great voice into a Gaelic song.

I have translated it here, not in verse (of which I have no skill) but at least in the king's English. He sang it often afterwards, and the thing became popular; so that I have heard it, and had it explained to me, many's the time.

This is the song of the sword of Alan;
The smith made it,
The fire set it;
Now it shines in the hand of Alan Breck.
Their eyes were many and bright,
Swift were they to behold,
Many the hands they guided:
The sword was alone.
The dun deer troop over the hill,
They are many, the hill is one;
The dun deer vanish,
The hill remains.

Come to me from the hills of heather,
Come from the isles of the sea.
O far-beholding eagles,
Here is your meat.

Now this song which he made (both words and music) in the hour of our victory, is something less than just to me, who stood beside him in the tussle. Mr. Shuan and five more were either killed outright or thoroughly disabled; but of these, two fell by my hand, the two that came by the skylight. Four more were hurt, and of that number, one (and he not the least important) got his hurt from me. So that, altogether, I did my fair share both of the killing and the wounding, and might have claimed a place in Alan's verses. But poets have to think upon their rhymes; and in good prose talk, Alan always did me more than justice.

In the meanwhile, I was innocent of any wrong being done me. For not only I knew no word of the Gaelic; but what with the long suspense of the waiting, and the scurry and strain of our two spirits of fighting, and more than all, the horror I had of some of my own share in it, the thing was no sooner over than I was glad to stagger to a seat. There was that tightness on my chest that I could hardly breathe; the thought of the two men I had shot sat upon me like a nightmare; and all upon a sudden, and before I had a guess of what was coming, I began to sob and cry like any child.

Alan clapped my shoulder, and said I was a brave lad and wanted nothing but a sleep. "I'll take the first watch," said he. "Ye've done well by me, David, first and last; and I wouldn't lose you for all Appin—no, nor for Breadalbane."

So I made up my bed on the floor; and he took the first spell, pistol in hand and sword on knee, three hours by the captain's watch upon the wall. Then he roused me up, and I took my turn of three hours; before the end of which it was broad day, and a very quiet morning, with a smooth, rolling sea that tossed the ship and made the blood run to and fro on the round-house floor, and a heavy rain that drummed upon the roof. All my watch there was nothing stirring; and by the banging of the helm, I knew they had even no one at the tiller. Indeed (as I learned afterwards) there were so many of them hurt or dead, and the rest in so ill a temper, that Mr. Riach and the captain had to take turn and turn like Alan and me, or the brig might have gone ashore and nobody the wiser. It was a mercy the night had fallen so still, for the wind had gone down as soon as the rain began. Even as it was, I judged by the wailing of a great number of gulls that went crying and fishing round the ship, that she must have drifted pretty near the coast of one of the islands of the Hebrides; and at last, looking out of the door of the round-house, I saw the great stone hills of Skye on the right hand, and, a little more astern, the strange isle of Rum.

Twenty-Seven

Mal de Mer

Jerome K. Jerome

We sat there for half an hour, describing to each other our maladies. I explained to George and William Harris how I felt when I got up in the morning, and William Harris told us how he felt when he went to bed; and George stood on the hearth-rug, and gave us a clever and powerful piece of acting, illustrative of how he felt in the night.

George fancies he is ill; but there's never anything really the matter with him, you know.

At this point, Mrs. Poppets knocked at the door to know if we were ready for supper. We smiled sadly at one another, and said we supposed we had better try to swallow a bit. Harris said a little something in one's stomach often kept the disease in check; and Mrs. Poppets brought the tray in, and we drew up to the table, and toyed with a little steak and onions, and some rhubarb tart.

I must have been very weak at the time; because I know, after the first half hour or so, I seemed to take no interest whatever in my food—an unusual thing for me—and I didn't want any cheese.

This duty done, we refilled our glasses, lit our pipes, and resumed the discussion upon our state of health. What it was that was actually the matter with us, we none of us could be sure of; but the unanimous opinion was that it—whatever it was—had been brought on by overwork.

"What we want is rest," said Harris.

"Rest and a complete change," said George. "The overstrain upon our brains has produced a general depression throughout the system. Change of scene, and absence of the necessity for thought, will restore the mental equilibrium."

George has a cousin, who is usually described in the charge sheet as a medical student, so that he naturally has a somewhat family physicianary way of putting things.

I agreed with George, and suggested that we should seek out some retired and oldworld spot, far from the madding crowd, and dream away a sunny week among its drowsy lanes—some half forgotten nook, hidden away by the fairies, out of reach of the noisy world—some quaint-perched eyrie on the cliffs of Time, from whence the surging waves of the nineteenth century would sound far-off and faint.

Harris said he thought it would be humpy. He said he knew the sort of place I meant; where everybody went to bed at eight o'clock, and you couldn't get a REFEREE for love or money, and had to walk ten miles to get your baccy.

"No," said Harris, "if you want rest and change, you can't beat a sea trip."

I objected to the sea trip strongly. A sea trip does you good when you are going to have a couple of months of it, but, for a week, it is wicked. You start on Monday with the idea implanted in your bosom that you are going to enjoy yourself. You wave an airy adieu to the boys on shore, light your biggest pipe, and swagger about the deck as if you were Captain Cook, Sir Francis Drake, and Christopher Columbus all rolled into one. On Tuesday, you wish you hadn't come. On Wednesday, Thursday, and Friday, you wish you were dead. On Saturday, you are able to swallow a little beef tea, and to sit up on deck, and answer with a wan, sweet smile when kindhearted people ask you how you feel now. On Sunday, you begin to walk about again, and take solid food. And on Monday morning, as, with your bag and umbrella in your hand, you stand by the gunwale, waiting to step ashore, you begin to thoroughly like it.

I remember my brother-in-law going for a short sea trip once, for the benefit of his health. He took a return berth from London to Liverpool;

and when he got to Liverpool, the only thing he was anxious about was to sell that return ticket.

It was offered round the town at a tremendous reduction, so I am told; and was eventually sold for eighteen pence to a bilious-looking youth who had just been advised by his medical men to go to the sea-side, and take exercise.

"Sea-side!" said my brother-in-law, pressing the ticket affectionately into his hand; "why, you'll have enough to last you a lifetime; and as for exercise! why, you'll get more exercise, sitting down on that ship, than you would turning somersaults on dry land."

He himself—my brother-in-law—came back by train. He said the North-Western Railway was healthy enough for him.

Another fellow I knew went for a week's voyage round the coast, and, before they started, the steward came to him to ask whether he would pay for each meal as he had it, or arrange beforehand for the whole series.

The steward recommended the latter course, as it would come so much cheaper. He said they would do him for the whole week at two pounds five. He said for breakfast there would be fish, followed by a grill. Lunch was at one, and consisted of four courses. Dinner at six—soup, fish, entree, joint, poultry, salad, sweets, cheese, and dessert. And a light meat supper at ten. My friend thought he would close on the two-pound-five job (he is a hearty eater), and did so.

Lunch came just as they were off Sheerness. He didn't feel so hungry as he thought he should, and so contented himself with a bit of boiled beef, and some strawberries and cream. He pondered a good deal during the afternoon, and at one time it seemed to him that he had been eating nothing but boiled beef for weeks, and at other times it seemed that he must have been living on strawberries and cream for years.

Neither the beef nor the strawberries and cream seemed happy, either—seemed discontented like.

At six, they came and told him dinner was ready. The announcement aroused no enthusiasm within him, but he felt that there was some of that two-pound-five to be worked off, and he held on to ropes and things and went down. A pleasant odour of onions and hot ham, mingled with fried fish and greens, greeted him at the bottom of the ladder; and then the steward came up with an oily smile, and said:

"What can I get you, sir?"

"Get me out of this," was the feeble reply.

And they ran him up quick, and propped him up, over to leeward, and left him.

For the next four days he lived a simple and blameless life on thin captain's biscuits (I mean that the biscuits were thin, not the captain) and soda water; but, towards Saturday, he got uppish, and went in for weak tea and dry toast, and on Monday he was gorging himself on chicken broth. He left the ship on Tuesday, and as it steamed away from the landing-stage he gazed after it regretfully.

"There she goes," he said, "there she goes, with two pounds' worth of food on board that belongs to me, and that I haven't had."

He said that if they had given him another day he thought he could have put it straight.

So I set my face against the sea trip. Not, as I explained, upon my own account. I was never queer. But I was afraid for George. George said he should be all right, and would rather like it, but he would advise Harris and me not to think of it, as he felt sure we should both be ill. Harris said that, to himself, it was always a mystery how people managed to get sick at sea—said he thought people must do it on purpose, from affectation—said he had often wished to be, but had never been able.

Then he told us anecdotes of how he had gone across the Channel when it was so rough that the passengers had to be tied into their berths, and he and the captain were the only two living souls on board who were not ill. Sometimes it was he and the second mate who were not ill; but it was generally he and one other man. If not he and another man, then it was he by himself.

It is a curious fact, but nobody ever is sea-sick—on land. At sea, you come across plenty of people very bad indeed, whole boatloads of them; but I never met a man yet, on land, who had ever known at all what it was to be sea-sick. Where the thousands upon thousands of bad sailors that swarm in every ship hide themselves when they are on land is a mystery.

If most men were like a fellow I saw on the Yarmouth boat one day, I could account for the seeming enigma easily enough. It was just off Southend Pier, I recollect, and he was leaning out through one of the port-holes in a very dangerous position. I went up to him to try and save him.

"Hi! come further in," I said, shaking him by the shoulder. "You'll be overboard."

"Oh my! I wish I was," was the only answer I could get; and there I had to leave him.

Three weeks afterwards, I met him in the coffee room of a Bath hotel, talking about his voyages, and explaining, with enthusiasm, how he loved the sea.

"Good sailor!" he replied in answer to a mild young man's envious query; "well, I did feel a little queer ONCE, I confess. It was off Cape Horn. The vessel was wrecked the next morning." I said:

"Weren't you a little shaky by Southend Pier one day, and wanted to be thrown overboard?"

"Southend Pier!" he replied, with a puzzled expression.

"Yes; going down to Yarmouth, last Friday three weeks."

"Oh, ah—yes," he answered, brightening up; "I remember now. I did have a headache that afternoon. It was the pickles, you know. They were the most disgraceful pickles I ever tasted in a respectable boat. Did you have any?"

For myself, I have discovered an excellent preventive against sea-sickness, in balancing myself. You stand in the centre of the deck, and, as the ship heaves and pitches, you move your body about, so as to keep it always straight. When the front of the ship rises, you lean forward, till the deck almost touches your nose; and when its back end gets up, you lean backwards. This is all very well for an hour or two; but you can't balance yourself for a week. George said:

"Let's go up the river."

He said we should have fresh air, exercise and quiet; the constant change of scene would occupy our minds (including what there was of Harris's); and the hard work would give us a good appetite, and make us sleep well. Harris said he didn't think George ought to do anything that would have a tendency to make him sleepier than he always was, as it might be dangerous.

He said he didn't very well understand how George was going to sleep any more than he did now, seeing that there were only twenty-four hours in each day, summer and winter alike; but thought that if he DID sleep any more, he might just as well be dead, and so save his board and lodging.

Harris said, however, that the river would suit him to a "T." I don't know what a "T" is (except a six-penny one, which includes bread-and-butter and cake AD LIB., and is cheap at the price, if you haven't had any dinner). It seems to suit everybody, however, which is greatly to its credit.

It suited me to a "T" too, and Harris and I both said it was a good idea of George's; and we said it in a tone that seemed to somehow imply that we were surprised that George should have come out so sensible.

The only one who was not struck with the suggestion was Montmorency. He never did care for the river, did Montmorency.

"It's all very well for you fellows," he says; "you like it, but I don't. There's nothing for me to do. Scenery is not in my line, and I don't smoke. If I see a rat, you won't stop; and if I go to sleep, you get fooling about with the boat, and slop me overboard. If you ask me, I call the whole thing bally foolishness."

We were three to one, however, and the motion was carried.

Twenty-Eight

Escape from the Ice

Ernest Shackleton

On April 7 at daylight the long-desired peak of Clarence Island came into view; bearing nearly north from our camp. At first it had the appearance of a huge berg, but with the growing light we could see plainly the black lines of scree and the high, precipitous cliffs of the island, which were miraged up to some extent. The dark rocks in the white snow were a pleasant sight.

So long had our eyes looked on icebergs that apparently grew or dwindled according to the angles at which the shadows were cast by the sun; so often had we discovered rocky islands and brought in sight the peaks of Joinville Land, only to find them, after some change of wind or temperature, floating away as nebulous cloud or ordinary berg; that not until Worsley, Wild, and Hurley had unanimously confirmed my observation was I satisfied that I was really looking at Clarence Island. The land was still more than sixty miles away, but it had to our eyes something of the appearance of home, since we expected to find there our first solid footing after all the long months of drifting on the unstable ice. We had adjusted ourselves to the life on the floe, but our hopes had been fixed all the time on some possible landing place. As one hope failed to materialize, our anticipations fed themselves on another. Our drifting home had no rudder to guide it, no sail to give it speed. We were dependent upon the caprice of wind and current; we went whither those irresponsible forces listed. The longing to feel solid earth under our feet filled our hearts. In the full daylight Clarence Island ceased to look like land and had the appearance of a berg of more than eight or ten miles away, so deceptive are distances in the clear air of the Antarctic. The sharp white peaks of Elephant Island showed to the west of north a little later in the day. "I have stopped issuing sugar now, and our meals consist of seal meat and blubber only, with 7 ozs. of dried milk per day for the party," I wrote. "Each man receives a pinch of salt, and the milk is boiled up to make hot drinks for all hands. The diet suits us, since we cannot get much exercise on the floe and the blubber supplies heat. Fried slices of blubber seem to our taste to resemble crisp bacon. It certainly is no hardship to eat it, though persons living under civilized conditions probably would shudder at it. The hardship would come if we were unable to get it."

I think that the palate of the human animal can adjust itself to anything. Some creatures will die before accepting a strange diet if deprived of their natural food. The Yaks of the Himalayan uplands must feed from the growing grass, scanty and dry though it may be, and would starve even if allowed the best oats and corn.

"We still have the dark water-sky of the last week with us to the southwest and west, round to the northeast. We are leaving all the bergs to the west and there are few within our range of vision now. The swell is more marked today, and I feel sure we are at the verge of the floe-ice. One strong gale, followed by a calm would scatter the pack, I think, and then we could push through. I have been thinking much of our prospects. The appearance of Clarence Island after our long drift seems, somehow, to convey an ultimatum. The island is the last outpost of the south and our final chance of a landing place. Beyond it lies the broad Atlantic. Our little boats may be compelled any day now to sail unsheltered over the open sea with a thousand leagues of ocean separating them from the land to the north and east. It seems vital that we shall land on Clarence Island or its neighbour, Elephant Island.

"The latter island has attraction for us, although as far as I know nobody has ever landed there. Its name suggests the presence of the plump and succulent sea elephant. We have an increasing desire in any case to get firm ground under our feet. The floe has been a good friend to us, but it is reaching the end of its journey, and it is liable at any time now to break up and fling us into the unplumbed sea."

A little later, after reviewing the whole situation in the light of our circumstances, I made up my mind that we should try to reach Deception Island. The relative positions of Clarence, Elephant, and Deception Islands can be seen on the chart. The two islands first named lay comparatively near to us and were separated by some eighty miles of water from Prince George Island, which was about 160 miles away from our camp on the berg. From this island a chain of similar islands extends westward, terminating in Deception Island. The channels separating these desolate patches of rock and ice are from ten to fifteen miles wide. But we knew from the Admiralty sailing directions that there were stores for the use of shipwrecked mariners on Deception Island, and it was possible that the summer whalers had not yet deserted its harbour.

Also we had learned from our scanty records that a small church had been erected there for the benefit of the transient whalers. The existence of this building would mean to us a supply of timber, from which, if dire necessity urged us, we could construct a reasonably seaworthy boat. We had discussed this point during our drift on the floe. Two of our boats were fairly strong, but the third, the *James Caird*, was light, although a little longer than the others. All of them were small for the navigation of these notoriously stormy seas, and they would be heavily loaded, so a voyage in open water would be a serious undertaking. I fear that the carpenter's fingers were already itching to convert pews into topsides and decks. In any case, the worst that could befall us when we had reached Deception Island would be a wait until the whalers returned about the middle of November.

Another bit of information gathered from the records of the west side of the Weddell Sea related to Prince George Island. The Admiralty "Sailing Directions," referring to the South Shetlands, mentioned a cave on this island. None of us had seen that cave or could say if it was large or small, wet or dry; but as we drifted on our floe and later, when navigating the treacherous leads and making our uneasy night camps, that cave seemed to my fancy to be a palace which in contrast would dim the splendours of Versailles.

The swell increased that night, and the movement of the ice became more pronounced. Occasionally a neighbouring floe would hammer against the ice on which we were camped, and the lesson of these blows was plain to read. We must get solid ground under our feet quickly. When the vibration ceased after a heavy surge, my thoughts flew round to the problem ahead. If the party had not numbered more than six men a solution would not have been so hard to find; but obviously the transportation of the whole party to a place of safety, with the limited means at our disposal, was going to be a matter of extreme difficulty. There were twenty-eight men on our floating cake of ice, which was steadily dwindling under the influence of wind, weather, charging floes, and heavy swell. I confess that I felt the burden of responsibility sit heavily on my shoulders; but, on the other hand, I was stimulated and cheered by the attitude of the men. Loneliness is the penalty of leadership, but the man who has to make the decisions is assisted greatly if he feels that there is no uncertainty in the minds of those who follow him, and that his orders will be carried out confidently and in expectation of success.

The sun was shining in the blue sky on the following morning (April 8). Clarence Island showed clearly on the horizon, and Elephant Island could also be distinguished. The single snow-clad peak of Clarence Island stood up as a beacon of safety, though the most optimistic imagination could not make an easy path of the ice and ocean that separated us from that giant, white and austere.

"The pack was much looser this morning, and the long rolling swell from the northeast is more pronounced than it was yesterday. The floes rise and fall with the surge of the sea. We evidently are drifting with the surface current, for all the heavier masses of floe, bergs, and hummocks are being left behind. There has been some discussion in the camp as to the advisability of making one of the bergs our home for the time being and drifting with it to the west. The idea is not sound. I cannot be sure that the berg would drift in the right direction. If it did move west and carried us into the open water, what would be our fate when we tried to launch the boats—down the steep sides of the berg in the sea-swell after the surrounding floes had left us? One must reckon, too, the chance of the berg splitting or even overturning during our stay. It is not possible to gauge the condition of a big mass of ice by surface appearance. The ice may have a fault, and when the wind, current, and swell set up strains and tensions, the line of weakness may reveal itself suddenly and disastrously. No, I do not like the idea of drifting on a berg. We must stay on our floe till conditions improve and then make another attempt to advance towards the land."

At 6.30 p.m. a particularly heavy shock went through our floe. The watchman and other members of the party made an immediate inspection and found a crack right under the *James Caird* and between the other two boats and the main camp. Within five minutes the boats were over the crack and close to the tents. The trouble was not caused by a blow from another floe. We could see that the piece of ice we occupied had slewed and now presented its long axis towards the oncoming swell. The floe, therefore, was pitching in the manner of a ship, and it had cracked across when the swell lifted the centre, leaving the two ends comparatively unsupported. We were now on a triangular raft of ice, the three sides measuring, roughly, 90, 100, and 120 yds. Night came down dull and overcast, and before midnight the wind had freshened from the west. We could see that the pack was opening under the influence of wind, wave, and current; and I felt that the time for launching the boats was near at hand. Indeed, it was obvious that even if the conditions were unfavourable for a start during the coming day, we could not safely stay on the floe many hours longer. The movement of the ice in the swell was increasing, and the floe might split right under our camp. We had made preparations for quick action if anything of the kind occurred. Our case would be desperate if the ice broke into small pieces not large enough to support our party and not loose enough to permit the use of the boats.

The following day was Sunday (April 9), but it proved no day of rest for us. Many of the important events of our Expedition occurred on Sundays, and this particular day was to see our forced departure from the floe on which we had lived for nearly six months, and the start of our journeyings in the boats.

"This has been an eventful day. The morning was fine, though somewhat overcast by stratus and cumulus clouds; moderate south-south-westerly and south-easterly breezes. We hoped that with this wind the ice would drift nearer to Clarence Island. At 7 a.m. lanes of water and leads could be seen on the horizon to the west. The ice separating us from the lanes was loose, but did not appear to be workable for the boats. The long swell from the north-west was coming in more freely than on the previous day and was driving the floes together in the utmost confusion. The loose brash between the masses of ice was being churned to mudlike consistency, and no boat could have lived in the channels that opened and closed around us. Our own floe was suffering in the general disturbance, and after breakfast I ordered the tents to be struck and everything prepared for an immediate start when the boats could be launched."

I had decided to take the *James Caird* myself, with Wild and eleven men. This was the largest of our boats, and in addition to her human complement she carried the major portion of the stores. Worsley had charge of the *Dudley Docker* with nine men, and Hudson and Crean were the senior men on the *Stancomb Wills*.

Soon after breakfast the ice closed again. We were standing by, with our preparations as complete as they could be made, when at 11 a.m. our floe suddenly split right across under the boats. We rushed our gear on to the larger of the two pieces and watched with strained attention for the next development. The crack had cut through the site of my tent. I stood on the edge of the new fracture, and, looking across the widening channel of water, could see the spot where for many months my head and shoulders had rested when I was in my sleeping bag. The depression formed by my body and legs was on our side of the crack. The ice had sunk under my weight during the months of waiting in the tent, and I had many times put snow under the bag to fill the hollow. The lines of stratification showed clearly the different layers of snow. How fragile and precarious had been our resting place! Yet usage had dulled our sense of danger. The floe had become our home, and during the early months of the drift we had almost ceased to realize that it was but a sheet of ice floating on unfathomed seas. Now our home was being shattered under our feet, and we had a sense of loss and incompleteness hard to describe.

The fragments of our floe came together again a little later, and we had our lunch of seal meat, all hands eating their fill. I thought that a good meal would be the best possible preparation for the journey that now seemed imminent, and as we would not be able to take all our meat with us when we finally moved, we could regard every pound eaten as a pound rescued. The call to action came at 1 p.m. The pack opened well and the channels became navigable. The conditions were not all one could have desired, but it was best not to wait any longer. The *Dudley Docker* and the *Stancomb Wills* were launched quickly. Stores were thrown in, and the two boats were pulled clear of the immediate floes towards a pool of open water three miles broad, in which floated a lone and mighty berg. The *James Caird* was the last boat to leave, heavily loaded with stores and odds and ends of camp equipment. Many things regarded by us as essentials at that time were to be discarded a little later as the pressure of the primitive became more severe. Man can sustain life with very scanty means. The trappings of civilization are soon cast aside in the face of stern realities, and given the barest opportunity of winning food and shelter, man can live and even find his laughter ringing true.

The three boats were a mile away from our floe home at 2 p.m. We had made our way through the channels and had entered the big pool when we saw a rush of foam-clad water and tossing ice approaching us, like the tidal bore of a river. The pack was being impelled to the east by a tide-rip, and two huge masses of ice were driving down upon us on converging courses. The *James Caird* was leading.

Starboarding the helm and bending strongly to the oars, we managed to get clear. The two other boats followed us, though from their position astern at first they had not realized the immediate danger. The *Stancomb Wills* was the last boat and she was very nearly caught; but by great exertion she was kept just ahead of the driving ice. It was an unusual and startling experience. The effect of tidal action on ice is not often as marked as it was that day. The advancing ice, accompanied by a large wave, appeared to be travelling at about three knots; and if we had not succeeded in pulling clear we would certainly have been swamped.

We pulled hard for an hour to windward of the berg that lay in the open water. The swell was crashing on its perpendicular sides and throwing spray to a height of sixty feet. Evidently there was an ice-foot at the east end, for the swell broke before it reached the bergface and flung its white spray on to the blue ice wall. We might have paused to have admired the spectacle under other conditions; but night was coming on apace, and we needed a camping place. As we steered north-west, still amid the ice floes, the *Dudley Docker* got jammed between two masses while attempting to make a short cut. The old adage about a short cut being the longest way round is often as true in the Antarctic as it is in the peaceful countryside. The *James Caird* got a line aboard the *Dudley Docker*, and after some hauling the boat was brought clear of the ice again. We hastened forward in the twilight in search of a flat, old floe, and presently found a fairly large piece rocking in the swell. It was not an ideal camping place by any means, but darkness had overtaken us. We hauled the boats up, and by 8 p.m. had the tents pitched and the blubberstove burning cheerily. Soon all hands were well fed and happy in their tents, and snatches of song came to me as I wrote up my log. Some intangible feeling of uneasiness made me leave my tent about 11 p.m. that night and glance around the quiet camp. The stars between the snow-flurries showed that the floe had swung round and was end on to the swell, a position exposing it to sudden strains. I started to walk across the floe in order to warn the watchman to look carefully for cracks, and as I was passing the men's tent the floe lifted on the crest of a swell and cracked right under my feet. The men were in one of the dome-shaped tents, and it began to stretch apart as the ice opened. A muffled sound, suggestive of suffocation, came from beneath the stretching tent. I rushed forward, helped some emerging men from under the canvas, and called out,

"Are you all right?"

"There are two in the water," somebody answered. The crack had widened to about four feet, and as I threw myself down at the edge, I saw a whitish object floating in the water. It was a sleeping bag with a man inside. I was able to grasp it, and with a heave lifted man and bag on to the floe. A few seconds later the ice edges came together again with tremendous force. Fortunately, there had been but one man in the water, or the incident might have been a tragedy. The rescued bag contained Holness, who was wet down to the waist but otherwise unscathed. The crack was now opening again. The *James Caird* and my tent were on one side of the opening and the remaining two boats and the rest of the camp on the other side. With two or three men to help me I struck my tent;

then all hands manned the painter and rushed the *James Caird* across the opening crack. We held to the rope while, one by one, the men left on our side of the floe jumped the channel or scrambled over by means of the boat. Finally I was left alone. The night had swallowed all the others and the rapid movement of the ice forced me to let go the painter. For a moment I felt that my piece of rocking floe was the loneliest place in the world. Peering into the darkness, I could just see the dark figures on the other floe. I hailed Wild, ordering him to launch the *Stancomb Wills*, but I need not have troubled. His quick brain had anticipated the order and already the boat was being manned and hauled to the ice edge. Two or three minutes later she reached me, and I was ferried across to the Camp.

We were now on a piece of flat ice about 200 ft. long and 100 ft. wide. There was no more sleep for any of us that night. The killers were blowing in the lanes around, and we waited for daylight and watched for signs of another crack in the ice. The hours passed with laggard feet as we stood huddled together or walked to and fro in the effort to keep some warmth in our bodies. We lit the blubber-stove at 3 a.m., and with pipes going and a cup of hot milk for each man, we were able to discover some bright spots in our outlook. At any rate, we were on the move at last, and if dangers and difficulties lay ahead we could meet and overcome them. No longer were we drifting helplessly at the mercy of wind and current.

The first glimmerings of dawn came at 6 a.m., and I waited anxiously for the full daylight. The swell was growing, and at times our ice was surrounded closely by similar pieces. At 6.30 a.m. we had hot hoosh, and then stood by waiting for the pack to open. Our chance came at 8, when we launched the boats, loaded them, and started to make our way through the lanes in a northerly direction; the *James Caird* was in the lead, with the *Stancomb Wills* next and the *Dudley Docker* bringing up the rear. In order to make the boats more seaworthy we had left some of our shovels, picks, and dried vegetables on the floe, and for a long time we could see the abandoned stores forming a dark spot on the ice. The boats were still heavily loaded. We got out of the lanes, and entered a stretch of open water at 11 a.m. A strong easterly breeze was blowing, but the fringe of pack lying outside protected us from the full force of the swell, just as the coral reef of a tropical island checks the rollers of the

Pacific. Our way was across the open sea, and soon after noon we swung round the north end of the pack and laid a course to the westward, the *James Caird* still in the lead. Immediately our deeply laden boats began to make heavy weather. They shipped sprays, which, freezing as they fell, covered men and gear with ice, and soon it was clear that we could not safely proceed. I put the *James Caird* round and ran for the shelter of the pack again, the other boats following. Back inside the outer line of ice the sea was not breaking. This was at 3 p.m., and all hands were tired and cold. A big floeberg resting peacefully ahead caught my eye, and half an hour later we had hauled up the boats and pitched camp for the night. It was a fine, big, blue berg with an attractively solid appearance, and from our camp we could get a good view of the surrounding sea and ice. The highest point was about 15 ft. above sea level. After a hot meal all hands, except the watchman, turned in. Every one was in need of rest after the troubles of the previous night and the unaccustomed strain of the last thirty-six hours at the oars. The berg appeared well able to withstand the battering of the sea, and too deep and massive to be seriously affected by the swell; but it was not as safe as it looked. About midnight the watchman called me and showed me that the heavy northwesterly swell was undermining the ice. A great piece had broken off within eight feet of my tent. We made what inspection was possible in the darkness, and found that on the westward side of the berg the thick snow covering was yielding rapidly to the attacks of the sea. An ice-foot had formed just under the surface of the water. I decided that there was no immediate danger and did not call the men. The north-westerly wind strengthened during the night. The morning of April 11 was overcast and misty. There was a haze on the horizon, and daylight showed that the pack had closed round our berg, making it impossible in the heavy swell to launch the boats. We could see no sign of the water. Numerous whales and killers were blowing between the floes, and Cape pigeons, petrels, and fulmars were circling round our berg. The scene from our camp as the daylight brightened was magnificent beyond description, though I must admit that we viewed it with anxiety. Heaving hills of pack and floe were sweeping towards us in long undulations, later to be broken here and there by the dark lines that indicated open water. As each swell lifted around our rapidly dissolving berg it drove floe-ice on to the icefoot, shearing off more of the top snow covering and reducing the size of our camp. When the floes retreated to attack again the water swirled over the icefoot, which was rapidly increasing in width. The launching of the boats under such conditions would be difficult. Time after time, so often that a track was formed, Worsley, Wild, and I, climbed to the highest point of the berg and stared out to the horizon in search of a break in the pack. After long hours had dragged past, far away on the lift of the swell then appeared a dark break in the tossing field of ice. Aeons seemed to pass, so slowly it approached. I noticed enviously the calm, peaceful attitudes of two seals which lolled lazily on a rocking floe. They were at home and had no reason for worry or cause for fear. If they thought at all, I suppose they counted it an ideal day for a joyous journey on the tumbling ice. To us it was a day that seemed likely to lead to no more days. I do not think I had ever before felt the anxiety that belongs to leadership quite so keenly. When I looked down at the camp to rest my eyes from the strain of watching the wide white expanse broken by that one black ribbon of open water, I could see that my companions were waiting with more than ordinary interest to learn what I thought about it all. After one particularly heavy collision somebody shouted sharply, "She has cracked in the middle." I jumped off the lookout station and ran to the place the men were examining. There was a crack, but investigation showed it to be a mere surface-break in the snow with no indication of a split in the berg itself. The carpenter mentioned calmly that earlier in the day he had actually gone adrift on a fragment of ice. He was standing near the edge of our camping-ground when the ice under his feet parted from the parent mass. A quick jump over the widening gap saved him.

The hours dragged on. One of the anxieties in my mind was the possibility that we would be driven by the current through the eighty-mile gap between Clarence Island and Prince George Island into the open Atlantic; but slowly the open water came nearer, and at noon it had almost reached us. A long lane, narrow but navigable, stretched out to the southwest horizon. Our chance came a little later. We rushed our boats over the edge of the reeling berg and swung them clear of the ice-foot as it rose beneath them. The *James Caird* was nearly capsized by a blow from below as the berg rolled away, but she got into deep water. We flung stores and gear aboard and within a few minutes were away. The *James Caird* and *Dudley Docker* had good sails and with a favourable breeze could make progress along the lane, with the rolling fields of ice on either side. The swell was heavy and spray was breaking over the ice floes. An attempt to set a little rag of sail on the *Stancomb Wills* resulted in serious delay. The area of sail was too small to be of much assistance, and while the men were engaged in this work the boat drifted down towards the ice-floe, where her position was likely to be perilous. Seeing her plight, I sent the *Dudley Docker* back for her and tied the *James Caird* up to a piece of ice. The *Dudley Docker* had to tow the *Stancomb Wills*, and the delay cost us two hours of valuable daylight. When I had the three boats together again we continued down the lane, and soon saw a wider stretch of water to the west; it appeared to offer us release from the grip of the pack. At the head of an ice tongue that nearly closed the gap through which we might enter the open space was a wave-worn berg shaped like some curious antediluvian monster, an icy Cerberus guarding the way. It had head and eyes and rolled so heavily that it almost overturned. Its sides dipped deep in the sea, and as it rose again the water seemed to be streaming from its eyes, as though it were weeping at our escape from the clutch of the floes. This may seem fanciful to the reader, but the impression was real to us at the time. People living under civilized conditions, surrounded by Nature's varied forms of life and by all the familiar work of their own hands, may scarcely realize how quickly the mind, influenced by the eyes, responds to the unusual and weaves about it curious imaginings like the firelight fancies of our childhood days. We had lived long amid the ice, and we half-unconsciously strove to see resemblances to human faces and living forms in the fantastic contours and massively uncouth shapes of berg and floe.

At dusk we made fast to a heavy floe, each boat having its painter fastened to a separate hummock in order to avoid collisions in the swell. We landed the blubber-stove, boiled some water in order to provide hot milk, and served cold rations. I also landed the dome tents and stripped the coverings from the hoops. Our experience of the previous day in the open sea had shown us that the tents must be packed tightly. The spray had dashed over the bows and turned to ice on the cloth, which had soon grown dangerously heavy. Other articles of our scanty equipment had to go that night. We were carrying only the things that had seemed essential, but we stripped now to the barest limit of safety. We had hoped for a quiet night, but presently we were forced to cast off, since pieces of loose ice began to work round the floe. Drift ice is always attracted to the lee side of a heavy floe, where it bumps and presses under the influence of the current. I had determined not to risk a repetition of the last night's experience and so had not pulled the boats up. We spent the hours of darkness keeping an offing from the main line of pack under the lee of the smaller pieces. Constant rain and snow squalls blotted out the stars and soaked us through, and at times it was only by shouting to each other that we managed to keep the boats together. There was no sleep for anybody owing to the severe cold, and we dared not pull fast enough to keep ourselves warm since we were unable to see more than a few yards ahead. Occasionally the ghostly shadows of silver, snow, and fulmar petrels flashed close to us, and all around we could hear the killers blowing, their short, sharp hisses sounding like sudden escapes of steam. The killers were a source of anxiety, for a boat could easily have been capsized by one of them coming up to blow. They would throw aside in a nonchalant fashion pieces of ice much bigger than our boats when they rose to the surface, and we had an uneasy feeling that the white bottoms of the boats would look like ice from below. Shipwrecked mariners drifting in the Antarctic seas would be things not dreamed of in the killers' philosophy, and might appear on closer examination to be tasty substitutes for seal and penguin. We certainly regarded the killers with misgivings.

Early in the morning of April 12 the weather improved and the wind dropped. Dawn came with a clear sky, cold and fearless. I looked around at the faces of my companions in the *James Caird* and saw pinched and drawn features. The strain was beginning to tell. Wild sat at the rudder with the same calm, confident expression that he would have worn under happier conditions; his steel-blue eyes looked out to the day ahead. All the people, though evidently suffering, were doing their best to be cheerful, and the prospect of a hot breakfast was inspiriting. I told all the boats that immediately we could find a suitable floe the cooker would be started and hot milk and Bovril would soon fix everybody up. Away we rowed to the westward through open pack, floes of all shapes and sizes on every side of us, and every man not engaged in pulling looking eagerly for a suitable camping place. I could gauge the desire for food of the different members by the eagerness they displayed in pointing out to me the floes they considered exactly suited to our purpose. The temperature was about 10 degrees Fahr., and the Burberry suits of the rowers crackled as the men bent to the oars. I noticed little fragments of ice and frost falling from arms and bodies. At eight o'clock a decent floe appeared ahead and we pulled up to it. The galley was landed, and soon the welcome steam rose from the cooking food as the blubber-stove flared and smoked. Never did a cook work under more anxious scrutiny. Worsley, Crean, and I stayed in our respective boats to keep them steady and prevent collisions with the floe, since the swell was still running strong, but the other men were able to stretch their cramped limbs and run to and fro "in the kitchen," as somebody put it. The sun was now rising gloriously. The Burberry suits were drying and the ice was melting off our beards. The steaming food gave us new vigour, and within three quarters of an hour were off again to the west with all sails set. We had given an additional sail to the *Stancomb Wills* and she was able to keep up pretty well. We could see that we were on the true pack-edge, with the blue, rolling sea just outside the fringe of ice to the north. White-capped waves vied with the glittering floes in the setting of blue water, and countless seals basked and rolled on every piece of ice big enough to form a raft.

We had been making westward with oars and sails since April 9, and fair easterly winds had prevailed. Hopes were running high as to the noon observation for position. The optimists thought that we had done sixty miles towards our goal, and the most cautious guess gave us at least thirty miles. The bright sunshine and the brilliant scene around us may have influenced our anticipations. As noon approached I saw Worsley, as navigating officer, balancing himself on the gunwale of the *Dudley Docker* with his arm around the mast, ready to snap the sun. He got his observation and we waited eagerly while he worked out the sight. Then the *Dudley Docker* ranged up alongside the *James Caird* and I jumped into Worsley's boat in order to see the result. It was a grievous disappointment. Instead of making a good run to the westward we had made a big drift to the south-east. We were actually thirty miles to the east of the position we had occupied when we left the floe on the 9th. It has been noted by sealers operating in this area that there are often heavy sets to the east in the Belgica Straits, and no doubt it was one of these sets that we had experienced. The originating cause would be a north-westerly gale off Cape Horn, producing the swell that had already caused us so much trouble. After a whispered consultation with Worsley and Wild, I announced that we had not made as much progress as we expected, but I did not inform the hands of our retrograde movement.

The question of our course now demanded further consideration. Deception Island seemed to be beyond our reach. The wind was foul for Elephant Island, and as the sea was clear to the southwest, I discussed with Worsley and Wild the advisability of proceeding to Hope Bay on the mainland of the Antarctic Continent, now only eighty miles distant. Elephant Island was the nearest land, but it lay outside the main body of pack, and even if the wind had been fair we would have hesitated at that particular time to face the high sea that was running in the open. We laid a course roughly for Hope Bay, and the boats moved on again. I gave Worsley a line for a berg ahead and told him, if possible, to make fast before darkness set in. This was about three o'clock in the afternoon. We had set sail, and as the *Stancomb Wills* could not keep up with the other two boats I took her in tow, not being anxious to repeat the experience of the day we left the reeling berg. The *Dudley Docker* went ahead, but came beating down towards us at dusk. Worsley had been close to the berg, and he reported that it was unapproachable. It was rolling in the swell and displaying an ugly ice-foot. The news was bad. In the failing light we turned towards a line of pack, and found it so tossed and churned by the sea that no fragment remained big enough to give us an anchorage and shelter. Two miles away we could see a larger piece of ice, and to it we managed, after some trouble, to secure the boats. I brought my boat bow on to the floe, whilst Howe, with the painter in his hand, stood ready to jump. Standing up to watch our chance, while the oars were held ready to back the moment Howe had made his leap, I could see that there would be no possibility of getting the galley ashore that night. Howe just managed to get a footing on the edge of the floe, and then made the painter fast to a hummock. The other two boats were fastened alongside the *James Caird*. They could not lie astern of us in a line, since cakes of ice came drifting round the floe and gathering under its lee. As it was we spent the next two hours poling off the drifting ice that surged towards us. The blubberstove could not be used, so we started the Primus lamps. There was a rough, choppy sea, and the *Dudley Docker* could not get her Primus under way, something being adrift. The men in that boat had to wait until the cook on the *James Caird* had boiled up the first pot of milk.

The boats were bumping so heavily that I had to slack away the painter of the *Stancomb Wills* and put her astern. Much ice was coming round the floe and had to be poled off. Then the *Dudley Docker*, being the heavier boat, began to damage the *James Caird*, and I slacked the *Dudley Docker* away. The *James Caird* remained moored to the ice, with the *Dudley Docker* and the *Stancomb Wills* in line behind her. The darkness had become complete, and we strained our eye to see the fragments of ice that threatened us. Presently we thought we saw a great berg bearing down upon us, its form outlined against the sky, but this startling spectacle resolved itself into a low-lying cloud in front of the rising moon. The moon appeared in a clear sky. The wind shifted to the southeast as the light improved and drove the boats broadside on towards the jagged edge of the floe. We had to cut the painter of the *James Caird* and pole her off, thus losing much valuable rope. There was no time to cast off. Then we pushed away from the floe, and all night long we lay in the open, freezing sea, the *Dudley Docker* now ahead, the *James Caird* astern of her, and the *Stancomb Wills* third in the line. The boats were attached to one another by their painters. Most of the time the *Dudley Docker* kept the *James Caird* and the *Stancomb Wills* up to the swell, and the men who were rowing were in better pass than those in the other boats, waiting inactive for the dawn. The temperature was down to 4 degrees below zero, and a film of ice formed on the surface of the sea. When we were not on watch we lay in each other's arms for warmth. Our frozen suits thawed where our bodies met, and as the slightest movement exposed these comparatively warm spots to the biting air, we clung motionless, whispering each to his companion our hopes and thoughts. Occasionally from an almost clear sky came snow-showers, falling silently on the sea and laying a thin shroud of white over our bodies and our boats.

The dawn of April 13 came clear and bright, with occasional passing clouds. Most of the men were now looking seriously worn and strained. Their lips were cracked and their eyes and eyelids showed red in their salt-encrusted faces. The beards even of the younger men might have been those of patriarchs, for the frost and the salt spray had made them white. I called the *Dudley Docker* alongside and found the condition of the people there was no better than in the *James Caird*. Obviously we must make land quickly, and I decided to run for Elephant Island. The wind had shifted fair for that rocky isle, then about one hundred miles away, and the pack that separated us from Hope Bay had closed up during the night from the south. At 6 p.m. we made a distribution of stores among the three boats, in view of the possibility of their being separated. The preparation of a hot breakfast was out of the question. The breeze was strong and the sea was running high in the loose pack around us. We had a cold meal, and I gave orders that all hands might eat as much as they pleased, this concession being due partly to a realization that we would have to jettison some of our stores when we reached open sea in order to lighten the boats. I hoped, moreover, that a full meal of cold rations would compensate to some extent for the lack of warm food and shelter. Unfortunately, some of the men were unable to take advantage of the extra food owing to sea-sickness. Poor fellows, it was bad enough to be huddled in the deeply laden, spray-swept boats, frost-bitten and half-frozen, without having the pangs of sea-sickness added to the list of their woes. But some smiles were caused even then by the plight of one man, who had a habit of accumulating bits of food against the day of starvation that he seemed always to think was at hand, and who was condemned now to watch impotently while hungry comrades with undisturbed stomachs made biscuits, rations, and sugar disappear with extraordinary rapidity.

We ran before the wind through the loose pack, a man in the bow of each boat trying to pole off with a broken oar the lumps of ice that could not be avoided. I regarded speed as essential. Sometimes collisions were not averted. The *James Caird* was in the lead, where she bore the brunt of the encounter with lurking fragments, and she was holed above the water-line by a sharp spur of ice, but this mishap did not stay us. Later the wind became stronger and we had to reef sails, so as not to strike the ice too heavily. The *Dudley Docker* came next to the *James Caird* and the *Stancomb Wills* followed. I had given orders that the boats should keep 30 or 40 yds. apart, so as to reduce the danger of a collision if one boat was checked by the ice. The pack was thinning, and we came to occasional open areas where thin ice had formed during the night. When we encountered this new ice we had to shake the reef out of the sails in order to force a way through. Outside of the pack the wind must have been of hurricane force. Thousands of small dead fish were to be seen, killed probably by a cold current and the heavy weather. They floated in the water and lay on the ice, where they had been cast by the waves. The petrels and skua-gulls were swooping down and picking them up like sardines off toast.

We made our way through the lanes till at noon we were suddenly spewed out of the pack into the open ocean. Dark blue and sapphire green ran the seas. Our sails were soon up; and with a fair wind we moved over the waves like three Viking ships on the quest of a lost Atlantis. With the sheet well out and the sun shining bright above, we enjoyed for a few hours a sense of the freedom and magic of the sea, compensating us for pain and trouble in the days that had passed. At last we were free from the ice, in water that our boats could navigate. Thoughts of home, stifled by the deadening weight of anxious days and nights, came to birth once more, and the difficulties that had still to be overcome dwindled in fancy almost to nothing.

During the afternoon we had to take a second reef in the sails, for the wind freshened and the deeply laden boats were shipping much water and steering badly in the rising sea. I had laid the course for Elephant Island and we were making good progress. The *Dudley Docker* ran down to me at dusk and Worsley suggested that we should stand on all night; but already the *Stancomb Wills* was barely discernible among the rollers in the gathering dusk, and I decided that it would be safer to heave to and wait for the daylight. It would never have done for the boats to have become separated from one another during the night. The party must be kept together, and, moreover, I thought it possible that we might overrun our goal in the darkness and not be able to return. So we made a sea anchor of oars and hove to, the *Dudley Docker* in the lead, since she had the longest painter. The *James Caird* swung astern of the *Dudley Docker* and the *Stancomb Wills* again had the third place. We ate a cold meal and did what little we could to make things comfortable for the hours of darkness. Rest was not for us. During the greater part of the night the sprays broke over the boats and froze in masses of ice, especially at the stern and bows. This ice had to be broken away in order to prevent the boats' growing too heavy. The temperature was below zero and the wind penetrated our clothes and chilled us almost unbearably. I doubted if all the men would survive that night. One of our troubles was lack of water. We had emerged so suddenly from the pack into the open sea that we had not had time to take aboard ice for melting in the cookers, and without ice we could not have hot food. The *Dudley Docker* had one lump of ice weighing about ten pounds, and this was shared out among all hands. We sucked small pieces and got a little relief from thirst engendered by the salt spray, but at the same time we reduced our bodily heat. The condition of most of the men was pitiable. All of us had swollen mouths and we could hardly touch the food. I longed intensely for the dawn. I called out to the other boats at intervals during the night, asking how things were with them. The men always managed to reply cheerfully. One of the people on the *Stancomb Wills* shouted, "We are doing all right, but I would like some dry mits." The jest brought a smile to cracked lips. He might as well have asked for the moon. The only dry things aboard the boats were swollen mouths and burning tongues. Thirst is one of the troubles that confront the traveller in polar regions. Ice may be plentiful on every hand, but it does not become drinkable until it is melted, and the amount that may be dissolved in the mouth is limited. We had been thirsty during the days of heavy pulling in the pack, and our condition was aggravated quickly by the salt spray. Our sleeping bags would have given us some warmth, but they were not within our reach. They were packed under the tents in the bows, where a mail-like coating of ice enclosed them, and we were so cramped that we could not pull them out.

At last daylight came, and with the dawn the weather cleared and the wind fell to a gentle southwesterly breeze. A magnificent sunrise heralded in what we hoped would be our last day in the boats. Rose-pink in the growing light, the lofty peak of Clarence Island told of the coming glory of the sun. The sky grew blue above us and the crests of the waves sparkled cheerfully. As soon as it was light enough we chipped and scraped the ice off the bows and sterns. The rudders had been unshipped during the night in order to avoid the painters catching them. We cast off our iceanchor and pulled the oars aboard. They had grown during the night to the thickness of telegraph poles while rising and falling in the freezing seas, and had to be chipped clear before they could be brought inboard.

We were dreadfully thirsty now. We found that we could get momentary relief by chewing pieces of raw seal meat and swallowing the blood, but thirst came back with redoubled force owing to the saltiness of the flesh. I gave orders, therefore, that meat was to be served out only at stated intervals during the day or when thirst seemed to threaten the reason of any particular individual. In the full daylight Elephant Island showed cold and severe to the north-northwest. The island was on the bearings that Worsley had laid down, and I congratulated him on the accuracy of his navigation under difficult circumstances, with two days dead reckoning while following a devious course through the pack-ice and after drifting during two nights at the mercy of wind and waves. The *Stancomb Wills* came up and McIlroy reported that Blackborrow's feet were very badly frost-bitten. This was unfortunate, but nothing could be done. Most of the people were frost-bitten to some extent, and it was interesting to notice that the "old timers," Wild, Crean, Hurley, and I, were all right. Apparently we were acclimatized to ordinary Antarctic temperature, though we learned later that we were not immune.

All day, with a gentle breeze on our port bow, we sailed and pulled through a clear sea. We would have given all the tea in China for a lump of ice to melt into water, but no ice was within our reach. Three bergs were in sight and we pulled towards them, hoping that a trail of brash would be floating on the sea to leeward; but they were hard and blue, devoid of any sign of cleavage, and the swell that surged around them as they rose and fell made it impossible for us to approach closely. The wind was gradually hauling ahead, and as the day wore on the rays of the sun beat fiercely down from a cloudless sky on painracked men. Progress was slow, but gradually Elephant Island came nearer. Always while I attended to the other boats, signalling and ordering, Wild sat at the tiller of the *James Caird*. He seemed unmoved by fatigue and unshaken by privation. About four o'clock in the afternoon a stiff breeze came up ahead and, blowing against the current, soon produced a choppy sea. During the next hour of hard pulling we seemed to make no progress at all. The *James Caird* and the *Dudley Docker* had been towing the *Stancomb Wills* in turn, but my boat now took the *Stancomb Wills* in tow permanently, as the *James Caird* could carry more sail than the *Dudley Docker* in the freshening wind.

We were making up for the southeast side of Elephant Island, the wind being between northwest and west. The boats, held as close to the wind as possible, moved slowly, and when darkness set in our goal was still some miles away. A heavy sea was running. We soon lost sight of the *Stancomb Wills*, astern of the *James Caird* at the length of the painter, but occasionally the white gleam of broken water revealed her presence. When the darkness was complete I sat in the stern with my hand on the painter, so that I might know if the other boat broke away, and I kept that position during the night. The rope grew heavy with the ice as the unseen seas surged past us and our little craft tossed to the motion of the waters. Just at dusk I had told the men on the *Stancomb Wills* that if their boat broke away during the night and they were unable to pull against the wind, they could run for the east side of Clarence Island and await our coming there. Even though we could not land on Elephant Island, it would not do to have the third boat adrift. It was a stern night. The men, except the watch, crouched and huddled in the bottom of the boat, getting what little warmth they could from the soaking sleeping bags and each other's bodies. Harder and harder blew the wind and fiercer and fiercer grew the sea. The boat plunged heavily through the squalls and came up to the wind, the sail shaking in the stiffest gusts. Every now and then, as the night wore on, the moon would shine down through a rift in the driving clouds, and in the momentary light I could see the ghostly faces of men, sitting up to trim the boat as she heeled over to the wind. When the moon was hidden its presence was revealed still by the light reflected on the streaming glaciers of the island. The temperature had fallen very low, and it seemed that the general discomfort of our situation could scarcely have been increased; but the land looming ahead was a beacon of safety, and I think we were all buoyed up by the hope that the coming day would see the end of our immediate troubles. At least we would get firm land under our feet. While the painter of the *Stancomb Wills* tightened and drooped under my hand, my thoughts were busy with plans for the future.

Towards midnight the wind shifted to the south-west, and this change enabled us to bear up closer to the island. A little later the *Dudley Docker* ran down to the *James Caird*, and Worsley shouted a suggestion that he should go ahead and search for a landing-place. His boat had the heels of the *James Caird*, with the *Stancomb Wills* in tow. I told him he could try, but he must not lose sight of the *James Caird*. Just as he left me a heavy snow squall came down, and in the darkness the boats parted. I saw the *Dudley Docker* no more. This separation caused me some anxiety during the remaining hours of the night. A cross sea was running and I could not feel sure that all was well with the missing boat. The waves could not be seen in the darkness, though the direction and force of the wind could be felt, and under such conditions, in an open boat, disaster might overtake the most experienced navigator. I flashed our compass lamp on the sail in the hope that the signal would be visible on board the *Dudley Docker*, but could see no reply. We strained our eyes to windward in the darkness in the hope of catching a return signal and repeated our flashes at intervals.

My anxiety, as a matter of fact, was groundless. I will quote Worsley's own account of what happened to the *Dudley Docker*:

"About midnight we lost sight of the *James Caird* with the *Stancomb Wills* in tow, but not long after saw the light of the *James Caird*'s compasslamp, which Sir Ernest was flashing on their sail as a guide to us. We answered by lighting our candle under the tent and letting the light shine through. At the same time we got the direction of the wind and how we were hauling from my little pocket compass, the boat's compass being smashed. With this candle our poor fellows lit their pipes, their only solace, as our raging thirst prevented us from eating anything. By this time we had got into a bad tide rip, which, combined with the heavy, lumpy sea, made it almost impossible to keep the *Dudley Docker* from swamping. As it was we shipped several bad seas over the stern as well as abeam and over the bows, although we were 'on a wind.' Lees, who owned himself to be a rotten oarsman, made good here by strenuous bailing, in which he was well seconded by Cheetham. Greenstreet, a splendid fellow, relieved me at the tiller and helped generally. He and Macklin were my right and left bowers as stroke-oars throughout. McLeod and Cheetham were two good sailors and oars, the former a typical old deepsea salt and growler, the latter a pirate to his fingertips. In the height of the gale that night Cheetham was buying matches from me for bottles of champagne, one bottle per match (too cheap; I should have charged him two bottles). The champagne is to be paid when he opens his pub in Hull and I am able to call that way. . . . We had now had one hundred and eight hours of toil, tumbling, freezing, and soaking, with little or no sleep. I think Sir Ernest, Wild, Greenstreet, and I could say that we had no sleep at all. Although it was sixteen months since we had been in a rough sea, only four men were actually sea-sick, but several others were off colour.

"The temperature was 20 degrees below freezing point; fortunately, we were spared the bitterly low temperature of the previous night. Greenstreet's right foot got badly frost-bitten, but Lees restored it by holding it in his sweater against his stomach. Other men had minor frost-bite, due principally to the fact that their clothes were soaked through with salt water. . . . We were close to the land as the morning approached, but could see nothing of it through the snow and spindrift. My eyes began to fail me. Constant peering to windward, watching for seas to strike us, appeared to have given me a cold in the eyes. I could not see or judge distance properly, and found myself falling asleep momentarily at the tiller. At 3 a.m. Greenstreet relieved me there. I was so cramped from long hours, cold, and wet, in the constrained position one was forced to assume on top of the gear and stores at the tiller, that the other men had to pull me amidships and straighten me out like a jackknife, first rubbing my thighs, groin, and stomach.

"At daylight we found ourselves close alongside the land, but the weather was so thick that we could not see where to make for a landing. Having taken the tiller again after an hour's rest under the shelter (save the mark!) of the dripping tent, I ran the *Dudley Docker* off before the gale, following the coast around to the north. This course for the first hour was fairly risky, the heavy sea before which we were running threatening to swamp the boat, but by 8 a.m. we had obtained a slight lee from the land. Then I was able to keep her very close in, along a glacier front, with the object of picking up lumps of freshwater ice as we sailed through them. Our thirst was intense. We soon had some ice aboard, and for the next hour and a half we sucked and chewed fragments of ice with greedy relish.

"All this time we were coasting along beneath towering rocky cliffs and sheer glacier faces, which offered not the slightest possibility of landing anywhere. At 9:30 a.m. we spied a narrow, rocky beach at the base of some very high crags and cliffs, and made for it. To our joy, we sighted the *James Caird* and the *Stancomb Wills* sailing into the same haven just ahead of us. We were so delighted that we gave three cheers, which were not heard aboard the other boats owing to the roar of the surf. However, we soon joined them and were able to exchange experiences on the beach."

Our experiences on the *James Caird* had been similar, although we had not been able to keep up to windward as well as the *Dudley Docker* had done. This was fortunate as events proved, for the *James Caird* and *Stancomb Wills* went to leeward of the big bight the *Dudley Docker* entered and from which she had to turn out with the sea astern. We thus avoided the risk of having the *Stancomb Wills* swamped in the following sea. The weather was very thick in the morning. Indeed at 7 a.m. we were right under the cliffs, which plunged sheer into the sea, before we saw them. We followed the coast towards the north, and ever the precipitous cliffs and glacier faces presented themselves to our searching eyes. The sea broke heavily against these walls and a landing would have been impossible under any conditions. We picked up pieces of ice and sucked them eagerly. At 9 a.m. at the northwest end of the island we saw a narrow beach at the foot of the cliffs. Outside lay a fringe of rocks heavily beaten by the surf but with a narrow channel showing as a break in the foaming water. I decided that we must face the hazards of this unattractive landing-place. Two days and nights without drink or hot food had played havoc with most of the men, and we could not assume that any safer haven lay within our reach. The *Stancomb Wills* was the lighter and handier boat—and I called her alongside with the intention of taking her through the gap first and ascertaining the possibilities of a landing before the *James Caird* made the venture. I was just climbing into the *Stancomb Wills* when I saw the *Dudley Docker* coming up astern under sail. The sight took a great load off my mind.

Rowing carefully and avoiding the blind rollers which showed where sunken rocks lay, we brought the *Stancomb Wills* towards the opening in the reef. Then, with a few strong strokes we shot through on the top of a swell and ran the boat on to a stony beach. The next swell lifted her a little farther. This was the first landing ever made on Elephant Island, and a thought came to me that the honour should belong to the youngest member of the Expedition, so I told Blackborrow to jump over. He seemed to be in a state almost of coma, and in order to avoid delay I helped him, perhaps a little roughly, over the side of the boat. He promptly sat down in the surf and did not move. Then I suddenly realized what I had forgotten, that both his feet were frost-bitten badly. Some of us jumped over and pulled him into a dry place. It was a rather rough experience for Blackborrow, but, anyhow, he is now able to say that he was the first man to sit on Elephant Island. Possibly at the time he would have been willing to forgo any distinction of the kind. We landed the cook with his blubber-stove, a supply of fuel and some packets of dried milk, and also several of the men. Then the rest of us pulled out again to pilot the other boats through the channel. The *James Caird* was too heavy to be beached directly, so after landing most of the men from the *Dudley Docker* and the *Stancomb Wills* I superintended the transhipment of the *James Caird*'s gear outside the reef. Then we all made the passage, and within a few minutes the three boats were aground. A curious spectacle met my eyes when I landed the second time. Some of the men were reeling about the beach as if they had found an unlimited supply of alcoholic liquor on the desolate shore. They were laughing uproariously, picking up stones and letting handfuls of pebbles trickle between their fingers like misers gloating over hoarded gold. The smiles and laughter, which caused cracked lips to bleed afresh, and the gleeful exclamations at the sight of two live seals on the beach made me think for a moment of that glittering hour of childhood when the door is open at last and the Christmas-tree in all its wonder bursts upon the vision. I remember that Wild, who always rose superior to fortune, bad and good, came ashore as I was looking at the men and stood beside me as easy and unconcerned as if he had stepped out of his car for a stroll in the park.

Soon half a dozen of us had the stores ashore. Our strength was nearly exhausted and it was heavy work carrying our goods over the rough pebbles and rocks to the foot of the cliff, but we dared not leave anything within reach of the tide. We had to wade knee-deep in the icy water in order to lift the gear from the boats. When the work was done we pulled the three boats a little higher on the beach and turned gratefully to enjoy the hot drink the cook had prepared. Those of us who were comparatively fit had to wait until the weaker members of the party had been supplied; but every man had his pannikin of hot milk in the end, and never did anything taste better. Seal steak and blubber followed, for the seals that had been careless enough to await our arrival on the beach had already given up their lives. There was no rest for the cook. The blubber-stove flared and spluttered fiercely as he cooked, not one meal, but many meals, which merged into a daylong bout of eating. We drank water and ate seal meat until every man had reached the limit of his capacity.

The tents were pitched with oars for supports, and by 3 p.m. our camp was in order. The original framework of the tents had been cast adrift on one of the floes in order to save weight. Most of the men turned in early for a safe and glorious sleep, to be broken only by the call to take a turn on watch. The chief duty of the watchman was to keep the blubber-stove alight, and each man on duty appeared to find it necessary to cook himself a meal during his watch, and a supper before he turned in again.

Wild, Worsley, and Hurley accompanied me on an inspection of our beach before getting into the tents. I almost wished then that I had postponed the examination until after sleep, but the sense of caution that the uncertainties of polar travel implant in one's mind had made me uneasy. The outlook we found to be anything but cheering. Obvious signs showed that at spring tides the little beach would be covered by the water right up to the foot of the cliffs. In a strong northeasterly gale, such as we might expect to experience at any time, the waves would pound over the scant barrier of the reef and break against the sheer sides of the rocky wall behind us. Well-marked terraces showed the effect of other gales, and right at the back of the beach was a small bit of wreckage not more than three feet long, rounded by the constant chafing it had endured. Obviously we must find some better resting place. I decided not to share with the men the knowledge of the uncertainties of our situation until they had enjoyed the full sweetness of rest untroubled by the thought that at any minute they might be called to face peril again. The threat of the sea had been our portion during many, many days, and a respite meant much to weary bodies and jaded minds.

The cliffs at the back of the beach were inaccessible except at two points where there were steep snow slopes. We were not worried now about food, for, apart from our own rations, there were seals on the beach and we could see others in the water outside the reef. Every now and then one of the animals would rise in the shallows and crawl up on the beach, which evidently was a recognized place of resort for its kind. A small rocky island which protected us to some extent from the northwesterly wind carried a ringed penguin rookery. These birds were of migratory habit and might be expected to leave us before the winter set in fully, but in the meantime they were within our reach. These attractions, however, were overridden by the fact that the beach was open to the attack of wind and sea from the northeast and east. Easterly gales are more prevalent than western in that area of the Antarctic during the winter. Before turning in that night I studied the whole position and weighed every chance of getting the boats and our stores into a place of safety out of reach of the water. We ourselves might have clambered a little way up the snow-slopes, but we could not have taken the boats with us. The interior of the island was quite inaccessible. We climbed up one of the slopes and found ourselves stopped soon by overhanging cliffs. The rocks behind the camp were much weathered, and we noticed the sharp, unworn boulders that had fallen from above. Clearly there was a danger from overhead if we camped at the back of the beach. We must move on. With that thought in mind I reached my tent and fell asleep on the rubbly ground, which gave a comforting sense of stability. The fairy princess who would not rest on her seven downy mattresses because a pea lay underneath the pile might not have understood the pleasure we all derived from the irregularities of the stones, which could not possibly break beneath us or drift away; the very searching lumps were sweet reminders of our safety.

Early next morning (April 15) all hands were astir. The sun soon shone brightly and we spread out our wet gear to dry, till the beach looked like a particularly disreputable gipsy camp. The boots and clothing had suffered considerably during our travels. I had decided to send Wild along the coast in the *Stancomb Wills* to look for a new camping ground, and he and I discussed the details of the journey while eating our breakfast of hot seal steak and blubber. The camp I wished to find was one where the party could live for weeks or even months in safety, without danger from sea or wind in the heaviest winter gale. Wild was to proceed westwards along the coast and was to take with him four of the fittest men, Marston, Crean, Vincent, and McCarthy. If he did not return before dark we were to light a flare, which would serve him as a guide to the entrance of the channel. The *Stancomb Wills* pushed off at 11 a.m. and quickly passed out of sight around the island. Then Hurley and I walked along the beach towards the west, climbing through a gap between the cliff and a great detached pillar of basalt. The narrow strip of beach was cumbered with masses of rock that had fallen from the cliffs. We struggled along for two miles or more in the search for a place where we could get the boats ashore and make a permanent camp in the event of Wild's search proving fruitless, but after three hours' vain toil we had to turn back. We had found on the far side of the pillar of basalt a crevice in the rocks beyond the reach of all but the heaviest gales. Rounded pebbles showed that the seas reached the spot on occasions. Here I decided to depot ten cases of Bovril sledging ration in case of our having to move away quickly. We could come back for the food at a later date if opportunity offered.

Returning to the camp, we found the men resting or attending to their gear. Clark had tried angling in the shallows off the rocks and had secured one or two small fish. The day passed quietly. Rusty needles were rubbed bright on the rocks and clothes were mended and darned. A feeling of tiredness—due, I suppose, to reaction after the strain of the preceding days—overtook us, but the rising tide, coming farther up the beach than it had done on the day before, forced us to labour at the boats, which we hauled slowly to a higher ledge. We found it necessary to move our makeshift camp nearer the cliff. I portioned out the available ground for the tents, the galley, and other purposes, as every foot was of value. When night arrived the *Stancomb Wills* was still away, so I had a blubber-flare lit at the head of the channel.

About 8 p.m. we heard a hail in the distance. We could see nothing, but soon like a pale ghost out of the darkness came the boat, the faces of the men showing white in the glare of the fire. Wild ran her on the beach with the swell, and within a couple of minutes we had dragged her to a place of safety. I was waiting Wild's report with keen anxiety, and my relief was great when he told me that he had discovered a sandy spit seven miles to the west, about 200 yds. long, running out at right angles to the coast and terminating at the seaward end in a mass of rock. A long snow slope joined the spit at the shore end, and it seemed possible that a "dugout" could be made in the snow. The spit, in any case, would be a great improvement on our narrow beach. Wild added that the place he described was the only possible camping-ground he had seen. Beyond, to the west and southwest, lay a frowning line of cliffs and glaciers, sheer to the water's edge. He thought that in very heavy gales either from the south-west or east the spit would be spray-blown, but that the seas would not actually break over it. The boats could be run up on a shelving beach. After hearing this good news I was eager to get away from the beach camp. The wind when blowing was favourable for the run along the coast. The weather had been fine for two days and a change might come at any hour. I told all hands that we would make a start early on the following morning. A newly killed seal provided a luxurious supper of steak and blubber, and then we slept comfortably till the dawn.

The morning of April 17 came fine and clear. The sea was smooth, but in the offing we could see a line of pack, which seemed to be approaching. We had noticed already pack and bergs being driven by the current to the east and then sometimes coming back with a rush to the west. The current ran as fast as five miles an hour, and it was a set of this kind that had delayed Wild on his return from the spit. The rise and fall of the tide was only about five feet at this time, but the moon was making for full and the tides were increasing. The appearance of ice emphasized the importance of getting away promptly. It would be a serious matter to be prisoned on the beach by the pack. The boats were soon afloat in the shallows, and after a hurried breakfast all hands worked hard getting our gear and stores aboard. A mishap befell us when we were launching the boats. We were using oars as rollers, and three of these were broken, leaving us short for the journey that had still to be undertaken. The preparations took longer than I had expected; indeed, there seemed to be some reluctance on the part of several men to leave the barren safety of the little beach and venture once more on the ocean. But the move was imperative, and by 11 a.m. we were away, the *James Caird* leading. Just as we rounded the small island occupied by the ringed penguins the "willywaw" swooped down from the 2000-ft. cliffs behind us, a herald of the southerly gale that was to spring up within half an hour.

Soon we were straining at the oars with the gale on our bows. Never had we found a more severe task. The wind shifted from the south to the southwest, and the shortage of oars became a serious matter. The *James Caird*, being the heaviest boat, had to keep a full complement of rowers, while the *Dudley Docker* and the *Stancomb Wills* went short and took turns using the odd oar. A big swell was thundering against the cliffs and at times we were almost driven on to the rocks by swirling green waters. We had to keep close inshore in order to avoid being embroiled in the raging sea, which was lashed snow-white and quickened by the furious squalls into a living mass of sprays. After two hours of strenuous labour we were almost exhausted, but we were fortunate enough to find comparative shelter behind a point of rock. Overhead towered the sheer cliffs for hundreds of feet, the sea-birds that fluttered from the crannies of the rock dwarfed by the height. The boats rose and fell in the big swell, but the sea was not breaking in our little haven, and we rested there while we ate our cold ration. Some of the men had to stand by the oars in order to pole the boats off the cliff-face.

After half an hour's pause I gave the order to start again. The *Dudley Docker* was pulling with three oars, as the *Stancomb Wills* had the odd one, and she fell away to leeward in a particularly heavy squall. I anxiously watched her battling up against wind and sea. It would have been useless to take the *James Caird* back to the assistance of the *Dudley Docker* since we were hard pressed to make any progress ourselves in the heavier boat. The only thing was to go ahead and hope for the best. All hands were wet to the skin again and many men were feeling the cold severely. We forged on slowly and passed inside a great pillar of rock standing out to sea and towering to a height of about 2400 ft. A line of reef stretched between the shore and this pillar, and I thought as we approached that we would have to face the raging sea outside; but a break in the white surf revealed a gap in the reef and we laboured through, with the wind driving clouds of spray on our port beam. The *Stancomb Wills* followed safely. In the stinging spray I lost sight of the *Dudley Docker* altogether. It was obvious she would have to go outside the pillar as she was making so much leeway, but I could not see what happened to her and I dared not pause. It was a bad time. At last, about 5 p.m., the *James Caird* and the *Stancomb Wills* reached comparatively calm water and we saw Wild's beach just ahead of us. I looked back vainly for the *Dudley Docker*.

Rocks studded the shallow water round the spit and the sea surged amongst them. I ordered the *Stancomb Wills* to run on to the beach at the place that looked smoothest, and in a few moments the first boat was ashore, the men jumping out and holding her against the receding wave. Immediately I saw she was safe I ran the *James Caird* in. Some of us scrambled up the beach through the fringe of the surf and slipped the painter round a rock, so as to hold the boat against the backwash. Then we began to get the stores and gear out, working like men possessed, for the boats could not be pulled up till they had been emptied. The blubber-stove was quickly alight and the cook began to prepare a hot drink. We were labouring at the boats when I noticed Rickenson turn white and stagger in the surf. I pulled him out of reach of the water and sent him up to the stove, which had been placed in the shelter of some rocks. McIlroy went to him and found that his heart had been temporarily unequal to the strain placed upon it. He was in a bad way and needed prompt medical attention. There are some men who will do more than their share of work and who will attempt more than they are physically able to accomplish. Rickenson was one of these eager souls. He was suffering, like many other members of the Expedition, from bad salt-water boils. Our wrists, arms, and legs were attacked. Apparently this infliction was due to constant soaking with sea-water, the chafing of wet clothes, and exposure.

I was very anxious about the *Dudley Docker*, and my eyes as well as my thoughts were turned eastward as we carried the stores ashore; but within half an hour the missing boat appeared, labouring through the spume-white sea, and presently she reached the comparative calm of the bay. We watched her coming with that sense of relief that the mariner feels when he crosses the harbour-bar. The tide was going out rapidly, and Worsley lightened the *Dudley Docker* by placing some cases on an outer rock, where they were retrieved subsequently. Then he beached his boat, and with many hands at work we soon had our belongings ashore and our three craft above high-water mark. The spit was by no means an ideal camping ground; it was rough, bleak, and inhospitable—just an acre or two of rock and shingle, with the sea foaming around it except where the snow slope, running up to a glacier, formed the landward boundary. But some of the larger rocks provided a measure of shelter from the wind, and as we clustered round the blubber-stove, with the acrid smoke blowing into our faces, we were quite a cheerful company. After all, another stage of the homeward journey had been accomplished and we could afford to forget for an hour the problems of the future. Life was not so bad. We ate our evening meal while the snow drifted down from the surface of the glacier, and our chilled bodies grew warm. Then we dried a little tobacco at the stove and enjoyed our pipes before we crawled into our tents. The snow had made it impossible for us to find the tideline and we were uncertain how far the sea was going to encroach upon our beach. I pitched my tent on the seaward side of the camp so that I might have early warning of danger, and, sure enough, about 2 a.m. a little wave forced its way under the tent cloth. This was a practical demonstration that we had not gone far enough back from the sea, but in the semi-darkness it was difficult to see where we could find safety. Perhaps it was fortunate that experience had inured us to the unpleasantness of sudden forced changes of camp. We took down the tents and repitched them close against the high rocks at the seaward end of the spit, where large boulders made an uncomfortable resting place. Snow was falling heavily. Then all hands had to assist in pulling the boats farther up the beach, and at this task we suffered a serious misfortune. Two of our four bags of clothing had been placed under the bilge of the *James Caird*, and before we realized the danger a wave had lifted the boat and carried the two bags back into the surf. We had no chance of recovering them. This accident did not complete the tale of the night's misfortunes. The big eight-man tent was blown to pieces in the early morning. Some of the men who had occupied it took refuge in other tents, but several remained in their sleeping-bags under the fragments of cloth until it was time to turn out.

A southerly gale was blowing on the morning of April 18 and the drifting snow was covering everything. The outlook was cheerless indeed, but much work had to be done and we could not yield to the desire to remain in the sleeping bags. Some sea-elephants were lying about the beach above high-water mark, and we killed several of the younger ones for their meat and blubber. The big tent could not be replaced, and in order to provide shelter for the men we turned the *Dudley Docker* upside down and wedged up the weather side with boulders. We also lashed the painter and stern-rope round the heaviest rocks we could find, so as to guard against the danger of the boat being moved by the wind. The two bags of clothing were bobbing about amid the brash and glacier ice to the windward side of the spit, and it did not seem possible to reach them. The gale continued all day, and the fine drift from the surface of the glacier was added to the big flakes of snow falling from the sky. I made a careful examination of the spit with the object of ascertaining its possibilities as a camping-ground. Apparently, some of the beach lay above high-water mark and the rocks that stood above the shingle gave a measure of shelter. It would be possible to mount the snow slope towards the glacier in fine weather, but I did not push my exploration in that direction during the gale. At the seaward end of the spit was the mass of rock already mentioned. A few thousand ringed penguins, with some gentoos, were on these rocks, and we had noted this fact with a great deal of satisfaction at the time of our landing. The ringed penguin is by no means the best of the penguins from the point of view of the hungry traveller, but it represents food. At 8 a.m. that morning I noticed the ringed penguins mustering in orderly fashion close to the water's edge, and thought that they were preparing for the daily fishing excursion; but presently it became apparent that some important move was on foot. They were going to migrate, and with their departure much valuable food would pass beyond our reach. Hurriedly we armed ourselves with pieces of sledge-runner and other improvised clubs, and started towards the rookery. We were too late. The leaders gave their squawk of command and the columns took to the sea in unbroken ranks. Following their leaders, the penguins dived through the surf and reappeared in the heaving water beyond. A very few of the weaker birds took fright and made their way back to the beach, where they fell victims later to our needs; but the main army went northwards and we saw them no more. We feared that the gentoo penguins might follow the example of their ringed cousins, but they stayed with us; apparently they had not the migratory habit. They were comparatively few in number, but from time to time they would come in from the sea and walk up our beach. The gentoo is the most strongly marked of all the smaller varieties of penguins as far as colouring is concerned, and it far surpasses the adelie in weight of legs and breast, the points that particularly appealed to us.

The deserted rookery was sure to be above high-water mark at all times; and we mounted the rocky ledge in search of a place to pitch our tents. The penguins knew better than to rest where the sea could reach them even when the highest tide was supported by the strongest gale. The disadvantages of a camp on the rookery were obvious. The smell was strong, to put it mildly, and was not likely to grow less pronounced when the warmth of our bodies thawed the surface. But our choice of places was not wide, and that afternoon we dug out a site for two tents in the debris of the rookery, levelling it off with snow and rocks. My tent, No. 1, was pitched close under the cliff, and there during my stay on Elephant Island I lived. Crean's tent was close by, and the other three tents, which had fairly clean snow under them, were some yards away. The fifth tent was a ramshackle affair. The material of the torn eightman tent had been drawn over a rough framework of oars, and shelter of a kind provided for the men who occupied it.

The arrangement of our camp, the checking of our gear, the killing and skinning of seals and sea-elephants occupied us during the day, and we took to our sleeping-bags early. I and my companions in No. 1 tent were not destined to spend a pleasant night. The heat of our bodies soon melted the snow and refuse beneath us and the floor of the tent became an evil smelling yellow mud. The snow drifting from the cliff above us weighted the sides of the tent, and during the night a particularly stormy gust brought our little home down on top of us. We stayed underneath the snow-laden cloth till the morning, for it seemed a hopeless business to set about repitching the tent amid the storm that was raging in the darkness of the night.

The weather was still bad on the morning of April 19. Some of the men were showing signs of demoralization. They were disinclined to leave the tents when the hour came for turning out, and it was apparent they were thinking more of the discomforts of the moment than of the good fortune that had brought us to sound ground and comparative safety. The condition of the gloves and headgear shown me by some discouraged men illustrated the proverbial carelessness of the sailor. The articles had frozen stiff during the night, and the owners considered, it appeared, that this state of affairs provided them with a grievance, or at any rate gave them the right to grumble. They said they wanted dry clothes and that their health would not admit of their doing any work. Only by rather drastic methods were they induced to turn to. Frozen gloves and helmets undoubtedly are very uncomfortable, and the proper thing is to keep these articles thawed by placing them inside one's shirt during the night. The southerly gale, bringing with it much snow, was so severe that as I went along the beach to kill a seal I was blown down by a gust. The cooking pots from No. 2 tent took a flying run into the sea at the same moment. A case of provisions which had been placed on them to keep them safe had been capsized by a squall. These pots, fortunately, were not essential, since nearly all our cooking was done over the blubber-stove. The galley was set up by the rocks close to my tent, in a hole we had dug through the debris of the penguin rookery. Cases of stores gave some shelter from the wind and a spread sail kept some of the snow off the cook when he was at work. He had not much idle time. The amount of seal and sea-elephant steak and blubber consumed by our hungry party was almost incredible. He did not lack assistance—the neighbourhood of the blubber-stove had attractions for every member of the party; but he earned everybody's gratitude by his unflagging energy in preparing meals that to us at least were savoury and satisfying. Frankly, we needed all the comfort that the hot food could give us. The icy fingers of the gale searched every cranny of our beach and pushed relentlessly through our worn garments and tattered tents. The snow, drifting from the glacier and falling from the skies, swathed us and our gear and set traps for our stumbling feet. The rising sea beat against the rocks and shingle and tossed fragments of floe ice within a few feet of our boats. Once during the morning the sun shone through the racing clouds and we had a glimpse of blue sky; but the promise of fair weather was not redeemed. The consoling feature of the situation was that our camp was safe. We could endure the discomforts, and I felt that all hands would be benefited by the opportunity for rest and recuperation.

Sources

"An American Sealer in the Russian Sea," from *Stories of Ships and the Sea*, Jack London, 1922.

"The Capture: A True Event," from *The Atrocities of the Pirates*, Aaron Smith, 1824.

"The Proud Narragansett's Escape," from *Yankee Ships and Yankee Sailors*, James Barnes, 1897.

"The Loss of the *Indianapolis*," from *The Tragic Fate of the U.S.S.* Indianapolis, Raymond B. Lech. New York: Cooper Square Press, a division of the Rowman & Littlefield Publishing Group, 1982.

"Mutiny on Board the Ship *Globe* of Nantucket," from *Narrative of the Mutiny, On Board the Ship* Globe, William Lay and Cyrus M. Hussey, 1828.

"The Savage Sea," from *Survive the Savage Sea*, Dougal Robertson. Lanham, MD: Sheridan House, a division of the Rowman & Littlefield Publishing Group, 1994.

"The Shetland Bus," from *The Shetland Bus*, David Howarth. Guilford, CT: The Lyons Press, a division of the Rowman & Littlefield Publishing Group, 2001.

"On the Grand Banks," from *Captains Courageous*, Rudyard Kipling, 1897.

"Loss of the Whaleship *Essex*," from *Shipwreck of the Whaleship* Essex, Owen Chase, 1821.

"Yammerschooner," from *Sailing Alone Around the World*, Joshua Slocum, 1900.

"The Wreck of the *Medusa*," from *Perils and Captivity*, Charlotte-Adélaïde Dard. Edinburgh: Constable and Co., 1827.

"Young Ironsides," from *Old Ironsides*, James Fenimore Cooper, 1853.

"Rounding Cape Horn," from *White-Jacket*, Herman Melville, 1850.

"The Last Cruise of the *Saginaw*," from *The Last Cruise of the* Saginaw, George H. Read. Boston and New York: Houghton Mifflin Company, 1912.

"Loss of a Man—Superstition," from *Two Years Before the Mast*, Richard Henry Dana, 1840.

"Treacherous Passage," from *Eight Survived: The Harrowing Story of the USS* Flier, Douglas A. Campbell. Guilford, CT: The Lyons Press, a division of the Rowman & Littlefield Publishing Group, 2010.

"The Merchant's Cup," from *Broken Stowage*, David W. Bone, 1922.

"Seventy Days in an Open Boat," from *Two Survived: The Timeless WWII Epic of Seventy Days at Sea in an Open Boat*, Guy Pearce Jones. Guilford, CT: The Lyons Press, a division of the Rowman & Littlefield Publishing Group, 2001.

"Loss of the *Pequod*," from *Moby Dick*, Herman Melville, 1851.

"Blueskin, the Pirate," from *Howard Pyle's Book of Pirates*, Merle Johnson, 1921.

"The Adventures of Captain Horn," from *The Adventures of Captain Horn*, Frank Stockton, 1895.

"The Wreck of the *Citizen*," from *The Arctic Whaleman*, Lewis Holmes, 1857.

"The Cruise of the *Wasp*," from *Hero Tales of American History*, Henry Cabot Lodge and Theodore Roosevelt, 1895.

"Dirty Weather," from *Typhoon*, Joseph Conrad, 1902.

"An Introduction to Informality," from *The Riddle of the Sands*, Erskine Childers, 1903.

"The Siege of the Round-House," from *Kidnapped*, Robert Louis Stevenson, 1886.

"Mal de Mer," from *Three Men in a Boat (To Say Nothing of the Dog)*, Jerome K. Jerome, 1889.

"Escape from the Ice," from *South*, Ernest Shackleton, 1919.